The Biology and Medicine of Rabbits and Rodents

The Biology and Medicine of Rabbits and Rodents

JOHN E. HARKNESS, D.V.M., M.S., M.Ed.

College of Veterinary Medicine
Mississippi State University
Mississippi State, Mississippi

JOSEPH E. WAGNER, D.V.M., M.P.H., Ph.D

College of Veterinary Medicine
University of Missouri-Columbia
Columbia, Missouri

3rd Edition

7 6 75

Illustrations by Donald L. Connor

Lea & Febiger

1989

Philadelphia • London

Lea & Febiger
600 Washington Square
Philadelphia, PA 19106
U.S.A.

This book contains much information about disease therapy, particularly by drugs. The authors, editors, and publisher have taken reasonable care to ensure the accuracy of the drugs, dosages, and schedules suggested. However, if a drug is used in a species or manner other than according to label directions, adverse reactions or lack of efficacy impart risk and even liability to the user. Since withdrawal times are not known for most drugs in rabbits, users will be liable if illegal drug residues are caused to be present in rabbit meat.

First Edition, 1977
    Translations:
        Spanish Edition by Editorial Acribia, Zaragoza, Spain, 1980

Second Edition, 1983
    Translations:
        Japanese Edition by Yokendo, LTD, Tokyo, Japan, 1985

Library of Congress Cataloging-in-Publication Data

Harkness, John E.
    The biology and medicine of rabbits and rodents / John E.
Harkness, Joseph E. Wagner; illustrations by Donald L. Connor.—
3rd ed.
        p. cm.
    Includes bibliographies and index.
    ISBN 0-8121-1176-1
    1. Laboratory animals—Diseases. 2. Rodents—Diseases.
3. Rabbits—Diseases. 4. Rodents as laboratory animals. 5. Rabbits
as laboratory animals. I. Wagner, Joseph E. II. Title.
SF996.5.H37 1989
636′.932—dc19   88-13330

PRINTED IN THE UNITED STATES OF AMERICA

Print Number 3 2 1

Preface

The purpose of *The Biology and Medicine of Rabbits and Rodents* is to provide people professionally concerned with the use and maintenance of rabbits and rodents with a concise, up-to-date, reasonably comprehensive source of needed information about their care and health. The book is intended for students and other people with a background in the biological sciences, including veterinarians and scientists. A basic knowledge of biology and an interest in rabbits and rodents as pets or laboratory animals is assumed. Beyond this, the emphasis throughout is on the practical aspects of rabbit and rodent care and health; however, substantial detail is provided about many aspects of biology and husbandry, clinical signs and procedures, specific diseases, and their diagnoses. This book is intended to bridge the gap between the hardcover, rather expensive reference works in the American College of Laboratory Animal Medicine-Academic Press series of monographs on various species of laboratory animals (rat, mouse, rabbit, guinea pig, and hamster) and a wide variety of class notes, handbooks, autotutorial materials, and other publications used as references or in teaching about rabbits and rodents. By publishing in paperback instead of hardcover, we aim to make this book affordable for students, veterinary practitioners, and institutional representatives. This is done in the interest of the well-being of the rabbits and rodents of the world, which are the focus of this effort, and ultimately, mankind, which is the beneficiary of their existence.

The response to the first two editions of this book (1977 and 1983) was extremely gratifying. To the more than 20,000 people who ordered these editions, we say thanks. We also thank our international readers, including those who ordered copies of the Japanese and Spanish translations. Our publisher informs us that this book has a following, i.e., many of those who bought the first edition also bought the second. Because the third edition has been substantially updated and expanded, we expect to continue to earn our following. We are already planning a future edition, and we are particularly anxious to hear readers' suggestions for improving and updating the next edition.

The major changes in the third edition occur in Chapter 5, where new diseases and conditions have been added (cryptosporidiosis, clostridial enterotoxemia, and mouse polio) and others have been substantially updated. In Chapter 3 we added a review of drugs approved by the United States Food and Drug Administration for specific uses in rabbits and rodents. A new section on analgesia has been added,

v

Preface

and information on diagnostic testing has been expanded and updated. In Chapter 6 new replacement case reports have been added. Throughout this edition, outdated references have been eliminated and replaced with more recent references.

Columbia, Missouri   Joseph E. Wagner
Mississippi State, Mississippi   John E. Harkness

Acknowledgments

Although many parts of this book have been substantially rewritten and updated from the second edition, and we ourselves have learned much since the appearance of the first edition, we are grateful to our colleagues in veterinary practice and animal care facilities for many of the changes incorporated into this edition. We thank our families for tolerating the eccentricities of book authors; the highly competent staff of Lea & Febiger; Sylvia Bradfield and Brian Carter for typing the manuscript; Karen Boillot and Tracy Swartz for their many trips to the library; and Drs. Nephi M. Patton, Susan Gibson, and Cynthia Williford for their reviews and ideas. Cheers to you all! We sincerely appreciate your help.

Portions of the text were developed with support from DHHS Grant RR00471.

J.E.W.
J.E.H.

Contents

CHAPTER 1 • GENERAL HUSBANDRY

Major Concerns in Husbandry 1
Factors Predisposing to Disease 3
Facility Cleaning, Disinfection, and Fumigation 4
Rabbits and Rodents in Research 5
General References 6

CHAPTER 2 • BIOLOGY AND HUSBANDRY

The Rabbit 9
The Guinea Pig 19
The Hamster 27
The Gerbil 34
The Mouse 40
The Rat 47

CHAPTER 3 • CLINICAL PROCEDURES

Drugs 55
Anesthesia 61
Surgical Procedures 68
Analgesia 69
Radiography 70
Euthanasia 71
Blood Collection 73
Serologic Testing 76
References 76

CHAPTER 4 • CLINICAL SIGNS AND DIFFERENTIAL DIAGNOSES

History Protocol 85
The Rabbit 86
The Guinea Pig 90
The Hamster 93
The Gerbil 95
The Mouse 98
The Rat 100
References 103

CHAPTER 5 • SPECIFIC DISEASES AND CONDITIONS

Acariasis 111
Allergies to Laboratory Animals 115
Anorexia 117
Bordetella bronchiseptica Infection 117
Cestodiasis 119
Coccidiosis (Hepatic) 122
Coccidiosis (Intestinal) 124
Colibacillosis 126
Corynebacterium kutscheri Infection 127
Cryptosporidiosis 129
Dermatophytosis 131
Encephalitozoonosis (Nosematosis) 132
Enteropathy (Nonspecific) 134
Enterotoxemia (Clostridial) 137
Epilepsy in Gerbils 139
Heat Stroke 140
Hemorrhagic Fever with Renal Syndrome 141
Hypovitaminosis C (Scurvy) 142
Lymphocytic Choriomeningitis 144
Malocclusion 145
Mastitis 146
Metabolic Toxemias of Pregnancy 147
Moist Dermatitis 148
Mouse Hepatitis Virus Infection 149
Mousepox (Ectromelia) 151
Mucoid Enteropathy 154
Murine Encephalomyelitis (Mouse Polio) 155
Murine Mycoplasmosis 156
Myxomatosis 160
Neoplasia 161
Nephrosis 166
Oxyuriasis (Pinworms) 168
Pasteurella multocida Infection 171
Pasteurella pneumotropica Infection 174
Pediculosis 175
Pneumocystosis 176
Proliferative Ileitis 178
Reovirus Type 3 Infection of Mice 179
Rotavirus Infection 180
Salmonellosis 182
Sendai Virus Infection 184
Sialodacryoadenitis 186
Spironucleosis (Hexamitiasis) 188
Staphylococcus aureus Infection 190
Streptococcus pneumoniae Infection 192
Streptococcus zooepidemicus Infection 194
Transmissible Murine Colonic Hyperplasia 195
Trichobezoars 197
Tularemia 197

CONTENTS

Tyzzer's Disease (*Bacillus piliformis* Infection) 198
Urolithiasis 201
Venereal Spirochetosis 202

CHAPTER 6 • CASE REPORTS

Rabbits 205
Guinea Pigs 208
Hamsters 209
Gerbils 210
Mice 210
Rats 212
Suggested Answers 213

INDEX 221

Chapter 1
General Husbandry

Emphasis in this text is on the biology, husbandry, medicine, and diseases of rabbits and the commonly used rodent species—guinea pigs, mice, rats, hamsters, and gerbils. Populations of rabbit and rodent pets are difficult to establish; however, a 1984 review of the veterinary services market indicated that around 2.1% of United States households own a total of 9 million such pets.

The content of this chapter on general husbandry, among the most important information in the entire text, will be referred to repeatedly in subsequent chapters on biology, syndromes, and diseases. If the client has a continuing regard for the principles of disease prevention, then discussions of syndromes, treatments, and specific diseases become less important.

Specific, detailed guidelines for establishing an acceptable level of research animal care and for meeting the legal regulations of the United States Department of Agriculture [Public Law 89-544 (1966), commonly referred to as "The Animal Welfare Act" and amended by Public Laws JL91-579 (1971), 94-279 (1976) and 99-198 (1985)] and the policies of the National Institutes of Health are contained in these publications:

Title 9: *Animals and Animal Products,* Chapter 1, Subchapter A—Animal Welfare, available upon request from the Deputy Administrator, Veterinary Services, Animal and Plant Health Inspection Service, U.S. Department of Agriculture, Federal Building, 6505 Belcrest Road, Hyattsville, MD 20782.

Guide for the Care and Use of Laboratory Animals. DHHS-PHS, NIH Publication No. 85-23, 1985. Published by the National Research Council, 2101 Constitution Ave, NW, Washington DC, 20418 and available from the Superintendent of Documents, Government Printing Office, Washington, DC 20402-9325. This publication is commonly referred to as "The Guide (item #S/N017-040-00498-2)."

MAJOR CONCERNS IN HUSBANDRY

Housing

The major considerations in preventive management include general husbandry factors, circumstances predisposing to disease, and methods of facility sanitation. These concerns extend to both pet and research animal housing.

Housing should be designed to provide for the psychosocial well-being and physical comfort of the animals. Primary enclosures (cages, pens, and runs) should be structurally sound, appropriate for the species housed, in good repair, free of sharp or abrasive surfaces, easily cleaned, constructed to prevent escape and intrusion, and large enough to provide for freedom of movement and normal postural adjustments, eating, and breeding. Housing areas should receive reliable electric power and potable water supply and should be uncluttered, dry, clean, easily sanitized, properly drained, well-lighted, well-ventilated, and heated or cooled as necessary. Unpainted wood, rusted metal, and other porous or deteriorating materials that are difficult to sanitize should not be used for housing.

Physical Comfort

The provision of physical comfort extends to animal care personnel as well as to animals. Housed animals should be dry, clean, away from excessive noise, and maintained within an ambient temperature range of 18° to 29° C (65° to 84° F). According to *The Guide*, temperature should not vary dramatically from an average level, usually between 18° and 26° C (64° to 79° F) for mice, rats, hamsters, and guinea pigs and 16° and 21° C (61° to 70° F) for rabbits. Hairless or nude animals require higher ambient temperatures. The relative humidity in the cage should be maintained between 40% and 70% for mice, rats, hamsters, and guinea pigs and 40% and 60% for rabbits.

A light intensity of 30 foot candles (323 lumens/m^2) at 1 m above floor level is adequate for routine animal care activities. At this illumination level, a rat in the front of an upper cage held in a cage rack would receive the equivalent of 3 to 3.7 foot candles (32 to 40 lumens/m^2). Illumination of excessive intensity and duration may cause retinal lesions in albino rats and mice. Room air changes, with fresh or filtered air, generally should not be fewer than 10 complete air changes per hour. The spatial dimensions of the room and the number of animals present also affect ventilation requirements. Animals housed outdoors must be in well-ventilated areas away from excessive drafts, dampness, direct sunlight, temperature extremes, vermin, and predators.

Health

Facilities and caging should be physically cleaned and sanitized when necessary (usually 1 to 3 bedding changes per week). Ammonia gas, which damages the respiratory epithelium, is reduced by lowering population density and by providing adequate ventilation, good sanitation, and frequent bedding changes. Vermin must be excluded from animal quarters. Different species and diseased animals should be housed separately, preferably in different rooms or cubicles. Isolation and quarantine facilities must be available and utilized. Professional and technical personnel should regularly inspect animals for injury and disease. Stock and replacement animals should be obtained from reputable dealers. Rodent producers should provide a recent health monitoring report.

Nutrition

Rabbits and rodents should be fed a wholesome, fresh, clean, nutritious, palatable diet on a regular basis and in adequate quantity. Feeding and watering devices should be clean, designed to prevent contamination by wastes, appropriate for the species and age of animal housed, accessible, and functional. Water should be fresh, clean, and available *ad libitum* in sipper-tube watering devices. Food should be stored in closed containers, kept at room temperature or below, observed for mold or vermin, and used while fresh.

Provision of an adequate diet should underlie all efforts to maintain the health of laboratory animals. Shortcut, discount, outdated, or improperly formulated diets and vitamin formulations should always be suspect. Although the nutritional requirements of rabbits and rodents have been extensively investigated and reported, optimal nutrient levels remain uncertain. Requirements known at present are available from feed company publications or from the publications of the National Academy of Sciences listed in the references.

With the exception of ascorbic acid deficiency in guinea pigs and net energy, water, and protein deficiencies in all species, malnutrition is uncommon in rabbits and rodents. Nutritional problems uncommonly encountered include vitamin D, calcium, and phosphorus imbalances in rabbits (atherosclerosis), guinea pigs (metastatic calcification), and young rodents (rickets); vitamin A (hydrocephalus, prenatal death) and vitamin E (muscular dystrophy, prenatal mortality, seminiferous tubule degeneration) deficiencies in rabbits and rodents; and certain mineral or amino acid deficiencies in several species.

Subclinical nutritional deficiencies, excesses, or imbalances may be more common than suspected, as the signs are often obscured by secondary bacterial infections or metabolic disorders. The importance of nutritional imbalances lies more in the predisposing role than in the causation of primary deficiency disease. Because the majority of research animals are involved in long-term studies, more attention will be focused on the role of laboratory animal diets in the causation of degenerative processes associated with aging, such as renal disease in rats and hamsters.

When nutritional studies are not involved, the small animal species should be fed a complete, fresh, wholesome, palatable, clean, pelleted diet. Supplementation with straw, greens, salt blocks, and antibiotics should be undertaken only with an understanding of possible benefits and adverse consequences. Also, the diet that might be preferred by the animal may not be necessarily the proper diet for that species.

Primary nutritional imbalances may be demonstrated as weight loss or failure to gain, increased susceptibility to disease, hair loss, prenatal mortality, agalactia, infertility, anemia, deformed bones, central nervous system abnormalities, or a reluctance to move. Individual pet animals, more frequently than laboratory animals, experience nutritional deficiencies because fresh, complete diets are less available to the pet owner than to the large research laboratory.

Fresh water from sipper-tube waterers and a complete, fresh, pelleted diet with more than 14% protein are adequate for a rodent pet. Vitamin drops, salt blocks, seeds, and various treats are usually not necessary.

Identification

Animals used in research should be prop-

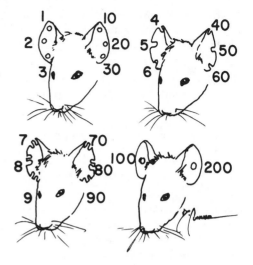

Fig. 1. Ear notch-punch code for identification of rodents. These number codes are used in various combinations to produce the desired number.

erly and clearly identified. Identification methods vary, but animals may be identified by cage cards, individual coat pattern, ear punch or notch (mice, rats, and hamsters), toe clip (mice), ear tag or stud (hamsters, guinea pigs, and rabbits), dye staining (on light-colored fur), or tattooing (ear, tail, foot pad, or shaved flank). An ear notch-punch code is shown in Figure 1.

Generally, permanent individual animal identification is used with cage cards. Cage card identification alone is often inadequate, because animals are easily mixed during cage cleaning and experimental manipulations.

FACTORS PREDISPOSING TO DISEASE

Certain organic or environmental factors increase the exposure or reduce the resistance of the animal to disease. These factors must be considered in disease prevention efforts. Factors influencing disease susceptibility include environmental, genetic, metabolic, experimental, and dietary variables.

Environmental Factors
  Climatic extremes
  Climatic changes
  Inadequate ventilation
  High ammonia levels
  Excessive drafts
  Dampness

Personnel changes
Single vs. group housing
Wire vs. plastic cages
Crowding
Improper bedding
Social hierarchies
Excessive noise
Improper lighting
Waste accumulation
Exposure to animal vectors of disease
High or low humidity

Genetic Factors
Sex differences
Inherited mutations
Strain differences
Congenital abnormalities
Immune system deficiencies
Inbreeding depression

Metabolic Factors
Age
Obesity
Concurrent disease
Anorexia
Lack of exercise
Lactation
Pregnancy
Nonspecific stressors

Experimental Procedures
Restraint
Surgery
Drug effects
Neoplastic induction
Radiation effects
Pathogen inoculation
Bleeding

Dietary Factors
Insufficient quantity of feed or water
  Insufficient amount supplied
  Feeders or waterers inaccessible, not recognized, not working, or not workable by the age group or species involved
  Feed supplied in a form (hard pellets or soft mash) that cannot be eaten by age group or species involved
  Increased demand for nutrients because of pregnancy, lactation, heat, cold, diet composition, or disease
  Competition for feed
  Frozen water
Inadequate feed or water quality
  Improper formulation
  Intended for another species
  Unpalatable
  Deteriorated
  Contaminated by insects, mold, bacteria, parasite ova, urine, feces
Dietary alterations
  Feed not recognized by the animal

Alteration of intestinal flora due to dietary change or antibiotic administration
Change in gastrointestinal flora at weaning

FACILITY CLEANING, DISINFECTION, AND FUMIGATION

Sanitation is a key operation in laboratory animal care. Clean cages are particularly important during pregnancy, lactation, and weaning, after the removal of sick animals, and preceding the introduction of new animals into the colony.

Clean water and a brush or automatic washer are used to remove accumulated bedding, feed, urine, and feces from the cages, feeders, and waterers. This step is extremely important, because organic materials remaining on the cage will retain microbes and inhibit the antimicrobial activity of disinfectants.

After gross waste accumulations have been removed, the cage is washed with 82° C (180° F) water for at least 3 minutes, or a disinfectant solution is applied to all surfaces. Disinfectant solutions (phenolic disinfectants, quaternary ammonium compounds, and halogens) are available in farm supply or feed stores and from manufacturers. Instructions for use are on the labels of the bottled concentrates. In all cases detergents and disinfectants must be thoroughly rinsed off the cleaned cages and feeders.

Disinfectants should be selected for broad-spectrum activity, rapid kill effect, cleaning capacity, solubility, stability, residual activity, and lack of odor, toxicity, and irritability. Disinfectants should be effective in the presence of organic materials, detergents, hard water, and varying pH levels and on porous, rough, or cracked surfaces. Some disinfectant preparations should be avoided because they cloud clear plastic cages.

Unfortunately, no single disinfectant meets all these criteria, and selection must be based on specific requirements. The effects of disinfectants vary with time of exposure, temperature and concentration of solution, and ionic content of the diluent. Important categories of microorganisms weakly or not affected by standard disinfectant solutions are bacterial spores, coccidial oocysts, and *Pseudomonas*. Viruses and fungi also vary in susceptibility to disinfectants.

Halogen-bearing disinfectants, the hypochlorites and iodophors, are effective in acidic solutions, but they may stain or damage fabrics and have reduced activity in the presence of organic matter, soap, or detergent residues. A good, practical, and safe disinfectant for pet animal cages is a solution of 30 ml of a 5% sodium hypochlorite solution (laundry bleach) in 1 liter of water (1 oz per quart). A fresh mixture should be prepared daily and used only on clean cages.

Phenol derivative compounds, which are the disinfectants least affected by environmental agents, kill the vegetative forms of both gram-positive and gram-negative bacteria (excepting *Pseudomonas* spp, which require longer exposures and higher concentrations) in approximately 30 minutes. The germicidal activity is increased with the increased concentration and temperature of the solution. Phenolic compounds, emulsified at 1% to 5% in weakly acidic, soapy water, have some antifungal, sporicidal, and virucidal activity. Because of a residual odor and toxicity, phenolic derivatives are not used to disinfect feeders and waterers. Synthetic phenolic compounds, including pine oils, have greater activity and less odor than do natural coal tar derivatives; but, as with other phenolic compounds, they are particularly toxic for cats.

Quaternary ammonium compounds are effective against gram-positive bacteria but are considerably less effective in the presence of organic matter, soaps, and an acidic pH. These compounds are useful for general purpose disinfection and for cleaning feeders and waterers. Residues of these compounds on the nest box have been implicated as the cause of death among suckling rabbits. Other disinfecting substances include 2% lye solution, formalin, ethylene oxide gas, and 10% ammonia solution.

Alkaline rabbit urine (pH 8.2) and guinea pig and hamster urine contain phosphate and carbonate crystals. When accumulated on the caging, these crystals form a scale that is difficult to remove. Acidic products (pH 2) are available for removing the scale. A flame may be used to remove accumulated hair and manure from rabbit cages. Polycarbonate plastic is affected by alkaline detergents, which cause the transparent plastic cages to become cloudy and brittle. Acid detergent preparations are less destructive.

Details for water acidification as a measure for reducing *Pseudomonas* contamination are discussed in the section on the biology and husbandry of the mouse.

Formaldehyde gas fumigation, if preceded by thorough mechanical cleaning of the room and caging, is an effective method for eliminating parasites and vegetative bacterial forms. Spores, oocysts, and parasitic ova are better eliminated by mechanical cleaning and the application of a solution of formaldehyde, ammonia, and activated glutaraldehyde or aldehyde-alcohol.

Before fumigation is attempted, the room must be free of animals, airtight, warmed to at least 21° C (70° F), and wetted to raise the relative humidity to 80% or more. Formaldehyde gas is generated by heating paraformaldehyde crystals on a hot plate. Provisions should be made for exhausting fumes from the room without the entry of personnel. Formaldehyde gas has been linked to nasal cancers in rats. Alcide (chlorine dioxide or ABQ—Alcide Corporation, Norwalk, CT 06851) and peracetic acid are powerful microbiocides used in maintaining gnotobiotic facilities. Interestingly, these products are not very effective against parasite ova and cocidial oocysts. For example it is not unusual to see pinworm infections in viral free barrier sustained mice or rats.

RABBITS AND RODENTS IN RESEARCH

Estimates of numbers of animals used in research are difficult to derive because of the limitations and outdated nature of applicable surveys and estimates. Based on USDA-APHIS data, a 1978 ILAR survey, commercial producer production data, and a 1986 U.S. Congressional Office of Technology Assessment report, approximately 10.5 million mice, 4.5 million rats, 475,000 hamsters, 525,000 guinea pigs, 525,000 rabbits, and 110,000 gerbils were used in research in 1988. Nearly all these animals are produced for the purpose of research and testing. Most mice and rats are reared pathogen-free behind barriers that preclude introduction of disease agents and are then maintained disease-free at the research and testing laboratories. Because of the highly sophisticated nature of the research in which these animals are used, they must be maximally defined physiologically, genetically, and microbiologically. Most facility managers work with diagnostic laboratories to maintain an awareness of contagious disease (if any) operant in their institutional facilities.

GENERAL REFERENCES

Altman, P.L., and Katz, D.D. (eds.): Inbred and Genetically Defined Strains of Laboratory Animals. Part 1. Mouse and Rat. Part 2. Hamster, Guinea Pig, Rabbit and Chicken. Bethesda, Federated American Societies of Experimental Biology, 1979.

Andrews, E.J., Ward, B.C., and Altman, N.H. (eds.): Spontaneous Animal Models of Human Disease. Vols. 1 and 2. Orlando, Academic Press, 1979.

Archibald, J., Ditchfield, J., and Rowsell, H.C. (eds.): The Contribution of Laboratory Animal Science to the Welfare of Man and Animals: Past, Present and Future. New York, Gustav Fischer Verlag, 1985.

Arrington, L.R.: Introductory Laboratory Animal Science: The Breeding, Care and Management of Experimental Animals, 2nd Ed. Danville, IL, The Interstate Printers & Publishers, Inc., 1978.

Balk, M.W., and Melby, E.C., Jr.: Importance of Laboratory Animal Genetics: Health and Environment in Biomedical Research. Orlando, Academic Press, 1984.

Bellhorn, R.W.: Lighting in the animal environment. Lab. Anim. Sci., 30:440–450, 1980.

Benirschke, K., Garner, F.M., and Jones, T.C. (eds.): Pathology of Laboratory Animals. 2 Vols. Springer-Verlag New York, Inc., 1978.

Biosafety in Microbiological and Biomedical Laboratories. Washington, DC, U.S. Government Printing Office, 1984.

Block, S.: Disinfection, Sterilization, and Preservation. Philadelphia, Lea & Febiger, 1983.

Comfortable Quarters for Laboratory Animals, 7th Ed. Washington, DC, Animal Welfare Institute, 1979.

Diner, J.: Physical and Mental Suffering of Experimental Animals. Washington, DC, Animal Welfare Institute, 1979.

Disinfectants: Their Chemistry, Use, and Evaluation. Publication 74–5. Cordova, TN, American Association for Laboratory Animal Science, 1974.

Experimental Animals/Jikken Dobutsu. Edited by the Japanese Association for Laboratory Animal Science and available from Kinokuniya Company, Ltd., 3-17-7 Shinjuku, Shinjuku-ku, Tokyo 160-91, Japan.

Fox, J.G., Cohen, J.B., and Loew, F.M.: Laboratory Animal Medicine. Orlando, Academic Press, 1984.

Fox, M.A.: The Case for Animal Experimentation: An Evolutionary and Ethical Perspective. Berkeley, University of California Press, 1986.

Fox, M.C.: Laboratory Animal Husbandry: Ethology, Welfare and Experimental Variables. Albany, State University of New York Press, 1986.

Gay, W.I. (ed.): Health Benefits of Animal Research. Washington, DC, Foundation for Biomedical Research, 1987.

Gay, W.I. (ed.): Methods of Animal Experimentation. Vols. I, II, III, IV, V, and VI. New York, Academic Press, 1965–1981.

Greenman, D.L., et al.: Influence of cage shelf level on retinal atrophy in mice. Lab. Anim. Sci., 32:353–356, 1982.

Guide to the Care and Use of Experimental Animals. Vols. I and II. Ottawa, ONT, Canadian Council on Animal Care, 1980 and 1984.

Hafez, E.S.E. (ed.): Reproduction and Breeding Techniques for Laboratory Animals. Philadelphia, Lea & Febiger, 1970.

Hime, J.M., and O'Donoghue, P.N.: Handbook of Diseases of Laboratory Animals—Diagnosis and Treatment. London, Heinemann Veterinary Books, 1979.

Holmes, D.D.: Clinical Laboratory Animal Medicine: An Introduction. Ames, IA, Iowa State University Press, 1984.

Inglis, J.K.: Introduction to Laboratory Animal Science and Technology. New York, Pergamon Press, 1980.

Institute of Laboratory Animal Resources: Animals for Research, 10th Ed. Washington, DC, National Academy of Sciences, 1979.

Institute of Laboratory Animal Resources: Animal Housing. Washington, DC, National Academy of Sciences, 1979.

Institute of Laboratory Animal Resources: Animal Models for Biomedical Research. Vols., I, II, III, IV, and V. Washington, DC, National Academy of Sciences, 1968–1974.

Institute of Laboratory Animal Resources: ILAR News. Washington, DC, National Academy of Sciences, 2101 Constitution Avenue, N.W.

Lab Animal. United Business Publications, Inc., 475 Park Avenue South, New York, NY 10016.

Laboratory Animal Science. American Association for Laboratory Animal Science, 70 Timbercreek Cordova, TN 38018.

Laboratory Animals. Journal of the Laboratory Animal Science Association, Laboratory Animals Ltd., 7 Warwick Court, London, WCIR 5DP, UK.

Melby, E.C., and Altman, N.H. (eds.): Handbook of Laboratory Animal Science. Vols. I, II, and III. Cleveland, Chemical Rubber Company Press, 1974.

Mitruka, B.M., et al.: Animals for Medical Research: Models for the Study of Human Disease. Huntington, NY, Robert E. Krieger Pub. Co., Inc., 1980.

Nutrient Requirements of Laboratory Animals, 3rd Ed. Washington, DC, National Academy of Sciences, 1978.

Orcutt, R.P., Otis, A.P., and Alliger, H.: Alcide℗: An alternative sterilant to peracetic acid. S. Sasaki et al.

(eds.): Recent Advances in Germfree Research, pp. 79–81, Tokai University Press, 1981.

Owen, D.G.: The effect of 'Alcide' on 4 strains of rodent coccidial oocysts. Lab. Anims., 17:267–269, 1983.

Pratt, D.: Alternatives to Pain in Experiments on Animals. New York, Argus Archives, 1980.

Regan, T., and Singer, P. (eds.): Animal Rights and Human Obligation. Englewood Cliffs, NJ, Prentice-Hall, Inc., 1976.

Rowan, A.N.: Of Mice, Models, and Men: A Critical Evaluation of Animal Research. Albany, State University of New York Press, 1984.

Ruitenberg, E.J., and Peters, P.W.J. (eds.): Laboratory Animals: Laboratory Animal Models for Domestic Animal Production. World Animal Science Series. Vol. C2. New York, Elsevier Science Publishers, Inc., 1986.

Short, D.J., and Woodnott, D.P. (eds.): The I. A. T. Manual of Laboratory Animal Practice and Techniques. Springfield, IL, Charles C Thomas, 1969.

Silverman, P.: Animal Behavior in the Laboratory. New York, Universe Books, Inc., 1978.

Singer, P.: Animal Liberation. New York, Avon Books, 1977.

Sperlinger, D. (ed.): Animals in Research: New Perspectives in Animal Experimentation. New York, John Wiley & Sons, 1981.

Universities Federation for Animal Welfare: The UFAW Handbook on the Care and Management of Laboratory Animals, 6th Ed. New York, Churchill Livingstone, Inc., 1987.

Williams, C.S.F.: Practical Guide to Laboratory Animals. St. Louis, The C.V. Mosby Co., 1976.

Zeitscrhrift für Versuchstierkunde. VEB Gustav Fischer Verlag, Villengang 2, Postfach 176, 6900 Jena, GDR.

Chapter 2
Biology and Husbandry

Chapter 2 covers selected topics in the biology and husbandry of rabbits, guinea pigs, hamsters, gerbils, mice, and rats. The sections are divided into ten categories: Origin and Description; Anatomic and Physiologic Characteristics; The Animal as a Pet; Housing; Feeding and Watering; Breeding; Disease Prevention; Public Health Concerns; Uses in Research; and Sources of Information. The references listed in each section may be found in university and public libraries or ordered from bookstores or publishers. Works included are for both scientist and layman, adult and child, and they vary greatly in quality and in price. The items are included because they either are currently in print (1988) or are of special interest to raisers and users of small animals. Further information about the six species included in this chapter is given, often in considerable detail, in the general reference works listed in Chapter 1.

THE RABBIT

The domestic or European rabbit, which can be housed indoors or outdoors and fed a readily available, pelleted feed, can be a pet, meat producer, or research subject. If rabbits are raised in clean cages with self-cleaning wire floors, receive adequate water and a complete, pelleted diet from hoppers, and are protected from predators, drafts, and temperature extremes, they will grow rapidly, reproduce well and live long, healthy lives.

Origin and Description

The domestic rabbit, *Oryctolagus cuniculus,* is a lagomorph of the family Leporidae. Domestic rabbits are descended from wild rabbits of western Europe and northwestern Africa, where wild *Oryctolagus* still exist. Rabbits have become feral in other areas of the world, most notably Australia. Wild rabbits are gregarious, burrowing, herbivorous, nocturnal or crepuscular animals bearing distant kinship with the ungulates. The young are born hairless and blind into hair-lined nests.

Several size, shape, and color variations, derived from centuries of selective breeding, constitute the breeds recognized by the rabbit breeders' associations. Representatives of the large breeds (6.4 to 7.3 kg or 14 to 16 lb) are the Giant Chinchilla and the Flemish Giant. Among the medium-sized breeds (1.8 to 7.3 kg or 4 to 14 lb) are the Californian and New Zealand rabbits. Small breeds (0.9 to 1.8 kg or 2 to 4 lb) include the Dutch and Polish breeds. The albino New Zealand is popular for meat production and research, Rex and Angora are used for their pelts, and the smaller breeds are used both as pets and research animals.

Oryctolagus cuniculus is the only genus of European or domestic rabbit. Hares (*Lepus*) and cottontails (*Sylvilagus*) are in different genera. Hares and rabbits are often incorrectly named, e.g., the jackrabbit is a hare and the Belgium hare is a rabbit. Rabbits are born naked and helpless, whereas hares are born furred and with eyes open. Fertile, cross-genera matings do not occur. Orphaned young hares and cottontails are occasionally kept as pets, but injury and mortality are common consequences of caging wild species.

Anatomic and Physiologic Characteristics

Rabbits have a well-developed nictitating membrane or third eyelid. During sleep or anesthesia this fold moves from the medial canthus across the cornea. Rabbits have a wide field of vision that reaches 190° for each eyeball. The wide pupillary dilation results in a light sensitivity approximately eight times that of man. Rabbit ears are highly vascularized organs that serve in heat regulation as well as in sound gathering. The ears are fragile and sensitive and should not be used for restraint.

The thoracic cavity is relatively small compared with the capacious abdominal cavity. The intestine is approximately 10 times body length and includes a large cecum and glandular stomach. The lymphoid appendix and sacculus rotundus (ileocecal tonsil) are prominent lymphatic organs associated with the rabbit's intestine. The fragile, almost birdlike skeleton comprises only 8% of the animal's body weight, whereas the cat's skeleton is 13% of its body weight. The long bones and lumbar spine,

engulfed in powerful muscle masses, are particularly susceptible to fracture.

The rabbit has a comparatively small heart, and the right atrioventricular valve is bicuspid instead of tricuspid, as occurs in other mammals. Rabbits have delicate, thin-walled veins that tear easily.

The inguinal canals, leading to the inguinal pouches, remain open for the life of the rabbit. The testes descend at 12 weeks. Rabbits have 8 to 10 mammary glands lying in broad bands from the throat to the groin. The milk is high in fat and protein.

Rabbit teeth are all open rooted (continuously growing). There are 2/1 incisors, 0/0 canines, 3/2 premolars, and 3/3 molars. Lagomorphs are distinguishable from rodents by the second pair of incisors in the upper jaw. Malocclusion and overgrowth are most likely to occur with the incisors, which grow 10 to 12 cm a year throughout life.

The neutrophil, especially when observed in suppurative lesions, resembles an eosinophil because of the numerous intracytoplasmic, eosinophilic granules. These cells are also known as pseudoeosinophils, heterophils, or amphophils. Neutrophils and lymphocytes are present in approximately equal numbers (30% to 70%). Basophils are more common in rabbits (2% to 7%) than in other mammals.

Rabbits as Pets

Behavior.   Rabbits make good pets and can be house trained. Older bucks and primiparous does with strong territorial instincts may bite people, but such biting is uncommon. They can, however, if improp-

Rabbit—Physiologic Values

The values listed below are approximations only and may not represent the normal range in a given population. Sources consulted are included among the comprehensive texts listed in Chapter 1 and the publications on specific species following the respective sections of Chapter 2.

Adult body weight: male (buck)	2–5 kg
Adult body weight: female (doe)	2–6 kg
Birth weight	30–80 g

Rabbit—Physiologic Values (*Continued*)

Body surface area (cm²)	9.5 (wt. in grams)$^{2/3}$
(See Paget 1965 ref., Chapter 3 bibliography)	
Rectal temperature	38.5–40.0° C
Diploid number	44
Life span	5–6 yr or more
Food consumption	5 g/100 g/day
Water consumption	5–10 ml/100 g/day or more
GI transit time	4–5 hr
Breeding onset: male	6–10 mo
Breeding onset: female	4–9 mo
Cycle length	induced ovulator
Gestation period	29–35 days
Postpartum estrus	none
Litter size	4–10
Weaning age	4–6 wk
Breeding duration,	1–3 yr
commercial	7–11 litters
Young production	2–4/mo
Milk composition	12.2% fat, 10.4% protein, 1.8% lactose
Respiratory rate	30–60/min
Tidal volume	4–6 ml/kg
Oxygen use	0.47–0.85 ml/g/hr
Heart rate	130–325/min
Blood volume	57–65 ml/kg
Blood pressure	90–130/60–90 mm Hg
Erythrocytes	4–7 × 10⁶/mm³
Hematocrit	36–48%
Hemoglobin	10.0–15.5 mg/dl
Leukocytes	9–11 × 10³/mm³
Neutrophils	20–75%
Lymphocytes	30–85%
Eosinophils	0–4%
Monocytes	1–4%
Basophils	2–7%
Platelets	250–270 × 10³/mm³
Serum protein	5.4–7.5 g/dl
Albumin	2.7–4.6 g/dl
Globulin	1.5–2.8 g/dl
Serum glucose	75–150 mg/dl
Blood urea nitrogen	17.0–23.5 mg/dl
Creatinine	0.8–1.8 mg/dl
Total bilirubin	0.25–0.74 mg/dl
Serum lipids	280–350 mg/dl
Phospholipids	75–113 mg/dl
Triglycerides	124–156 mg/dl
Cholesterol	35–53 mg/dl
Serum calcium	5.6–12.5 mg/dl
Serum phosphate	4.0–6.2 mg/dl

erly restrained, inflict painful scratches with their powerful rear limbs.

Rabbits protect themselves, as they have for eons, through acute senses, burrowing, challenges, and thumping. Otherwise, aggressive behavior is confined to the breeding cage, where pursuit, tail flagging, urination, squealing, and combat are exhibited. Sexually maturing rabbits (older than 3 months) often attack one another and therefore should be individually housed. Because wounding, pseudopregnancies, and infertility may occur in groups housed together, mature rabbits should be paired only at mating.

Life span.   A rabbit has a breeding life from approximately 4½ months to 3 to 4 years of age, an average life span of 5 to 6 years, and a possible longevity of 15 years. The breeding life spans 7 to 11 litters before the number of young per litter (7 or 8 for medium breeds) declines. This onset of senescence is probably due to a progressive endometrial fibrosis and consequent failure of ova to be implanted.

Restraint.   When a rabbit is carried a short distance, the neck skin is grasped with one hand and the rear quarters are supported with the other (Fig. 2); the small young rabbit can be grasped over the back (Fig. 3). For longer distances, the rabbit is placed on the bearer's forearm with its head concealed in the bend of the elbow (Fig. 4). When rabbits are dropped or otherwise improperly handled, they struggle and may break their backs or scratch the handler. When rabbits must be carried long distances, the use of carrier boxes, with a door and handle, is recommended. The animal should not be lifted or restrained by its ears.

Housing

Plans for rabbit hutches are available from libraries, feed companies, and extension agents. Assembled wire cages, with ancillary equipment, are available from farm supply stores or mail order houses advertising in publications of rabbit breeder associations or at conventions. Caging is also advertised in professional and trade journals

Fig. 2.   Two-hand grip for restraining rabbit. Rabbits improperly restrained will struggle and may luxate or fracture the lumbar spine.

Fig. 3.   One-hand grip for short carry of young or small rabbit. A one-hand grip may cause subcutaneous hemorrhaging and back damage from straining against carry.

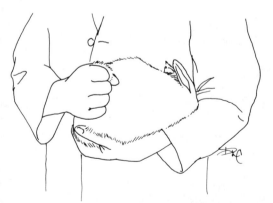

Fig. 4.   Method of restraint for carrying rabbit. The head and eyes are concealed in the elbow. (Adapted by permission from Reproduction and Breeding Techniques for Laboratory Animals by E. S. E. Hafez © 1970 by Lea & Febiger, Philadelphia, Pennsylvania.)

in laboratory animal science. Critical considerations in rabbit housing include structural strength, absence of sharp edges and rough surfaces, adequate size, wire construction (1 × 2.5 cm mesh on floor and 2.5 cm × 5 cm on sides), ease of cleaning, protection from climatic extremes, corrosion resistance, portableness, and provision of self-feeding "J" hoppers and sipper-tube or dewdrop water valves. Male rabbits are separated at weaning, whereas does are individually housed at 12 weeks of age.

Rabbits should be housed in areas commensurate with the animal's size and weight and in accordance with current regulations or guidelines. Rabbits up to 2 kg (4.4 lb) should have 0.14 m² (1.5 ft²) of floor space per animal; rabbits 2 to 4 kg (4.4 to 8.8 lb) require 0.28 m² (3 ft²), rabbits 4 to 5.4 kg (8.8 to 11.9 lb) require 0.37 m² (4 ft²), and rabbits over 5.4 kg (11.9 lb) require 0.46 m² (5 ft²) per animal. A doe with a litter requires an additional 0.19 m² (2 ft²) per cage.

Rabbits are housed indoors at temperatures between 4° C and 27° C (40° to 80° F), with a recommended average at a constant level between 16° C and 21° C (61° to 70° F), although much lower temperatures are well tolerated. Other characteristics of the rabbit's environment should include a moderate humidity (40% to 60%), draftless ventilation, regular manure removal or treatment to reduce ammonia levels, and a year-long light cycle of 14 to 16 hours for females, 8 to 10 hours for males, or an intermediate compromise. Shortening of daylight may bring on an autumnal sexual depression.

Hanging wire cages remain relatively clean for weeks, but hair and matted feces do accumulate. Dropping pans should be emptied and enclosed hutches should be cleaned weekly or as needed. Flies are attracted to rabbit droppings, and recently hatched fly larvae or maggots may give the false impression that the rabbits have "worms." Rabbit owners may bring these "worms" to experts to ask that they be identified and that treatment be recommended. Properly ventilated and dried manure

troughs can be left undisturbed for months. Cleaning of rabbit cages to remove infectious agents is particularly important before parturition, during weaning, before installing new stock, and after removing a sick animal.

Because of the plant diet, rabbit urine is turbid, has a pH of approximately 8.2, and varies from light yellow to deep orange or red brown, colors enhanced by dehydration and alkalinity. The carbonate and phosphate salt scale that accumulates on cage surfaces is difficult to remove.

Detergents, disinfectants, and lime-scale removers (acidic solutions such as vinegar or acids at pH 2) are applied with a stiff brush in routine cage cleaning. Acidic solutions damage floors and metals and should be used cautiously. Sodium hypochlorite (5%) mixed at 30 ml per liter of water (1 oz per qt) is an excellent disinfectant for small animal caging. Flaming may be necessary to remove hair and kill coccidial oocysts.

Rabbits can be housed outdoors if protected from cold (below 4° C or 40° F) by an enclosed area within the cage. In hot weather (above 29° C or 85° F), cages should be cooled by ventilation, shade, evaporator pads, or overhead sprinklers (care must be taken to prevent wetting of rabbits). Rabbits should be protected from rapid temperature changes (moving between outdoors and indoors), excessive drafts, predators, and insect and rodent vermin.

The overheating of bucks is a crucial consideration in the tropics or during the summer. At temperatures of 29° C and above, especially in heavily furred older rabbits in a high humidity, heat stress may lead to death or infertility. Does experience embryonic mortality under similar conditions. Heat lowers feed intake and increases water consumption, whereas cold increases both, if the water remains unfrozen.

Feeding and Watering

Rabbits should be fed a nutritious, wholesome, pelleted diet containing 12% to 22.5% crude fiber and 14% to 17% crude protein. Commercially available high-fiber diets (22.5%) are recommended for long periods of maintenance of rabbits during research or testing. Such diets allow for free choice feeding while still allowing control of weight gain and obesity. Higher-fiber diets reduce chances of hairball formation and diarrhea, but they are not recommended for use during periods of reproduction, lactation, and intended growth. The nutrient requirements of rabbits fed *ad libitum* are approximately 2500 kilocalories per kilogram (Kcal/kg) of diet during growth, gestation, or lactation, whereas maintenance requirements are 2100 Kcal/kg of diet. Growing and lactating rabbits, dwarf breeds, and rabbits in cold climates have higher caloric requirements. A once-daily feeding (limit feeding) of about 90 to 120 g (3 or 4 oz or $\frac{2}{3}$ cup) of pellets will maintain an adult medium-sized rabbit at a constant weight. Beginning with kindling, the feed quantity is increased approximately 30 g or 1 oz per day until, at peak lactation, the doe is consuming feed *ad libitum.* High energy levels may lead to increased susceptibility to mucoid enterotoxemia. Rabbits require dietary fiber levels that are greater than those required by other species. Rabbits do not digest more than 18% of dietary fiber in a single passage, but fiber has an important role as bulk. Rabbits are naturally coprophagous and, usually in the early morning, ingest large quantities of protein- and vitamin-rich feces directly from the anus. This soft, moist "night stool" may be seen in the stomachs of slaughtered rabbits. Wire floors do not inhibit coprophagy.

Medium-sized rabbits eat approximately 5 g of 2500 Kcal/kg feed per 100 g body weight per day. They should be supplied *ad libitum* with fresh, clean water. Does drink approximately 10 ml per 100 g body weight per day if nonpregnant and up to 90 ml per 100 g if lactating. Automatic waterers are commonly used with rabbits because they provide considerable savings in labor.

Supplementation of the pelleted diet with quality hay is practiced by some people with apparent success in reducing enteropathies and hair chewing. The provision of salt, table scraps, vitamins, antimicrobials, and such herbs as comfrey, pigweed, lamb's quarters, and tea without a medical reason (such as anorexia) is usually not advised.

Because the rabbit's intestinal microflora is sensitive to the intestinal milieu (e.g., osmolarity, pH, starch content), food changes, especially in the 4- to 12-week-old rabbit, should be graded over a 4- to 5-day period to allow adjustments in the balance of normal flora. Old and new feeds can be combined during the conversion.

Vitamin D toxicity may occur if rabbits are fed excessive levels of the vitamin. A rabbit suffering from vitamin D toxicity may exhibit impaired movement and appetite loss. Vitamin D is important in the regulation of calcium absorption and an excessive amount may result in soft tissue calcification. Vitamin A deficiency (under 10,000 IU/kg diet) and excess (over approximately 70,000 IU/kg diet) may lead to reproductive disturbances and hydrocephalus.

Breeding

Sexing Males have a rounded, protruding penile sheath with a rounded urethral opening; females have an elongated vulva with a slit opening. Stretching the perineum provides a clear view of the anogenital structures and everts the penis from its sheath. The anus is above the genitalia, and blind, perineal, or inguinal pouches containing scent glands lie immediately lateral to the anogenital line (Figs. 5 and 6). These scent glands are used to mark kits and also cage surfaces. In mature males, the hairless scrotal sacs are evident lateral and anterior to the penis. In immature males, the scrotal sacs are not readily apparent.

Other dimorphic sex characteristics, some more certain indicators than others, include the male's larger head and the heavy dewlap or fold of skin at the throat of many large females and of some males.

Estrous Cycle. Rabbits are induced ovulators releasing ova 9 to 13 hours following copulation. Does do not have an estrous cycle. Does do, however, have a 7- to 10-

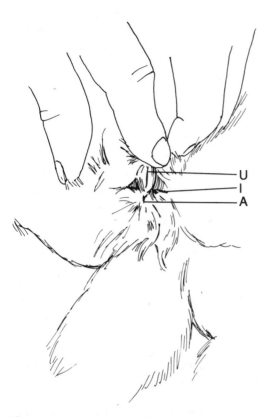

Fig. 5. External genitalia of young female rabbit. U = urogenital orifice; A = anus. Inguinal pouches (I) lie laterally to urogenital opening.

day period of receptivity followed by a short (1- to 2-day) period of inactivity during which a new "wave" of follicles replaces the atretic precursors. Receptivity is variably indicated by a swollen, reddened vulva and by a doe that stands and allows mounting. The vulvar sign is inconstant and should not be the sole criterion for pairing. Rabbits are receptive during periods of pregnancy and lactation with libido peaks about 26 and 39 days postpartum.

Ovarian activity, reflecting hypothalamic changes, is diminished as days shorten, which accounts for the proported loss of breeding activity and efficiency occurring in rabbits from late summer into winter. The use of an artificial 16:8 or 14:10 light:dark cycle, adequate caloric intake, and a slightly elevated ambient temperature may prevent or reduce this suppressive seasonal and environmental effect.

Small breeds (Dutch, Polish) may be mated at 4 to 5 months; medium breeds (New Zealand, California) at 4 to 6 months; and large breeds (Flemish, Checkered) at 5 to 7 months. Does mature earlier than bucks, which do not attain adult sperm levels until 6 to 7 months.

Artificial Insemination. Artificial insemination of does can be an economical and efficient method for rabbit breeding if practitioners are trained, equipment is correct, and strict hygiene practices are followed. A clean, lubricated artificial vagina is maintained at 37° C. Sperm are obtained from a practiced buck stimulated by a teaser doe. Such collections may be obtained 2 to 3 times each week. After obtaining the 0.3 to 1.0 ml of ejaculum, the gel plug is removed and the semen is kept at 37° C and used within 12 hours or is diluted in egg yolk-sodium citrate buffer. Insemination into a doe restrained on her back is effected through a smooth 15 cm × 0.6 cm plastic or glass tube inserted into the anterior vagina. Ten to 20 million live sperm are recommended for optimal conception. Additionally, several methods can be used to induce ovulation. Immediately following insemination, luteinizing hormone (1.25 mg in physiologic saline) can be given intra-

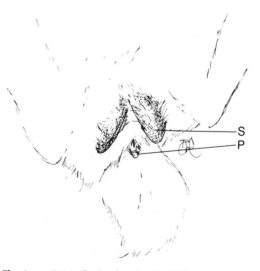

Fig. 6. External genitalia of male rabbit. S = scrotal sac; P = tip of penis. The penis overlies the anus.

venously into the marginal ear vein, a go-
nadotropin releasing hormone can be given
intramuscularly, or less desirably, a vasec-
tomized buck can be used for copulation and
the ovulatory stimulus.

Breeding Programs.   The optimal breeding
age for rabbits is between 4½ months and
3 years, although does may produce litters
for several more years. Breeding problems
are greatly reduced when the initial breeding occurs before the doe is 6 months old.
One buck is used for up to 25 to 30 does.
An intensive breeding schedule, requiring
good management, will result in up to 8
litters per doe per year, or approximately 55
kg (120 lb) of meat per year. Show rabbits
are usually limited to 3 or fewer litters per
year. In rabbitries where maximum production is important, does are rebred at 1
day to 4 weeks postpartum, and the young
are weaned when the doe's milk supply
drops at 4 to 5 weeks. Weaning at 5 to 8
weeks and rebreeding at 5 to 6 weeks are
commonly practiced in many commercial
rabbitries.

Receptivity is determined by carrying the
doe to the buck's cage and leaving the pair
together for 2 to 10 minutes. If mating does
not occur within several minutes, the doe
should be taken to another buck's cage. Care
should be taken to avoid injury to the buck.
Among healthy rabbits, approximately 85%
of copulations are followed by pregnancy;
when artificial insemination is used, the
percentage of successful breedings rises.

Pregnancy and Rearing.   The rabbit's gestation period lasts 29 to 35 days, with an
average of 31 or 32 days. Litters retained
longer than 34 days are usually stillborn and
may contain large or abnormal fetuses.
Pregnant rabbits are particularly prone to
fetal loss at 13 days, when placentation
changes from the yolk sac to the hemo-
chorial type, and at 23 days, when the tense,
rounded fetal structures are susceptible to
dislodgment.

Gentle palpation of the ventral abdomen
at 10 to 14 days into gestation will reveal
marble-sized swellings in the uterus (Fig.
7). The hand is placed between the hind
legs, just in front of the pelvis. With the
thumb on one side of the ventral abdomen
and the forefinger on the other, a light pressure is exerted as the hand is gently moved

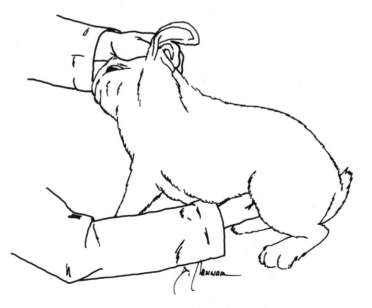

Fig. 7.   Palpation of rabbit for pregnancy. Rabbits are palpated for pregnancy between 10 and 14 days of
gestation.

back and forth in search of the uterine swellings.

During the last week of pregnancy (26 to 29 days), a clean nest box with a solid floor covered with bedding and a raised wire floor is placed into the doe's cage. The doe is then palpated once more. The nest box should be about 1 in. wider and longer than the doe's length and width in her normal sitting position (about 16 in. × 12 in. × 6 in. for a NZW doe). The wire overlay allows the young and the nest to remain dry. A doe may fail to use the nest box and may scatter newborn young on the wire if the bedding in the box is damp or contaminated with urine and feces or has an odor of disinfectants or predators. Subterranean or drop nest boxes with rims that are level with the main cage floor are preferred because fewer young are scattered. If scattering does occur, the young can return more easily to the safe and warm nest. Front loading nest boxes are also available.

A few days or hours before kindling, the doe removes hair from her forequarters, mammary region, and hips to construct a nest. Occasionally the nest is divided, scant, or too warm for hot weather, in which cases the nest should be reassembled or adjusted.

Rabbit litter size varies from 1 to 22 with an average of 7 or 8 in the medium-sized breeds. Smaller breeds and older does have fewer young. The young (kits) are usually born in the early morning within an hour or less, although litters divided by hours or days are known. Complications associated with delivery are rare, although some young may be stillborn or rejected from the nest. Difficult deliveries may be assisted by gentle manual pressure or with an intramuscular injection of 1 to 2 units oxytocin.

Cannibalism is rare in rabbits. When it occurs, it may be associated with dead or deformed young, a hyperexcitable, primiparous doe, placentophagy, environmental disturbances, or a low energy diet. Young rabbits may be abandoned if the nest is divided or inadequate, the young crawl out of the nest or are chilled, the doe has agalactia or mastitis, or the doe is disturbed.

Does do not retrieve young to the nest. Some producers raise rabbits over earthen pits stocked with earthworms to hasten decomposition of rabbit feces. The earthworms may attract snakes, opossums, skunks, raccoons, and other animals whose presence may disturb the doe and her young, causing death of the young. Toxins produced in association with uterine horn infections may decrease maternal instincts and cause the doe to scatter and neglect young.

Locking the doe out of the nest box except for a 15- to 30-minute period for morning nursing reduces litter mortality. The doe nurses the young once a day, usually in the early morning, and therefore she gives the erroneous impression of maternal neglect. Orphaned rabbits under 3 weeks of age can be fed warmed (29° C or 85° F) orphaned puppy milk formula by syringe, doll bottle, or gastric tube. The liquid is given slowly 2 or 3 times a day. Up to 5 ml per day are given the first few days; the amount is increased slowly to 15 ml the second week, and to 25 ml by the third week. Aspiration pneumonia, chilling, and diarrhea often result from hand feeding infant rabbits, and the mortality approaches 40% and more. The genitalia and anus of the young should be massaged daily with a wetted, warm cloth or cotton swab to stimulate defecation and urination.

Orphan or excess young may also be fostered to a lactating mother if the kits are under 2 weeks of age and are within 2 days of the age of the host litter. Masking odors during fostering is unnecessary. Mortality within the fostered litter is often high. Rabbit milk contains approximately 10% protein, 2.5% mineral, 12% fat, and 2% sugar. It can be roughly duplicated from 4% cow's milk by beating a whole egg in 250 ml (1 cup) of milk.

Pseudopregnancy may occasionally occur following sterile matings, excitement by nearby bucks, or mounting by other does. Ovulation is followed by a persistent corpus luteum, which lasts 15 to 17 days and secretes the progesterone that causes mammary development. Pseudopregnant does may build nests.

Disease Prevention

The general recommendations for husbandry and disease prevention listed in Chapter 1 should be observed, especially the removal of ammonia gas from the rabbit's environment. Common management practices with rabbits include periodic trimming of toenails and the exclusion of dogs, cats, and other animals, including insects, from the rabbitry. Catching and injuring limbs in the cage floor is a problem with rabbits and other small animals raised on wire. Rabbits trapped in this way may sit for hours or days with few signs of distress. There are no vaccines available at present for use in rabbits, although vaccines for the poxvirus disease myxomatosis are prepared as needed.

Special concerns are necessary for neonatal and weanling rabbits. The nest box should be sanitized prior to use and kept clean and dry during use. Ideally, the top of the nest box should be level with the floor of the doe's cage. Also, in contrast with common practice at weaning, the doe may be removed from the cage and the litter left in the familiar environment of nest box and maternal cage. This procedure may reduce the stress accompanying weaning.

Public Health Concerns

Diseases of major public health importance in domestic rabbits are rarely encountered. Such diseases include salmonellosis, tularemia, rabies, ringworm, tuberculosis, and toxoplasmosis. Rabies vaccinations (killed vaccine) are not given except in cases of high exposure probability, such as after a rabies outbreak in local skunks. Wounds caused by rabbit scratches on humans may be infected with *Pasteurella multocida*.

Uses in Research

Approximately 525,000 rabbits are used annually in research in the United States. Rabbits are used in a great variety of biomedical investigations, including studies of hydrocephalus, arteriosclerosis, hyperthermia, malignant lymphoma, teratology, cosmetics, ophthalmology, and reproductive physiology. Rabbits are frequently used as sources of hyperimmune sera and antibody.

Sources of Information

The American Rabbit Breeders Association, 1007 Morrisey Drive, Bloomington, IL 61701, has several publications on rabbits and cavies, including the magazine *Domestic Rabbits*.

Comprehensive works on the rabbit are *The Biology of the Laboratory Rabbit*, edited by S. Weisbroth, R. Flatt, and A. Kraus (1974, Academic Press, Inc., Orlando, FL 32887); *Rabbit Production*, 6th edition, by P.R. Cheeke, N.M. Patton, and G.S. Templeton (1987, Interstate Printers and Publishers, 19–27 N. Jackson Street, Danville, IL 61832-0594); *Domestic Rabbit Biology and Production* by L. Arrington and K. Kelley (1976, University Presses of Florida, 15 N.W. 15th Street, Gainesville, FL 32603); *The UFAW Handbook on the Care and Management of Laboratory Animals*, 6th edition, pp. 415–435, edited by T.B. Poole (1987, Churchill Livingstone Inc., 1560 Broadway, New York, NY 10036); and *The Guide to the Care and Use of Experimental Animals*, Vol. 2, Chapter 16, Rabbits, edited by H.C. Rowsell, et al. (1984, Canadian Council on Animal Care, 1105–151 Slater Street, Ottawa, Ontario K1P 5H3).

The Oregon State University (OSU) Rabbit Research Center, Oregon State University, Corvallis, OR 97331, publishes *The Journal of Applied Rabbit Research* and conducts studies of practical importance to rabbit raisers.

Other works on rabbits include:

Adams, C.E.: Artificial insemination in the rabbit: The technique and application to practice. J. Appl. Rabbit Res., 4:10–13, 1981.

Arvy, L., and More, J.: Atlas d'histologie du lapin (Histological Atlas of the Rabbit. *Oryctolagus cuniculus*). 1975. Librairie Maloine, 27 rue de l'Ecole de Médecine, 75006 Paris.

Barone, R., Pavaux, C., and Blin, P.C.: Atlas d'anatomie du lapin. Atlas of Rabbit Anatomy. 1973. Masson, 120 boulevard Saint-Germain, 75006, Paris.

Bennett, R.: Raising Rabbits Successfully. 1984. Williamson Publishing Co., Church Hill Road, P.O. Box 185, Charlotte, VT 05445.

Boy Scouts of America: Rabbit Raising. 1974. Boy Scouts of America, Eastern Distribution Center, 2109 Westinghouse Boulevard, P.O. Box 7143, Charlotte, NC 28217.

Brown, M.: Exhibition and Pet Rabbits. 1982. Saiga Publishing Co., Ltd., 51 Washington Street, Dover, NH 03820.

Casady, R.B., and Jawin, P.B.: Commercial Rabbit Raising. Shorey Publications, 110 Union Street, Seattle, WA 98111.

Cheeke, P.R. (ed.): Rabbit Feeding and Nutrition. 1987. Academic Press, Marketing Dept., 1250 6th Avenue, San Diego, CA 92101.

Collewijn, H.: The Oculomotor System of the Rabbit and Its Plasticity. (Studies of Brain Function Series: Vol. 5). 1981. Springer-Verlag New York, Inc., 175 Fifth Avenue, New York, NY 10010.

Committee On Animal Nutrition: Nutrient Requirements of Rabbits, 7th Ed. 1977. National Academy Press, Printing and Publishing Office, 2101 Constitution Avenue, Washington, DC 20418.

Craigie, E.H.: Laboratory Guide to the Anatomy of the Rabbit, 2nd Ed. 1966. University of Toronto Press, 33 East Tupper Street, Buffalo, NY 14203.

Downing, E.: Keeping Rabbits. 1984. Merrimack Publisher's Circle, 47 Pelham Road, Salem, NH 03079.

Faivre, M.I.: How to Raise Rabbits for Fun and Profit. 1973. Nelson-Hall, Inc., 111 North Canal Street, Chicago, IL 60606.

Hirschhorn, H.: All About Rabbits. 1974. T.F.H. Publications, 211 West Sylvania Avenue, Neptune, NJ 07753.

Kaissling, B., and Kriz, W.: Structural Analysis of the Rabbit Kidney (Advances in Anatomy, Embryology and Cell Biology: Vol. 56). 1979. Springer-Verlag New York, Inc., 175 Fifth Avenue, New York, NY 10010.

Kanable, A.: Raising Rabbits. 1977. Rodale Press, Inc., 33 East Minor Street, Emmaus, PA 18049.

Kaplan, H.M., and Timmons, E.H.: The Rabbit: A Model for the Principles of Mammalian Physiology and Surgery. 1979. Academic Press, Orlando, FL 32887.

Khera, K.D.: Maternal toxicity: A possible etiological factor in embryo fetal deaths and fetal malformation of rodent-rabbit species. Teratology, 31:129–153, 1985.

Longo, V.G.: Electroencephalographic Atlas for Pharmacological Research (Rabbit Brain Research: Vol. 2). 1962. Elsevier-North Holland Publishing Co., 52 Vanderbilt Avenue, New York, NY 10017.

Paget, G.E.: Clinical Testing of New Drugs, pg. 33. Revere Publishing Co., NY, 1965.

Paradise, P.R.: Rabbits. 1979. T.F.H. Publications, 211 West Sylvania Avenue, Neptune, NJ 07753.

Prince, J.H. (ed.): The Rabbit in Eye Research. 1964. Charles C Thomas, Publisher, 2600 South First Street, Springfield, IL 62717.

Robinson, D.: Encyclopedia of Pet Rabbits. 1979. T.F.H. Publications, 211 West Sylvania Avenue, Neptune, NJ 07753.

Sanchez, W.F., Cheeke, P.E., and Patton, N.M.: Effect of dietary crude protein level on the reproductive performance and growth of New Zealand White rabbits. J. Anim. Sci., 60:1029–1039, 1985.

Sawin, P.: Rabbit Raising. A.R. Harding Publishing Co., 2878 East Main Street, Columbus, OH 43209.

Scott, G.R.: Rabbit Keeping. 1979. David & Charles, Inc., P.O. Box 57, North Pomfret, VT 05053.

Shek, J.W., et al.: Atlas of the Rabbit Brain and Spinal Cord. 1985. S. Karger, 150 Fifth Avenue, Suite 1103, New York, NY 10011.

Silverstein, A., and Silverstein, V.: Rabbits: All About Them. 1973. Lothrop, Lee & Shepard Co., Division of William Morrow & Co., Inc., 105 Madison Avenue, New York, NY 10016.

Steinberg, P.: You and Your Pet: Rodents and Rabbits. 1978. Lerner Publishing Co., 241 First Avenue North, Minneapolis, MN 55401.

Urban, I., and Richard, P.: A Stereotaxic Atlas of the New Zealand Rabbit's Brain. 1972. Charles C Thomas, Publisher, 2600 South First Street, Springfield, IL 62717.

THE GUINEA PIG

The guinea pig or cavy is a docile rodent often encountered as a pet or research animal or, in its native Andes Mountains, as a culinary delicacy. Guinea pigs are notable for their fastidious eating habits, dependence on dietary vitamin C, long gestation period, and large and precocious young.

Origin and Description

The guinea pig, *Cavia porcellus,* is related to the wild and semidomesticated Caviae still found in the mountains and grasslands of Peru, Argentina, Brazil, and Uruguay. Andean Indians use the guinea pig as a special food much as North Americans use the turkey. Guinea pigs, which are hystricomorph (hedgehog-like) rodents related to chinchillas and porcupines, are gregarious, herbivorous, crepuscular animals. They were carried as curiosities to Europe in the sixteenth century, where they were selectively bred by fanciers.

The most common pet and laboratory variety is the English, American or short-haired guinea pig. The Duncan-Hartley and Hartley strains are representative lines of the English variety. Inbred strains 2 and 13 are popular research animals. Other varieties or breeds are the Abyssinian, with short, rough hair arranged in whorls or rosettes, and the Peruvian long hair or "rag mop" variety. Guinea pigs may be mono-colored, bicolored, or tricolored. The varieties interbreed, and a profusion of colors and hair coats is possible.

Anatomic and Physiologic Characteristics

This brief summary is intended only as an introduction to the several physical and functional aspects of guinea pigs.

Guinea pigs are tailless with compact stocky bodies and short legs. All guinea pig teeth are open rooted and erupt continuously. The teeth most commonly overgrown are the premolars. The dental formula of the guinea pig is 1/1 incisors, 0/0 canines, 1/1 premolars, and 3/3 molars.

In immature animals the functional thymus lies subcutaneously, on either side of the trachea, and within the neck, although in some individuals a portion extends caudally into the precardial mediastinum. The guinea pig's thymus is easily located and removed.

Dense populations of sebaceous glands are located circumanally and on the rump. Rodents, including guinea pigs, often walk or sit with these marking glands pressed on a surface.

Both male and female guinea pigs have two inguinal nipples, although mammary glands themselves are confined to the sow. Despite the apparent shortage of nipples, guinea pigs successfully raise litters of three, four, and more.

The sow has an intact, epithelial vaginal closure membrane except for the few days of estrus and at parturition. Both these events are signaled by the perforation of this membrane.

The large vesicular glands of the boar are bilateral, smooth, and transparent and extend approximately 10 cm into the abdominal cavity from their base in the accessory sex glands. To an uncertain observer these elongated organs resemble the uterus.

Kurloff's bodies are intracytoplasmic inclusions of glycoprotein composition found in mononuclear leukocytes. These cells proliferate during estrogenic stimulation and are found in highest numbers in the placenta, where they may function to protect fetal antigens from sensitized maternal lymphocytes and immune globulins. These cells are most prominent during pregnancy and may originate from the thymus gland.

The pubic symphysis, under the influence of the hormone relaxin, begins to separate during the latter half of gestation and continues parting until, at 48 hours prepartum, a palpable gap of about 15 mm is present. At parturition this opening may reach 22 mm. If first breeding for guinea pigs is delayed past 7 or 8 months, the symphysis separates less easily, and fat pads occlude the pelvic canal. Such impediments often lead to dystocia and death.

Guinea pig young are born fully furred with eyes open and teeth erupted. Within a few hours they are walking unsteadily and eating and drinking from pans. Food preferences are established within a few days of birth.

Guinea pigs are distinguished anatomically by large adrenal glands, a long colon (60% of the length of the small intestine versus only 16% in the rat), large tympanic bullae, and open inguinal canals.

Guinea pigs exhibit lethal bronchiolar smooth muscle contraction following histamine administration, and they are sensitive to the potentially lethal effects of bacterial proliferation in the cecum following stress or antibiotic administration. Mature sows are an excellent source of serum complement, the protein complex involved naturally in immune and endotoxin reactions and experimentally in serologic testing.

Cesarean derivation is relatively easy to accomplish with late-term guinea pigs. Neonates are precocious, nearly self-sufficient, and require little hand rearing if removed from the dam at birth. Although the young can eat or lap food from small bowls, bottle feeding for several days may be advantageous. Hand rearing requires regular stimulation of defecation and urination and the provision of a heat source for a week or so. Additionally, females allow young other than their own to nurse; thus, foster nursing can also be used to establish SPF colonies.

Endocrine control of gestation in the guinea pig is similar to that in the horse, monkey, and man in that the gestation period may be divided into trimesters of about 3 weeks' duration. In many instances pregnancy is maintained following ovariectomy at 25 or more days. The guinea pig is therefore an animal of choice for studying the effects of hormones and endocrine glands on pregnancy.

Guinea pigs produce antibodies to specific proteins, and production of anaphylaxis is an indicator of the presence or absence of small amounts of antigen. Unlike the rabbit or chicken, the guinea pig injected with antibodies is protected from anaphylaxis.

Guinea Pig—Physiologic Values

The values listed below are approximations only and may not represent the normal range in a given population. Sources consulted are included among the comprehensive texts listed in Chapter 1 and the publications on specific species following the respective sections of Chapter 2.

Adult body weight: male	900–1200 g
Adult body weight: female	700–900 g
Birth weight	70–100 g
Body surface area (cm²)	9.5 (wt. in grams)$^{2/3}$
Body temperature	37.2–39.5° C
Diploid number	64
Life span	4–5 yr
Food consumption	6 g/100 g/day
Water consumption	10 ml/100 g/day
GI transit time	13–30 hr
Breeding onset: male	600–700 g (3–4 mo)
Breeding onset: female	350–450 g (2–3 mo)
Cycle length	15–17 days
Gestation period	59–72 days
Postpartum estrus	fertile, 60–80% pregnancy
Litter size	2–5
Weaning age (lactation duration)	150–200 g 14–21 days
Breeding duration, commercial	18 mo to 4 years 4–5 litters
Young production (index per female)	0.7–1.4/mo
Milk composition	3.9% fat, 8.1% protein, 3.0% lactose
Respiratory rate	42–104/min
Tidal volume	2.3–5.3 ml/kg
Heart rate	230–380/min
Blood volume	69–75 ml/kg
Blood pressure	80–94/55–58 mm Hg
Erythrocytes	4.5–7.0 × 10⁶/mm³
Hematocrit	37–48%
Hemoglobin	11–15 g/dl

Guinea Pig—Physiologic Values (*Continued*)

Leukocytes	$7-18 \times 10^3/mm^3$
Neutrophils	28–44%
Lymphocytes	39–72%
Eosinophils	1–5%
Monocytes	3–12%
Basophils	0–3%
Platelets	$250-850 \times 10^3/mm^3$
Serum protein	4.6–6.2 g/dl
Albumin	2.1–3.9 g/dl
Globulin	1.7–2.6 g/dl
Serum glucose	60–125 g/dl
Blood urea nitrogen	9.0–31.5 mg/dl
Creatinine	0.6–2.2 mg/dl
Total bilirubin	0.3–0.9 mg/dl
Serum lipids	95–240 mg/dl
Phospholipids	25–75 mg/dl
Triglycerides	0–145 mg/dl
Cholesterol	20–43 mg/dl
Serum calcium	5.3–12 mg/dl
Serum phosphate	3.0–7.6 mg/dl

Guinea Pigs as Pets

Behavior. Guinea pigs, if gently handled, make good pets. They rarely bite or scratch but do scatter feed and bedding. Guinea pigs seldom climb or jump out of pens with open tops, and an 18-cm (7-in.) barrier will confine all but sexually mature males. They respond favorably to frequent handling and readily become conditioned to squeal before reward situations. They have a variety of vocalizations that correlate with different social behaviors.

Guinea pigs have certain behavioral characteristics that have important implications for both health and husbandry. Guinea pigs, as they mature, develop rigid habit patterns that must be accommodated if the animal is to thrive. Any changes in feed (taste, odor, texture, form), water, feeder, or waterer may cause the guinea pig to stop eating or drinking, sometimes, especially in pregnant females, with fatal consequences. Also, guinea pigs born onto wire learn to walk without mishap, but inexperienced animals placed onto wire often fall through the mesh and break or lacerate their limbs. Excited guinea pigs may stampede or "circle" and either trample the young or plunge to the floor. A rectangular cage or barriers within the cage reduce circling, but stampedes still occur.

Guinea pigs establish male-dominated hierarchies, and within these hierarchies subordinate animals are chewed and barbered. Although these hierarchical colony arrangements are usually stable, strange males placed together, especially in crowded conditions or in the presence of an estrous female, will fight.

Life span. The breeding life of laboratory housed guinea pigs is from 18 months to 4 years, and they have been known to live 8 years; however, they rarely survive in the home longer than 5 years, and their litter size is reduced to 1 or 2 per litter by 27 to 30 months of age.

Restraint. Guinea pigs are lifted by grasping under the trunk with one hand while supporting the rear quarters with the other hand (Fig. 8). Two-hand support is particularly important with adult and pregnant animals. The grasping hand should be

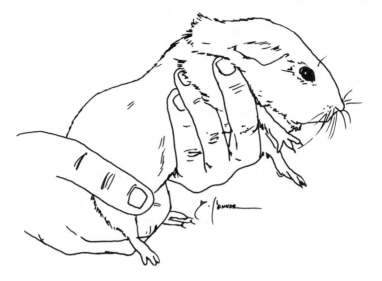

Fig. 8. Restraint of guinea pig. Hand placed under body supports animal without causing injury to thoracic viscera.

beneath the thorax and abdomen and the supporting hand under the rear feet or hind-quarters (Fig. 9). Injured lungs or liver may result from grabbing the animal around the thorax or abdomen.

Housing

Guinea pigs may be housed in colony pens on the floor, in tiered bins, or in large "shoe box" cages. Young and breeder animals will occasionally climb out of bins. Cages may be plastic, metal slat, or wire, but when guinea pigs not raised on wire are placed on wire, broken limbs often result. Guinea pigs on wire may also exhibit hair loss, foot pad ulcers, decreased production, and effects from stress.

Fig. 9. Restraint of pregnant guinea pig. Hand beneath rear quarters prevents struggling and supports heavy body.

Adult guinea pigs may be placed on "rat mesh" (0.85 cm² or 2 wires/in.), but some guinea pig breeders recommend a rectangular (1.2 × 3.8 cm) mesh to reduce leg injuries. Open bins should have sides 15 to 25 cm high, and covered cages should have sides at least 18 cm (7 in.) high. Adult animals should be provided with at least 652 cm² (101 in.²) and breeders with 1100 cm² (180 in.²) of floor space per animal.

Beddings of wood shavings, shredded paper, or other material of plant origin may be used.

Guinea pigs are housed at temperatures between 18° and 26° C (65° to 79° F), with the optimal ambient temperature at 21° C (70° F). Lower temperatures may lower the survival rate of newborns. High ambient temperatures without adequate air flow predispose to sterility and heat stress; low temperatures and wet bedding predispose to respiratory disease. The environmental humidity should be between 40% and 70%. A 12:12 light:dark cycle is commonly used with 10 to 15 air changes per hour.

Guinea pigs are messy housekeepers, and their urine is opaque, creamy yellow, and contains crystals. Cages, pens, and feeding receptacles should be cleaned and sanitized at least once a week. Satisfactory cleaning procedures include washing the cages with a detergent and hot water or with a disinfectant solution followed by a thorough rinse. A weakly acidic solution may be needed to remove the urine scale.

Feeding and Watering

Bowls, pans, and crocks should not be used for guinea pig waterers if solid-bottom cages with bedding are used because the waterers readily become contaminated with feces and bedding. Guinea pigs are notorious for chewing on, and playing with, sipper tubes or nipple valves and for quickly draining water bottles. Water then accumulates in cages. Although automatic watering can be used advantageously, guinea pigs may need to be shown how to drink from the valves. With guinea pigs it is important to use valves located outside the cage or other devices to divert spilled or leaked water to the outside of the cage. A further advantage of automatic waterers is that they ensure that guinea pigs are not without water for long periods of time, as may occur after they characteristically drain their water bottle. Automatic waterers must be checked frequently and cleaned and flushed periodically. Guinea pigs may block sipper-tube waterers by mixing dry feed and water in their mouths and passing the slurry into the sipper tube, thereby blocking the tube or causing it to drip. Guinea pigs readily learn to drink from sipper tubes; however, animals reared with water supplied in bowls or crocks may succumb to dehydration before "finding" water when it is provided by unfamiliar sipper tubes or nipple valves of automatic watering systems. Because guinea pigs may defecate or kick litter into their feed and water crocks, feeders and waterers should be suspended above the bedding.

Though herbivores, guinea pigs are fastidious eaters and may refuse to eat or drink if the feed or feeders are changed. They should receive a freshly milled, properly stored, pelleted, complete guinea pig diet. Whether this feed should be supplemented with hay or fresh greens is a matter of debate. Such supplement may improve production and growth, although pathogenic microorganisms may be introduced on the hay or greens. Adult guinea pigs will consume approximately 6 g feed and 10 to 40 ml water per 100 g body weight daily; these amounts, however, vary with ambient temperature, breeding status, food and water wastage, and relative humidity. Guinea pig chows are provided as small pellets that contain approximately 18% to 20% crude protein and 10% to 16% fiber. Large amounts of feed wastage are associated with wire-floor cages or use of feeders without inside lips.

Guinea pigs, like primates, lack an enzyme (L-gulonolactone oxidase) in the glucose-to-vitamin C pathway, and therefore they require dietary ascorbic acid at approximately 5 mg/kg body weight per day for maintenance and up to 30 mg/kg body weight per day for pregnancy. If ascorbic acid is not supplied in the feed, it can be added to the water (1 g/L), or each guinea pig may be fed approximately 1 small handful of cabbage or kale or ¼ orange daily. Carrots and lettuce are not good sources of vitamin C. The practice of supplementing complete pelleted rations with fruits, vegetables, and hays is to be discouraged because it may lead to digestive disturbances and bacterial contamination culminating in disease. The activity of vitamin C decreases as much as 50% in a 24-hour period in water in an open crock and even faster if metal or organic material is present or the room temperature is elevated. Feed containing ascorbic acid should be properly stored (cool and dry) and used within 90 days of milling. Guinea pigs should not be fed diets indicated for other species. Rabbit food, for example, contains no ascorbic acid and, for guinea pigs, excess levels of vitamin D.

Breeding

Sexing. The male guinea pig has no break in the ridge between the urethral orifice and the anus. The female has a shallow, U-shaped break (the vaginal membrane) in this ridge. The intact vaginal membrane is revealed by placing the thumb and forefinger on either side of the genital ridge. The fingers are drawn apart gently to reveal the vaginal closure membrane (Fig. 10). In males the penis and testes may be palpated or the penis may be extruded by gentle digital pressure (Fig. 11).

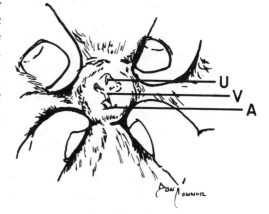

Fig. 10. External genitalia of female guinea pig. U = urethral orifice; V = vaginal closure membrane; A = anus.

Estrous Cycle. The estrous cycle in guinea pigs lasts 15 to 17 days. Most sows have an estrus 2 to 15 hours postpartum, at which time 60% to 80% will become pregnant if mated. The estrous period itself lasts about 24 to 48 hours, and the female accepts the male for 6 to 11 hours.

Sows reach puberty as early as 5 to 6 weeks of age and boars at 8 weeks, but 68 to 70 days is a more realistic average. They are not paired until the female weighs approximately 400 g (2 to 3 months) and the male 650 g (3 to 4 months). Because pairs are often housed together, detection of estrus is not necessary unless timed matings are desired. In estrus the sow exhibits lordosis and mounts other sows. If timed pregnancies are desired, mating can be detected by the observation of the perforated vaginal

Fig. 11. External genitalia of male guinea pig. Digital pressure will cause extrusion of penis. A = anus; P = penis; S = scrotal sac.

membrane, vaginal plug, or sperm in a vaginal smear.

Breeding Programs. One boar is housed in an adequate-sized pen (652 cm^2 or 101 in.^{2} per animal) with one to ten sows. Size requirements of the pen will limit the number of females housed with a boar. In intensive harem-style breeding systems, which utilize the postpartum estrus, the sows and boar remain together for their entire breeding life (approximately 18 months), and the young are removed at weaning. In a nonintensive system, the sows are removed for delivery and are rebred after weaning. Another variation is to return the sow alone to the harem for several hours to permit postpartum breeding. Pregnant sows may be separated or remain together, although in harems, a sow's milk may be stripped by the larger young of other sows.

Pregnancy and Rearing. The gestation period in guinea pigs lasts from 59 to 72 days, with an average between 63 and 68 days. The length of gestation depends on the size of the litter. Shorter gestation periods are associated with larger litters. Sows may double in weight during gestation. Fetuses may be palpated from 4 to 5 weeks to term, and beginning the last week of pregnancy, the pubic symphysis gradually separates up to 22 mm. A separation of 15 mm indicates parturition within 48 hours. Abortions and stillbirths are common.

No nests are built. Guinea pigs may deliver from 1 to 6 young, but 3 to 4 is an average litter. Litters of 5 or more retained over 72 days are usually born dead. Pseudopregnancy is rare in guinea pigs, but, when it does occur, it lasts approximately 17 days.

Newborn are precocious (haired, eyes open, teethed) at birth, weigh 60 to 100 g, and can begin eating solid food within a few days. Young under 50 or 60 g usually die. Young may also be trampled if the adults are crowded or become excited. If the sow abandons the young before they are a week old, or if the young are weaned early, mortality is high. The problems with ear chewing and trampling are reduced if the young and sow are removed to a nursery before or soon after birth. Young guinea pigs are weaned at 14 to 28 days when they weigh between 150 g and 200 g.

Because the young will begin eating solid food during the first few days postpartum, hand rearing is not as difficult as with the altricial species, i.e., rats, mice, and rabbits. One should not attempt to force-feed newborn orphaned guinea pigs during the first few hours of life, because the neonates will not be hungry until 12 to 24 hours of age. During the first few days, the neonates may be fed guinea pig chow softened with water or cow's milk. If given the opportunity, the young will nurse mothers other than their own, often to the detriment of their smaller cagemates.

Disease Prevention

The general husbandry and disease prevention admonitions listed in Chapter 1 should be followed for guinea pigs. Guinea pigs scatter their bedding into feeders and crocks, are susceptible to *Bordetella* pneumonia, have an absolute requirement for dietary ascorbic acid, and are fastidious eaters. If the food is changed, guinea pigs may refuse to eat. Because of the *Bordetella* susceptibility, guinea pigs should not be housed with rabbits, cats, dogs, and other species that carry *Bordetella* subclinically.

Public Health Concerns

Diseases of public health significance in guinea pigs are rare. Guinea pigs may carry bacteria (*Bordetella, Salmonella, Yersinia pseudotuberculosis, Streptococcus*) that are potential human pathogens, but transfer of these infections to humans would be most unusual. Allergies to guinea pig dander are known.

Uses in Research

Approximately 525,000 guinea pigs are used annually in research in the United States. Guinea pigs are used in projects involving anaphylaxis, delayed hypersensitivity, genetics, ketoacidosis, several aspects of nutrition, optic neuropathies, amoebiasis, scorbutus, leukemia, allergic encephalo-myelitis, ulcerative colitis, immunology, auditory function, and infectious diseases including tuberculosis. Complement commonly obtained from guinea pig serum is widely used in complement fixation tests to diagnose infectious diseases. Biochemical and pharmacologic tests are also commonly carried out in guinea pigs.

Sources of Information

Comprehensive works on the guinea pig are *The Biology of the Guinea Pig*, edited by J.E. Wagner and P.J. Manning (1976, Academic Press, 111 Fifth Avenue, New York, NY 10003); *The UFAW Handbook on the Care and Management of Laboratory Animals*, 6th edition, pp. 393–410, edited by T.B. Poole (1987, Churchill Livingstone Inc., 1560 Broadway, New York, NY 10036); and *The Guide to the Care and Use of Experimental Animals*, Vol. 2, Chapter 13, Guinea Pigs, edited by H.C. Rowsell, et al. (1984, Canadian Council on Animal Care, 1105–151 Slater Street, Ottawa, ONTARIO K1P 5H3).

Other works on the guinea pig include:

Axelrod J.: Breeding Guinea Pigs. 1980. T.F.H. Publications, 211 West Sylvania Avenue, Neptune, NJ 07753.

Bleier, R.: The Hypothalamus of the Guinea Pig: A Cytoarchitectronic Atlas. 1984. University of Wisconsin Press, Madison, WI.

Cooper, G., and Schiller, A.L.: Anatomy of the Guinea Pig. 1975. Harvard University Press, 79 Garden Street, Cambridge, MA 02138.

Elward, M., and Whiteway, C.E.: Encyclopedia of Guinea Pigs. 1980. T.F.H. Publications, 211 West Sylvania Avenue, Neptune, NJ 07753.

Paget, G.E.: *Clinical Testing of New Drugs*, pg. 33. Revere Publishing Co., NY, 1965.

Ragland, K.: Guinea Pigs. 1979. T.F.H. Publications, 211 West Sylvania Avenue, Neptune, NJ 07753.

THE HAMSTER

The golden or Syrian hamster is popular both as a pet and as a research animal. These short-tailed, stocky animals weigh about 100 g and are known for their short gestation period, ease of domestication, cheek pouches, immunologic tolerance, and ability to escape confinement. Hamsters do well on pelleted rodent chows or high-protein rodent diets and usually remain healthy and active throughout their short lives.

Origin and Description

The golden hamster, *Mesocricetus auratus*, is a rodent of the family Cricetidae. Wild golden hamsters occur within a limited range in the Middle East, where destruction of the territory and predation by owls are threatening the species' existence.

Hamsters in Syria, where they are called the Arabic equivalent of "originator of saddle bags" because of their distended cheek pouches, live independent lives in deep burrows, where several pounds of grain may be stored. Hamsters are nocturnal but have short periods of diurnal activity. Most hamsters used in research and all sold as pets are descended from 1 male and 2 female siblings that survived capture and domestication in 1930.

Color varieties of domestic hamsters include the wild type or reddish-brown, cinnamon, cream, white, piebald, and the long-haired "teddy bears." The smaller, dark brown Chinese hamster (*Cricetulus griseus*) and the larger common or European hamster (*Cricetus cricetus*) are used in biomedical research.

Anatomic and Physiologic Characteristics

The characteristics listed and briefly described in this section were obtained from Magalhaes' "Gross Anatomy" and Robinson's "General Aspects of Physiology" in *The Golden Hamster*, edited by R. Hoffman, et al. (The Iowa State University Press, 1968); and from Barnes' *Special Anatomy of Laboratory Mammals* (Department of Anatomy, School of Veterinary Medicine, University of California, Davis, 1971). This brief summary is intended only as an introduction to the several physical and functional aspects of hamsters.

Although hamsters are not true hibernators, with shortened day length and temperatures of 5° C to 15° C they gather food and pseudohibernate for extended periods of time.

Golden hamsters in pet or laboratory colonies exhibit a high degree of immunologic tolerance to both homografts and heterografts of both normal and neoplastic tissue. This tolerance is even more pronounced in the tissue of the cheek pouch.

Cheek pouches are oral cavity evaginations reaching alongside the head and neck to the scapulae. These pouches, lined with stratified squamous epithelium, are thin walled, highly distensible, well vascularized, and markedly deficient in lymphatics. The cheek pouches serve the hamster as a food transportation mechanism and man as immunologically privileged sites for research use.

On either flank, usually buried in hair and always more prominent in males, lie dark brown patches known as flank glands. These sebaceous glands are used to mark territory and contribute to the stimuli determining mating behaviors.

The hamster stomach is separated into squamous and glandular divisions by incisurae of the greater and lesser curvatures. The forestomach, anatomically linked to the esophagus, is the site of some pregastric fermentation.

At normal pH of 8, hamster urine is turbid and milky because of the large number of small crystals it contains.

Unlike those in other rodents, the adrenal glands in the male hamster are larger than those in the female. Hamsters exhibit the same often fatal reaction to antibiotics seen in guinea pigs, but they are far more shock resistant than that species. Newborn hamsters have fully erupted incisor teeth. The incisors are used in nursing to strip milk from the teats of the dam while suckling. Hamsters have 3/3 molars with fixed roots and 1/1 open rooted incisors. There are no canine or premolar teeth.

Hamster—Physiologic Values

The values listed below are approximations only and may not represent the normal range in a given population. Sources consulted are included among the comprehensive texts listed in Chapter 1 and the publications on specific species following the respective sections of Chapter 2.

Adult body weight: male	85–130 g
Adult body weight: female	95–150 g
Birth weight	2 g
Body surface area (cm^2)	10.5 (wt. in grams) $^{2/3}$
Body temperature	37–38° C
Diploid number	44
Life span	18–24 mo
Food consumption	> 15 g/100 g/day
Water consumption	> 20 ml/100 g/day
Breeding onset: male	10–14 wk
Breeding onset: female	6–10 wk
Cycle length	4 days
Gestation period	15–16 days
Postpartum estrus	infertile
Litter size	5–9
Weaning age	20–25 days
Breeding duration, commercial	10–12 mo 5–7 litters
Young production	3 mo
Milk composition	12.0% fat, 9.0% protein, 3.4% lactose
Respiratory rate	35–135/min

Hamster—Physiologic Values (*Continued*)

Tidal volume	0.6–1.4 ml
Oxygen use	0.6–1.4 ml/g/hr
Heart rate	250–500/min
Blood volume	78 ml/kg
Blood pressure	150/100 mm Hg
Erythrocytes	$6–10 \times 10^6/mm^3$
Hematocrit	36–55%
Hemoglobin	10–16 g/dl
Leukocytes	$3–11 \times 10^3/mm^3$
Neutrophils	10–42%
Lymphocytes	50–95%
Eosinophils	0–4.5%
Monocytes	0–3%
Basophils	0–1%
Platelets	$200–500 \times 10^3/mm^3$
Serum protein	4.5–7.5 g/dl
Albumin	2.6–4.1 g/dl
Globulin	2.7–4.2 g/dl
Serum glucose	60–150 mg/dl
Blood urea nitrogen	12–25 mg/dl
Creatinine	0.91–0.99 mg/dl
Total bilirubin	0.25–0.60 mg/dl
Cholesterol	25–135 mg/dl
Serum calcium	5–12 mg/dl
Serum phosphate	3.4–8.2 mg/dl

Hamsters as Pets

Behavior.   Hamsters have an undeserved reputation for pugnacity, a characteristic usually exhibited only after rough handling or a sudden disturbance. Hamsters, even those recently captured in the wild, are readily tamed, although they remain remarkably adept at chewing on and escaping from their cages. Escaped hamsters will not return to their cages, as will rats and gerbils. The use of live traps or a ramp leading into a bucket placed against a room wall may be necessary to catch an escaped hamster. Hamsters are most active during the dark and enjoy wheel-running activity. Pregnant females are reported to run up to 8 km per day.

Except for the few hours of estrus occurring once during the 4-day cycle, the sexually unreceptive female will usually attack a recently introduced male. Following copulation, the male is frequently removed from the breeding cage. Hamsters may be grouped by sex (males and females separate) in holding cages. Hamsters fight less often if housed together at the time of weaning or before sexual maturity, awakened simultaneously from anesthesia in a neutral arena, or provided with bolt holes or hiding places. Females may fight other females, and less often, males may fight other males.

The female hamster, when excited, is able to conceal her newborn litter in her cheek pouches. Later, when the perceived danger has passed, the mother returns the young to the nest; however, young may suffocate in the cheek pouch. Females may also conceal feed or bedding and other objects in their cheek pouches. Females with litters should be disturbed as little as possible.

Life span.   Female hamsters produce 5 to 6 large litters and then produce smaller litters for another 6 to 9 months, or until the dams are 12 to 15 months old. Males are used for commercial breeding until 12 months of age, but pet males may serve longer. Hamsters usually have a life span of 18 to 24 months, but older individuals are frequently reported.

Restraint. Hamsters are not naturally aggressive; however, they frequently feign biting. Nonetheless, care must be taken to avoid being bitten. Forceps should not be used to handle hamsters or other rodents. Use of forceps leads to aggressive tendencies, reduced reproduction, and abortion. Hamsters may be picked up in a small can (Fig. 12), by the loose skin of the neck (Fig. 13), by cupping the hands (Fig. 14), or by gripping over the back (Fig. 15). A protective glove is used if the animals are unaccustomed to handling or if the handler is afraid of or allergic to hamsters. Hamsters may bite if roughly handled, startled, injured, or awakened. When hamsters are restrained with a glove, extra care must be taken to avoid rough handling. Factors affecting hamster restraint are the loose skin, which must be gathered to immobilize the animal, the tendency to bite when startled, and the predisposition to sleep deeply and then awaken suddenly.

Housing

Several acceptable types of hamster cages are available, some equipped with wheels,

Fig. 13.   Scruff-of-the-neck grip for picking up and restraining hamster. Because of the cheek pouches, a hamster has ample loose skin about the neck.

Fig. 14.   Two-hand technique for picking up and restraining hamster. (Adapted by permission from The Golden Hamster by R. A. Hoffman, P. F. Robinson, and H. Magalhaes, © 1968 by the Iowa State University Press, Ames, IA.)

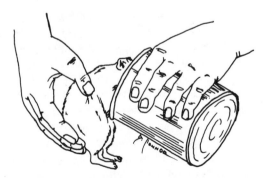

Fig. 12.   Picking up and restraining hamster with small container. (Adapted by permission from The Golden Hamster by R. A. Hoffman, P. F. Robinson, and H. Magalhaes, © 1968 by the Iowa State University Press, Ames, IA.)

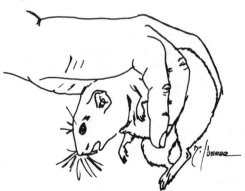

Fig. 15.   One-hand hold for restraining hamster. The thumb and third finger grasp the body. (Adapted by permission from The Golden Hamster by R. A. Hoffman, P. F. Robinson, and H. Magalhaes, © 1968 by the Iowa State University Press, Ames, IA.)

tunnels, and nest boxes. In research colonies, hamsters are usually housed in plastic, solid-bottom cages (Fig. 16) with deeply piled wood shavings for bedding. Such components are also desirable in pet hamster homes. Because hamsters have blunt noses, they may have difficulty eating from the slotted, sheet metal hoppers used for mice and rats and from wire mesh feeders. When the slots are large enough for the hamster to eat through comfortably, it will nevertheless pull the food pellets into the cage. Hamsters chew plastic, wood, and soft metals and will readily escape from poorly secured or constructed cages. Transport boxes are usually lined with mesh or light metal screening.

An adult hamster (over 100 g) requires a floor area of at least 123 cm² (19 in.²) and a cage height of at least 15 cm (6 in.). A female breeding hamster should have 790 cm² (150 in.²) of floor space. Adult hamsters are housed at temperatures between 18° and 29° C with a suggested level of 18° to 26° C (65° to 79° F). The young are maintained at 22° to 24° C (71° to 75° F). The relative humidity should be between 40% and 70%. The 12:12 light:dark cycle is most often used, but other lighting arrangements have been used successfully.

Hamsters produce little waste or odor, but in research colonies, hamster cages are changed once or twice weekly, except when neonates are present. Cages are sanitized with hot (82° C or 180° F) water with or without a detergent or with a nontoxic disinfectant and a thorough rinse. Bottles and hoppers are cleaned at the same time as the cages.

Feeding and Watering

Although the nutritional requirements of the omnivorous hamster have not been specifically determined, a pelleted rodent diet containing approximately 16% protein and 4 to 5% fat is conveniently used until better information is available. Protein deficiency may cause alopecia, while dietary fat above 7 to 9% may increase mortality. Laboratory rodent feeds are often repackaged in small, unlabeled bags and sold in pet stores. The low-protein but attractively wrapped "treats" may not be adequate diets for growth and reproduction.

Dams with litters should receive their feed directly on the floor, as preoccupation with hopper-bound pellets may result in neglect of the young. Young hamsters begin gnawing solid food and drinking water at 7 to 10 days of age; therefore, the sipper tube should extend low in the cage but not into the bedding. Feeder slots should be greater than 11 mm wide. Hamsters eat 5 to 12 g feed (about 2 tsp) and drink 10 ml (½ oz) water per 100 g body weight daily.

Breeding

Just as hamsters in the wild have a hibernal sexual quiescence, laboratory and pet hamsters also have a normal seasonal breeding quiescence in the winter months. During this time, fecundity drops and litter mortality increases. Provision of a constant temperature (22° to 24° C) and 12 to 14 hours of light reduces some of this effect.

Sexing. Viewed from above, the rear margin of the male hamster is rounded because of the scrotal sacs (Fig. 17), and the female posterior is pointed toward the tail (Fig. 18). Gentle pressure applied to the lower abdomen will cause the testes to protrude. Males have a greater anogenital dis-

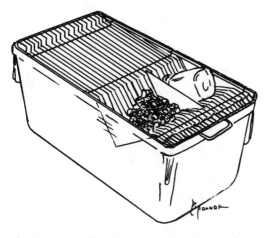

Fig. 16. Familiar "shoe-box" style of cage for housing laboratory rodents.

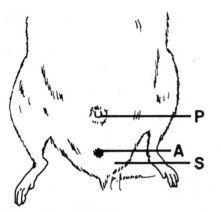

Fig. 17. External genitalia of male hamster. P = tip of penis; S = scrotal sac; A = anus. (Adapted by permission from The Golden Hamster by R. A. Hoffman, P. F. Robinson, and H. Magalhaes, © 1968 by the Iowa State University Press, Ames, IA.)

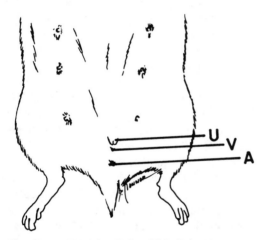

Fig. 18. External genitalia of female hamster. U = urethral orifice; V = vaginal orifice; A = anus. (Adapted by permission from The Golden Hamster by R. A. Hoffman, P. F. Robinson, and H. Magalhaes, © 1968 by the Iowa State University Press, Ames, IA.)

tance than females and a pointed genital papilla with a round opening. Male hamsters have prominent, dark marking glands in the skin of both flanks. These glands are difficult to see without shaving the animal.

Pregnancy and Rearing. The absence of estral discharge at 5 and 9 days postmating and the observation of a distended abdomen and rapid weight gain at 10 days are signs of pregnancy. The hamster's gestation period lasts 15½ to 16 days, unless there is dystocia or other difficulty. Before delivery,

the female becomes active, restless, and bleeds from the vulva. In good breeding colonies, litters usually contain from 5 to 9 young. Infertile matings or crowding of females may produce an occasional pseudo-pregnancy of 7 to 13 days.

Hamsters, pregnant or not, build nests, but more nests are built when temperatures are cool. Active nest building may enhance hormonal functions associated with parturition and lactation. Nest material, such as paper tissues, should be provided to pregnant hamsters several days prior to parturition. The opaque nesting material seems to give the dam with a newborn litter a certain amount of security and privacy when clear plastic cages are used. Transfer of the nest containing neonates reduces disturbances that may occur when changing cages. As neonatal hamsters are very immature and require a warm environment, the female utilizes the maternal nest.

Estrous Cycle. Golden hamsters become sexually mature at 32 to 42 days, but breeding is usually delayed until the female is 6 to 10 weeks and the male 10 to 14 weeks old (85–130 g).

Hamsters have an estrous cycle lasting about 4 days. Postpartum estrus is usually anovulatory. A fertile estrus follows weaning by 2 to 18 days.

As estrus approaches in a sexually mature female, a thin, stringy, translucent, cobweb-like mucus can be drawn from the vulva. On the morning following estrus, an opaque, tenacious, stringy mucus appears. A receptive, nonbelligerent approach of the female toward the male during early evening portends mating.

Breeding Programs. In a hand-mating scheme, the female hamster is placed into the male's cage about 1 hour before dark, and the pair is observed for mating or fighting. The male is removed following either outcome. This method is considered too labor intensive for large production colonies.

Monogamous pair mating involves placing a prepubital male and female together permanently. Although more males are kept with this system, it is labor efficient. Litter intervals are usually 35 to 40 days.

In sequential monogamy mating, 7 females are rotated through a male's cage in series at weekly intervals. With this system, after weaning of the litter, the female is returned to the male's cage for 1 week every seventh week. This method is labor efficient but results in injured or "burned-out" males that are either rested, replaced, retired, or destroyed.

With group or harem mating, 1 to 5 males are placed with a larger number of females. Seven to 12 days after pairing, the pregnant females are removed and individually housed until after weaning. Fighting occurs when females are returned to the cage.

Litter abandonment and cannibalism, common during the first pregnancy and the first week postpartum, may occur if the nest is scant, disturbed, or too visible; if the young are born onto wire, handled, or bite the mother while nursing; or if the diet is inadequate, the litter large or small, the mother is disturbed, or agalactia occurs. Unless efforts are made to disguise foreign odors, the young should not be handled until 7 days postpartum, especially if the female is primiparous. To avoid disturbing the neonates, the female should be supplied with a week's food and bedding at 13 days postmating. This allows the mother to nest undisturbed, especially if some feed is placed on the floor of the cage and the water bottle tube can be reached by the young. Weaning occurs at 20 to 25 days postpartum.

Fostering is rarely successful with hamsters, as both adopted and natural litters may be cannibalized. Cross-fostering onto dams of other rodent species is also impossible because of the extreme immaturity of hamster young and their sharp incisors. Hand rearing has not been successful. For these reasons hamsters have never been derived germ free.

Disease Prevention

The general husbandry and disease prevention measures mentioned in Chapter 1 should be observed. Among major husbandry problems are fighting, cannibalism of young, escape from cages, susceptibility to climatic changes, and food and water deprivation.

Public Health Concerns

Lymphocytic choriomeningitis (LCM) has received attention because of the 1974–1975 outbreak in man, but the severe forms of LCM are rare, even in humans at risk. Salmonellosis and hymenolepid tapeworm infections are hamster diseases of potential public health significance. The possibility of rabies occurring in pet or research hamsters is extremely unlikely as they are not apt to be exposed to wild animals in the rabies chain.

Uses in Research

Approximately 475,000 hamsters are used annually in research in the United States. Hamsters are used in investigations of congenital anomalies, dental caries, microcirculation, protozoal infections, gerontology, behavior, histocompatibility, infectious diseases, respiratory neoplasia, muscular dystrophy, cardiomyopathy, gallstones, amyloidosis, pancreatic neoplasia, cestodiasis, and, in the Chinese hamster, diabetes mellitus.

Sources of Information

Comprehensive works on the hamster are *The UFAW Handbook on the Care and Management of Laboratory Animals*, 6th edition, pp. 377–392, edited by T.B. Poole (1987, Churchill Livingstone Inc., 1560 Broadway, New York, NY 10036); *Guide to the Care and Use of Experimental Animals*, Vol. 2, Chapter 15, Hamsters, edited by H.C. Rowsell, et al. (1984, Canadian Council on Animal Care, 1105–151 Slater Street, Ottawa, ONTARIO K1P 5H3); *The Hamster: Reproduction and Behavior*, edited by H.I. Siegel (1985, Plenum Publishers, 233 Spring Street, New York, NY 10013); *The Golden Hamster: Its Biology and Use in Medical Research*, edited by R. Hoffman, et al. (1968, The Iowa State University Press, 2121 South State Avenue, 112C Press Office, Ames, IA 50010); and *The Hamster in Biomedical Research*, edited by G.L. Van Hoosier, et al. (1987, Academic Press, Inc., Orlando, FL 32887)

Other works on hamsters include:

Ostrow, M.: Breeding Hamsters. 1982. T.F.H. Publications, 211 West Sylvania Avenue, Neptune, NJ 07753.

Reznik, G., et al.: Clinical Anatomy of the European Hamster *Cricetus cricetus*. State Mutual Book and Periodical Service, Ltd., 521 Fifth Avenue, 17th Floor, New York, NY 10017.

Roberts, M.F.: Teddy Bear Hamsters. 1974. T.F.H. Publications, 211 West Sylvania Avenue, Neptune, NJ 07753.

Silverstein, A., and Silverstein, V.: Hamsters: All About Them. 1974. Lothrop, Lee & Shepard, Division of William Morrow & Co., Inc., 105 Madison Avenue, New York, NY 10016.

Streilein, J.W., et al., Editors. Hamster Immune Responses in Infectious and Oncologic Diseases (Advances in Experimental Medicine and Biology Series: Vol. 134). 1981. Plenum Publishers, 233 Spring Street, New York, NY 10013.

Symposium on the Syrian Hamster in Toxicology and Carcinogenesis Research, Boston, November 30–December 2, 1977. The Syrian Hamster in Toxicology and Carcinogenesis: Proceedings. F. Homburger, Editor (Progress in Experimental Tumor Research: Vol. 24). 1979. S. Karger, 150 Fifth Avenue, Suite 1103, New York, NY 10011.

Zim, H.S.: Golden Hamsters. 1951. William Morrow & Co., Inc., 105 Madison Avenue, New York, NY 10016.

Paget, G.E.: *Clinical Testing of New Drugs*, pg. 33. Revere Publishing Co., NY, 1965.

THE GERBIL

The Mongolian gerbil is a curious, nearly odorless, friendly rodent distinguished by its monogamous mating behavior, water conservation mechanisms, spontaneous epileptiform seizures, and relative freedom from spontaneous diseases.

depending on which taxonomist is consulted. Although these other species have rarely been used as research subjects in the United States, in other countries *M. libycus*, *M. shawii*, and *M. tristami* have been used.

Origin and Description

The Mongolian gerbil, *Meriones unguiculatus*, is a rodent of the family Cricetidae, although it has also been included among the Muridae. This gerbil is native to desert regions of Mongolia and northeastern China.

Gerbils are active, burrowing, social animals with cycles of activity and rest during day and night, although in captivity peak activity is in the middle of the dark cycle. They consume equal amounts of food day and night. In the wild, gerbils are generally predominately diurnal with crepuscular tendencies. Their burrows contain elaborate tunnels with multiple entrances, nesting rooms, and food chambers for storing seeds over the winter.

The agouti or mixed brown Mongolian gerbil is the variety commonly sold as pets or research animals, but black and also white gerbils are available. Black gerbils frequently have a white stripe or patch that extends from the chin to the chest and white spots on the feet. There are also dove, piebald, and cinnamon color mutations as well as a hairless mutation. Eleven to 15 genera or genera and species of other gerbils exist,

Anatomic and Physiologic Characteristics

Gerbils of both sexes have a distinct midventral abdominal pad composed of large sebaceous glands under control of gonadal hormones. Secretions from this gland are used in territorial marking and pup identification. The female gerbil has four pair of teats, two thoracic and two inguinal.

Spontaneous, convulsive (epileptiform) seizures occur in approximately half of pet or laboratory gerbils. These seizures, which begin at about 2 months of age, occur more often in families with a history of seizures. The seizures range from mild to severe and are elicited in unfamiliar surroundings or under a perceived threat.

Cerebral ischemia and infarction can be produced in some younger gerbils through unilateral carotid ligation. In such animals, the posterior communicating artery is reduced and the right and left cerebral vascular supplies are poorly connected, and therefore an ipsilateral "stroke" syndrome may be produced.

Gerbils are frequently reported to have lipemia. The lipemia is probably of dietary origin and would be accentuated by feeding

Gerbil—Physiologic Values

The values listed below are approximations only and may not represent the normal range in a given population. Sources consulted are included among the comprehensive texts listed in Chapter 1 and the publications on specific species following the respective sections of Chapter 2.

Adult body weight: male	65–100 g
Adult body weight: female	55–85 g
Birth weight (depends on litter size)	2.5–3.0
Body surface area (cm²)	10.5 (wt. in grams)$^{2/3}$
Body temperature	37.0–38.5° C
Diploid number-Karyotype	44
Life span	3–4 yr
Food consumption	5–8 g/100 g/day
Water consumption	4–7 ml/100 g/day
Vaginal opening	41 days or 28 g
Breeding onset: male	70–85 days
Breeding onset: female	65–85 days
Estrous cycle length	4–6 days
Gestation period (nonlactating)	24–26 days
Gestation period (concurrent lactation)	27–48 days
Postpartum estrus	Yes-fertile
Litter size	3–7; 5 avg.
Weaning age	20–26 days
Breeding duration,	12–17 mo
commercial	4–10 litters
Litters per year	7 avg.
Average production index	>1/wk per breeding pair
Respiratory rate	90/min
Oxygen use	1.4 ml/g/hr
Heart rate	360/min
Blood volume	66–78 ml/kg
Erythrocytes	8–9 × 10⁶/mm³
Reticulocytes	21–54/1000 rbc
Stippled rbc	2–16/1000 rbc
Polychromatophilic rbc	5–30/1000 rbc
Hematocrit	43–49%
Hemoglobin	12.6–16.2 mg/dl
Leukocytes	7–15 × 10³/mm³
Neutrophils	5–34%
Lymphocytes	60–95%
Eosinophils	0–4%
Monocytes	0–3%
Basophils	0–1%
Platelets	400–600 × 10³/mm³
Serum protein	4.3–12.5 mg/dl
Albumin	1.8–5.5 mg/dl
Globulin	1.2–6.0 mg/dl
Serum glucose	50–135 mg/dl

Gerbil—Physiologic Values (*Continued*)

Blood urea nitrogen	17–27 mg/dl
Creatinine	0.6–1.4 mg/dl
Total bilirubin	0.2–0.6 mg/dl
Cholesterol	90–150 mg/dl
Serum calcium	3.7–6.2 mg/dl
Serum phosphate	3.7–7.0 mg/dl

high-fat breeder chows or sunflower seeds. It is likely that gerbils develop lipemia more readily than do other rodent species. Even on diets with standard fat levels (4–6%), gerbils develop lipemia, and high serum cholesterol levels (e.g., 1590 mg/dl) have resulted from feeding rodent diets plus 1% cholesterol. This serum cholesterol is primarily associated with low density lipoproteins. Hepatic lipidosis and gallstones, but not atherosclerosis, developed in gerbils on the high-fat diet.

Gerbil blood has several characteristics not seen in the blood of other rodents. Male gerbils have higher packed red-cell volumes, hemoglobin levels, total leukocyte counts, and circulating lymphocyte counts than do females. Some erythrocytes of both sexes have a prominent polychromasia and basophilic stippling.

Gerbils have a large adrenal weight-to-body weight ratio, which contributes to their unique water conserving capability. The thymus is in the thorax and persists in adults.

Body weight of gerbils tends to be quite variable from animal to animal and between various colonies. Although such variability may be a genetic phenomenon, diet is a likely major contributing factor, i.e., higher fat content and diets of better nutritional quality lead to heavier body weights. Males are about 5% to 10% heavier than females. Cheal (1986) has published an excellent review of the unique physiologic attributes of gerbils and their use in research.

Gerbils as Pets

Behavior.   Gerbils are clean, friendly, cu-rious, and quiet, produce little waste and odor, rarely bite or fight, and are easily handled. Gerbils are more exploratory and less neophobic than are other small rodents. An escaped gerbil prefers midroom or its own cage to hiding. Gerbils often sit upright or, when threatened or excited, drum their rear legs on the floor. Although gerbils are social animals that live peacefully in either mixed or same-sex groups, they can be aggressive toward intruders. Gerbils may fight if crowded or mixed as adults, but the usual social interactions are grooming, wrestling, and communal sleeping. Incessant "digging" with the forepaws in the bottom corners of the cage is a characteristic of gerbils. Old gerbils are more inclined to bite than are young or breeding adults.

Gerbils paired after reaching puberty (10 weeks) or reunited after prolonged separation may fight. Fighting is reduced if the gerbils reach maturity in the same cage or if the animals are allowed to recover simultaneously from anesthesia in a neutral cage. It may not be possible to mix adults from established pairs without serious fighting.

Other behavioral characteristics of gerbils are their stable monogamous matings, rapid learning of avoidance responses but poor maze performances (too curious), and epileptiform seizures in some younger gerbils over 2 months of age following excitement, sudden noise, handling, or introduction to a strange environment.

Life Span.   The reproductive life of a female gerbil ranges from 15 to 20 months, during which time 4 to 10 litters may be produced, with 7 an average. The average life span is 3 years, but 4-year-old gerbils are common.

Restraint. The gerbil's tail is firmly grasped at the base, and the animal is lifted and cradled in the hand (Fig. 19). Extreme care should be taken not to pull on the tail, as the skin comes off easily. The gerbil may be more securely restrained with an over-the-back grip (Fig. 20). Gerbils resist being placed on their backs and, while struggling, may be dropped accidentally.

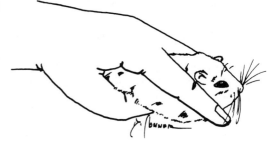

Fig. 20.   Over-the-back grip for restraining gerbil.

Housing

Plastic or metal rodent cages, with a solid floor and deep bedding, are the usual housing arrangement and are much preferred to wire cages without bedding. Bedding should be of sufficient depth (2 cm) to facilitate nest building; the bedding should be clean, dry, absorbent, and nonabrasive. Provision of an opaque cage, material for nest building, or a hiding place within the cage may improve reproductive and maternal performance. Gerbils are active gnawers and burrowers, and cages should be designed to prevent escape.

Mature gerbils over 12 weeks of age require a floor area of 230 cm² (36 in.²) per animal. A breeding pair requires a cage area of 1300 cm² (180 in.²). Cage sides should be at least 15 cm (6 in.) high. Cages suitable for rats and hamsters are generally satisfactory for gerbils.

Gerbils tolerate a wide range of temperatures but they become uncomfortable at temperatures above 35° C. They possess greater capacity for heat regulation than do other pets and laboratory rodents not of desert origin. Their tolerance of heat decreases, however, as humidity increases. Gerbils are usually housed at temperatures between 18° and 29° C (65° to 85° F) with a temperature at 22° C (72° F) being the usual compromise. The environmental humidity should be above 30%, although at approximately 50% the gerbil's hair coat becomes roughened. A 12:12 light:dark cycle is the usual laboratory arrangement.

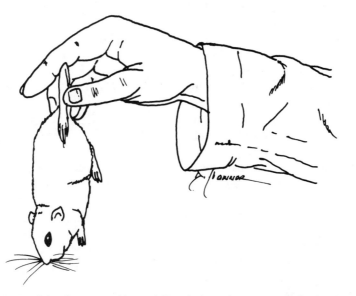

Fig. 19.   Picking up gerbil with grip at tail base. If the tail of a rodent is grasped distal to the base, the tail skin may be pulled off.

Gerbils produce only a few drops of urine daily, and fecal pellets are small, dry, and hard. Thus, their bedding remains odorless for several weeks. Gerbil cages require less cleaning than do the cages of other rodents. In gerbil colonies the bedding is usually changed when dirty or every 2 weeks. Nests of lactating females should be disturbed as little as possible. Cages should be washed with hot water and detergent, disinfected, and rinsed.

Feeding and Watering

Gerbils are granivorous or herbivorous. Gerbils should be fed *ad libitum* through feed hoppers a complete, freshly milled, pelleted rodent chow containing approximately 22% protein. Young begin eating at approximately 15 days of age. Mortality ensues when food in hoppers or water is inaccessible; therefore, supplementary food provided on the floor of cages may be necessary. Furthermore, it may be useful to soak hard pellets before feeding them to young. Quality hamster and gerbil feeds are available in pet stores. Adult gerbils consume 5 to 8 g of feed daily. Females tend to display more food hoarding behavior than do males. Gerbils are fond of sunflower seeds, preferring them to pelleted chows, but these seeds, with a low calcium and high fat content, are not a complete diet. Gerbils 2 to 5 weeks old may have difficulty opening the sunflower seeds or gnawing the hard pellets. Although gerbils in the wild, feeding on seeds and succulents, require little water and will stabilize at 86% of hydrated weight, caged gerbils should be supplied continuously with clean water. Contrary to popular opinion, gerbils maintained on dry food require water. Inadequate water consumption results in infertility, lower body weight, and possibly death. Although gerbils maintain weight when receiving as little as 2 ml of water daily, adult gerbils will consume approximately 4 to 10 ml water daily. Older males drink more water than do females or younger males. Sipper tubes should reach to and be operable by the smallest cage inhabitant.

Breeding

Sexing. Weanling and older males have a dark scrotum, and the anogenital distance is substantially greater in males (10 mm) than in females (5 mm), which makes sexing quite easy. Both sexes have a genital papilla (Figs. 21 and 22). The midventral marking (sebaceous) or scent gland is androgen-sensitive and therefore about twice as large in the male.

Estrous Cycle. Gerbils breed throughout the year. Gerbils are mated at 10 to 12 weeks of age when they weigh approximately 55 to 70 g; however, the testes may descend and the vagina may open as early as 28 and 40 days, respectively.

As gerbils are often permanently paired, the detection of estrus is rarely a factor in breeding. The gerbil's estrous cycle lasts 4 to 6 days, and a fertile, postpartum estrus occurs within 18 hours of parturition in approximately 60% of gerbils. As a rule, the gestation period of lactating gerbils is in-

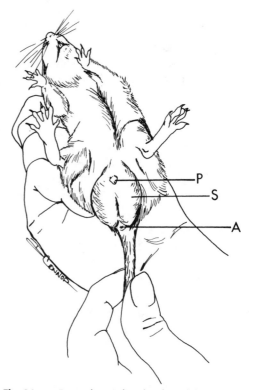

Fig. 21. External genitalia of male gerbil. P = tip of penis; S = scrotal sac; A = anus.

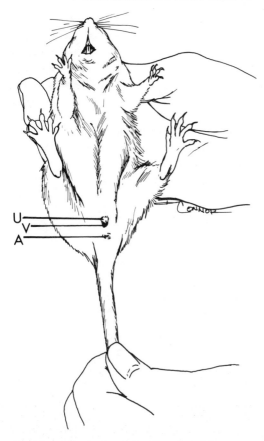

Fig. 22. External genitalia of female gerbil. U = urethral orifice; V = vaginal orifice; A = anus.

2 weeks postpartum to eliminate postpartum mating and reduce disturbance of the female and young. The 2-week period should not be lengthened, because the possibility of fighting when reunited is increased. Polygamous (trio) harem groups are also used successfully, but fighting may result.

Pregnancy and Rearing. During pregnancy a mature gerbil will gain between 10 and 30 g. The gestation period of non-lactating gerbils lasts 24 to 26 days, although both shorter (19 to 21 days) and longer periods have been noted. Utilization of the postpartum estrus will lead to concomitant pregnancy and lactation, a situation that leads to delayed implantation and therefore prolonged gestation. In this case, gestation may be as long as 48 days.

Gerbils, pregnant or not, build nests in the bedding. This activity is accentuated in cooler temperatures. Nests are constructed of a variety of paper or bedding materials, which gerbils are adept at shredding. Maternal nests are often covered. A wooden box 12 cm on a side placed in the cage facilitates nest building.

Usually 3 to 7 (1 to 12) gerbils are born, and delivery is almost always at night. Neonates weigh approximately 2.5 to 3 g at birth and are naked with sealed eyes and ears. Neonates from small litters tend to weigh more than those from large litters. Ears open in approximately 5 days, hair appears in 6 days, incisors erupt in 12 to 14 days, and eyes open in 16 to 17 days. If the litter is destroyed and lactation ceases, the female will resume cycling. There may be a selective advantage in the wild for destroying smaller litters and rebreeding.

Young are rarely abandoned or cannibalized. Factors contributing to abandonment include small litters (3 or fewer young), excessive handling of the young, lack of nesting material, and cages without provision for concealment.

Fostering is possible if the orphaned and host litters are born within a few days of each other. Hand feeding of neonatal rodents is difficult.

creased by up to 2 days for each young nursed. Estrus lasts about 24 hours, and gerbils in estrus are restless and may have a congested vulva. Estrus may be induced out of normal sequence by vaginal stimulation or by pairing adult males and females.

Breeding Programs. Gerbils breed readily in the home or laboratory environment. Breeding pairs established before sexual maturity (approximately 7 or 8 weeks) are usually stable, monogamous, lifelong arrangements in which one partner's removal or death may render the other sexually and socially incompatible in subsequent pairings or may induce pregnancy failure. The male participates in the care of young and is left in the cage at all times. Separation of breeding pairs should, however, follow fighting, rejection, or aggression toward the young.

Some breeders remove the male gerbil for

Disease Prevention

Observe the general husbandry practices listed in Chapter 1. Gerbils' fondness of sunflower seeds can produce the false impression of good dietary habits; however, sunflower seeds are high in fat, low in calcium, and are not a complete diet for rodents.

Public Health Concerns

Gerbils in captivity have few spontaneous diseases and even fewer of public health significance. *Salmonella* and *Hymenolepis* infections are potential health problems for man. Both conditions are rare in gerbil colonies. Gerbils could become feral in North America and cause extensive damage to crops. It is illegal to maintain gerbils as pets in California.

Uses in Research

Approximately 110,000 gerbils are used annually in the United States for studies of lead nephropathology, cerebral ischemia and stroke, auditory phenomena, parasite infections, epilepsy, infectious diseases, histocompatibility, dental disease, behavior, radiobiology, endocrinology, water conservation, and lipid metabolism. Gerbils are more resistant to radiation exposure than are other laboratory rodents.

Sources of Information

Comprehensive works on the gerbil include *The UFAW Handbook on the Care and Management of Laboratory Animals,* 6th edition, pp. 360–376, edited by T.B. Poole (1987, Churchill Livingstone Inc., 1560 Broadway, New York, NY 10036) and *The Guide to the Care and Use of Experimental Animals,* Vol. 2, Chapter 14, Mongolian Gerbils, edited by H.C. Rowsell, et al. (1984, Canadian Council on Animal Care, 1105–151 Slater Street, Ottawa, ONTARIO K1P 5H3). Additionally, *The Gerbil Digest* is published quarterly for biomedical researchers, supervisors, and technicians at their institutional address and is available from Tumblebrook Farm, Inc., West Brookfield, MA 01585.

Other works on gerbils include:

Agren, G.: Pair formation in the Mongolian gerbil. Anim. Behav., *32*:528–535, 1984.

Cheal, M.: Lifespan ontogeny of breeding and reproductive success in Mongolian gerbils. Lab. Anim., *17*:240–245, 1983.

Cheal, M.: The gerbil: A unique model for research in aging. Exper. Aging Res., *12*:3–21, 1986.

Clark, M.M., Spencer, C.A., and Galef, B.G., Jr.: Improving the productivity of breeding colonies of Mongolian gerbils (*Meriones unguiculatus*). Lab. Anims., *20*:313–315, 1986.

Dillon, W.G., and Glomski, C.A.: The Mongolian gerbil: Qualitative and quantitative aspects of the cellular blood picture. Lab. Anim., *9*:283–287, 1975.

How to Raise and Train Gerbils, T.F.H. Publications, 211 West Sylvania Avenue, Neptune, NJ 07753.

LeRoi, D.: Fancy Mice, Rats and Gerbils. 1976. Sportshelf & Soccer Associates, P.O. Box 634, New Rochelle, NY 10802.

Noms, M.L.: Disruption of pair bonding induces pregnancy failure in newly mated Mongolian gerbils (*Meriones unguiculatus*). J. Reprod. Fertil., *75*:43–47, 1985.

Ordken, C.C., and Scott, S.E.: Feeding characteristics of Mongolian gerbils (*Meriones unguiculatus*). Lab. Anim. Sci., *34*:181–184, 1984.

Ostrow, M.: The T.F.H. Book of Gerbils. 1984. T.F.H. Publications, 211 West Sylvania Avenue, Neptune, NJ 07753.

Paget, G.E.: *Clinical Testing of New Drugs,* pg. 33. Revere Publishing Co., NY, 1965.

Robinson, D.: Encyclopedia of Gerbils. 1980. T.F.H. Publications, 211 West Sylvania Avenue, Neptune, NJ 07753.

Tumblebrook Farm: An Indexed Bibliography of Gerbil Behavioral Studies. 1978. Tumblebrook Farm, Inc., West Brookfield, MA 01585.

Williams, W.M.: The Anatomy of the Mongolian Gerbil (*Meriones unguiculatus*). 1974. Tumblebrook Farm, Inc., West Brookfield, MA 01585.

THE MOUSE

Approximately 10.5 million mice are used annually in biomedical research in the United States. Commensal and wild mice have spread around the world, and the laboratory or house mouse is occasionally kept as a pet. They are secretive, burrowing, and nesting creatures. Because mice are small, prolific breeders; are easily and economically maintained in large populations; possess great genetic diversity; and are well characterized anatomically and physiologically, they are the most widely used ver-

Mouse—Physiologic Values

The values listed below are approximations only and may not represent the normal range in a given population. Sources consulted are included among the comprehensive texts listed in Chapter 1 and the publications on specific species following the respective sections of Chapter 2.

Adult body weight: male	20–40 g
Adult body weight: female	25–40 g
Birth weight	0.75–2.0 g
Body surface area (cm^2)	10.5 (wt. in grams)$^{2/3}$
Rectal temperature	36.5–38.0° C
Diploid number	40
Life span	1.5–3 yr
Food consumption	15 g/100 g/day
Water consumption	15 ml/100 g/day
GI transit time	8–14 hr
Breeding onset: male	50 days
Breeding onset: female	50–60 days
Cycle length	4–5 days
Gestation period	19–21 days
Postpartum estrus	fertile
Litter size	10–12
Weaning age	21–28 days
Breeding duration,	7–9 mo
commercial	6–10 litters
Young production	8/mo
Milk composition	12.1% fat, 9.0% protein, 3.2% lactose
Respiratory rate	60–220/min
Tidal volume	0.09–0.23 ml
Oxygen use	1.63–2.17 ml/g/hr
Heart rate	325–780/min
Blood volume	76–80 mg/kg
Blood pressure	113–147/81–106 mm Hg
Erythrocytes	7.0–12.5 × 10^6/mm^3
Hematocrit	39–49%
Hemoglobin	10.2–16.6 mg/dl
Leukocytes	6–15 × 10^3/mm^3
Neutrophils	10–40%
Lymphocytes	55–95%
Eosinophils	0–4%
Monocytes	0.1–3.5%
Basophils	0–0.3%
Platelets	160–410 × 10^3/mm^3
Serum protein	3.5–7.2 g/dl
Albumin	2.5–4.8 g/dl
Globulin	0.6 g/dl
Serum glucose	62–175 mg/dl
Blood urea nitrogen	12–28 mg/dl
Creatinine	0.3–1.0 mg/dl
Total bilirubin	0.1–0.9 mg/dl

Mouse—Physiologic Values (*Continued*)

Cholesterol	26–82 mg/dl
Serum calcium	3.2–8.5 mg/dl
Serum phosphate	2.3–9.2 mg/dl

tebrate animal in biomedical research and testing.

Origin and Description

The mouse, *Mus musculus,* is a rodent (order Rodentia) of the family Muridae. In recent times mice spread worldwide from a presumed original focus in temperate Asia stretching from modern day Turkey to China. Mice live permanently caged and managed by man, commensally in human habitations in cold weather and in the wild in warm weather, or entirely in the wild. These wilder mice undergo periodic population "explosions," but at all times they cause extensive damage to human food supplies.

The domestication and breeding of mice by fanciers led to the genetic diversity of mouse populations and to the research interest of nineteenth century scientists. The long-established Swiss albino mouse became the source of many non-inbred, white laboratory stocks, although there are several other sources both white and colored. Several genera of wild mice other than *Mus* are encountered in research colonies, including the field mouse *Microtus,* the grasshopper mouse *Onychomys,* and the white-footed mouse *Peromyscus.*

Research mice, as well as other research animals, may be included in two major categories based on ecologic and genetic characteristics. The ecologic category includes: (1) germfree (axenic) mice, which are free from detectable organisms; (2) defined flora (gnotobiotic) mice, which possess a specified flora and fauna; (3) specific pathogen-free mice, which are free of specified pathogenic microorganisms; and (4) conventional mice, which include almost all other mice housed under a variety of conditions.

The genetic category includes (1) random-bred mice, which are derived from mating unrelated mice; (2) inbred mice, which are the genetically homogeneous descendents of at least 20 consecutive brother-sister matings; and (3) the F1 hybrid, which is produced by mating mice of two different inbred strains. Several inbreeding schemes exist, but each brother-sister mating reduces existing heterozygosity by approximately 19%. Selective breedings and inbreedings have produced the great diversity of mice encountered as pets and research animals.

Inbred mice come in a variety of attractive, strain-specific coat colors. Consequently, these strains find their way into laboratories, schools, and homes where they are bred and raised as pets rather than as models of human disease, the feature for which they were originally selected. Regardless of their use, these mice are destined to develop those complex disease conditions characteristic of the strain. This diversity in disease profiles among strains presents the investigator, animal facility manager, or small animal practitioner with multiple differential diagnostic challenges.

Anatomic and Physiologic Characteristics

The conventional, random-bred white mouse is distinguished for its small size and the physiologic accommodations for that size. The rapid heart rate (500–600/min), oxygen consumption (1.7 ml O_2/g/hr), and fecundity (1 million descendants after 8 litters or 425 days) are some notable mouse characteristics.

Over 60 years of inbreeding and strain development have produced a considerable reservoir of mutants, which are utilized by the millions in the investigations of a vast array of normal and abnormal phenomena. The following characteristics provide a small sample of available mutants: anemic, athymic, bent tail, diabetic, frizzy, fuzzy, grizzled, jolting, naked, nervous, radiation sensitive, and tottering. A wide variety of spontaneous neoplasms occur in several inbred strains, listings of which can be obtained from the National Institutes of Health's *Catalogue of NIH Rodents*, NIH, Bethesda, MD 20014, or from the Roscoe B. Jackson Memorial Laboratory, Bar Harbor, ME 04609.

Some of the anatomic characteristics of mice include spleens in males 50% larger than those in females; dentition of 1/1 incisors, no canines or premolars, and 3/3 molars with no deciduous dentition and open rooted incisors; 3 pairs of thoracic mammary glands and 2 pairs of inguinal, all reaching from the ventral midline over the flanks, thorax, and portions of the neck; a stomach divided into squamous nonglandular, and glandular portions; and inguinal canals open for the life of the male.

Mice as Pets

Behavior. Mice, if gently handled, make good, albeit small, pets. Mice are timid, social, omnivorous, territorial, and escape-prone rodents that require little cage space and only small quantities of food and water. Although wild mice are distinctly nocturnal, laboratory mice and pet mice are active and resting alternately throughout the day and night. Although mice may live in compatible, single male dominated groups, the overall social organization is either loose or poorly defined and understood.

Adult male mice, especially those previously housed alone, newly assembled in the same cage usually fight, as do those existing in unstable hierarchies. The relative, hierarchical status of males in a cage can often be determined from the number and severity of bite wounds over the rump and back. Therefore, male mice should be housed separately to prevent fighting and the resulting abscesses, dermatitis, septicemia, and death. This male aggression is influenced by both testosterone and training, and both innate and learned components are necessary for the full demonstration of aggression. Inter-male aggression can be avoided by castration or destroying the sense of smell. Female mice seldom fight. Females with litters may fight to defend their nests. Mice housed alone tend to be more aggressive than their group-housed counterparts. If roughly handled or startled, mice will bite, or more likely pinch, with their teeth.

Human allergies to rodent dander or urinary proteins are common. Mice may have an undesirable odor, particularly if male mice are kept or the cages are excessively moist and dirty.

Life Span. Mice may live 2 to 3 years; however, there are major differences in longevity among strains, owing mainly to differential disease susceptibilities. For example, AKR (white) and C3H (agouti) mice live shorter lives than A (white) and C57BL (black) strains. Mice breed for approximately 7 to 18 months and produce 6 to 10 litters.

Restraint. Mice may be picked up by grasping the tail with the fingers (Fig. 23). If inspection or manipulation is intended, the mouse is lifted by the tail, placed on a rough, "toe-gripping" surface, grasped on the scruff of the neck by thumb and forefinger and inverted, and the tail is held between the palm and little finger (Fig. 24). Grasping and pulling the tip of the tail may result in stripping the skin from the tail.

Housing

Mice used in research or as pets are usually housed in either metal or plastic cages with wire mesh or slotted bar tops. A homemade shoe-box cage with a hardware cloth

Fig. 23. Restraint of mouse by grasping neck and tail base.

Fig. 24.   One-hand restraint of mouse for injection or bleeding procedures.

top is satisfactory for pet mice, as are the several wire and plastic cages available in pet stores. Large mice may have difficulty eating through 0.65 cm² mesh, and weanling mice may escape through larger meshes. Care should be taken to prevent escape, loss of young, and fractured limbs.

Mouse cages should provide at least 97 cm² (15 in.²) floor space per adult (25 g) mouse. A female with a litter should have 390 cm² (63 in.²) floor space. Mouse cages should be at least 13 cm (5 in.) high.

Bedding, which may be paper, sawdust, or soft pine, aspen, or cedar wood shavings, should be nonallergenic, dust-free, inedible, absorbent (but not cause dehydration of newborn mice), nontoxic, and free of pathogenic organisms. Soft pine and cedar wood shavings are widely used for pet rodent bedding because of their pleasant aroma. However, because they stimulate hepatic microsomal enzymes, they are generally avoided as a bedding material for research animals. Soft wood shavings and tissue paper make excellent rodent nesting material.

Temperatures in mouse rooms should be maintained between 18° and 26° C (64° to 79° F), with an approximate average of 22° C (72° F). The relative humidity in mouse rooms is usually between 40% and 70%. Cage filter covers, or covers of any type that retard diffusion of air, raise ambient cage temperature and humidity and make temperature control more difficult.

Mouse caging, like all animal caging and bedding, is cleaned or changed as often as necessary to prevent accumulation of odor and waste and to keep the animals dry and clean. Depending on the cage population, caging is usually cleaned at least twice a week.

Feeding and Watering

Mice are fed a clean, wholesome, and nutritious pelleted rodent diet *ad libitum* and are watered with an automatic watering system or water bottles with sipper tubes. Several varieties of special purpose rodent diets are available, including a wide variety of pet "treats" and feeds. Probably the best feed for rodents, including rats and mice, is the commercially available pelleted rodent laboratory diets sold in 50-lb bags or repackaged and sold in small boxes or bags in pet stores. Pelleted, balanced, high protein pet rodent diets are also satisfactory. Diets for maintenance usually contain 4 to 5% fat and about 14% protein. Diets for growth and reproduction contain 7 to 11% fat and 17 to 19% protein. Because these feeds vary greatly in price, freshness, and protein content and quality, careful shopping is necessary. Fresh water and a high-quality pelleted feed provide a complete diet; sweets, low protein "treats," vitamins, salt blocks, and vegetables are usually not necessary and may even lead to illness.

An adult mouse will consume approximately 15 g feed and 15 ml water per 100 g body weight per day. Water and feed consumption vary with the ambient temperature, humidity, water availability, dryness of the feed, breeding status, feed quality, and state of health.

Breeding

Sexing.   Neonatal males may be distinguished from females by a greater anogenital distance in males (1.5 to 2 ×), the pale testes visible through the abdominal wall, and the larger genital papilla. In females, the conspicuous row of nipples at 9 days is a distinguishing feature (Figs. 25 and 26).

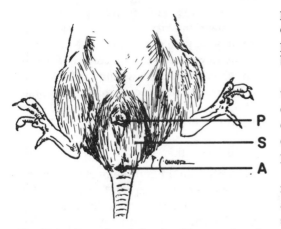

Fig. 25.   External genitalia of male mouse. P = tip of penis; S = scrotal sac; A = anus.

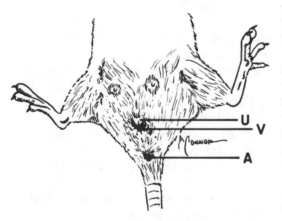

Fig. 26.   External genitalia of female mouse. U = urethral orifice; V = vaginal orifice; A = anus.

Estrous Cycle.   Although mice may have a first estrus at 28 to 40 days, they are first bred when 50 or more days of age (20 to 30 g). Mice bred too young or after 10 weeks have reduced fertility. The onset of sexual maturity varies with strain, season, growth rate, litter size, and level of nutrition.

The estrous cycle lasts 4 to 5 days, with an average evening estrous period of 12 hours. Except for postpartum estrus, estrus does not occur during lactation. Group-caged female mice may enter a period of continuous anestrus, which is terminated by the odor or presence of a male. Most of the females will then come into heat in ap-

proximately 72 hours, a synchronization of cycles known as the Whitten effect. Also, pseudopregnancy may prolong the interval between estrous periods.

During proestrus the vaginal smear contains epithelial cells with some cornified cells and leukocytes; in estrus, cornified cells predominate. In late metestrus and diestrus, cornification decreases and lymphocytes increase.

Breeding Programs.   Considerations in mouse breeding programs include space available, strain fecundity, inbreeding schemes, epidemiologic concerns, and production requirements. Mice are continuously polyestrus with minor seasonal variations. Mice have a fertile, postpartum estrus occurring 14 to 28 hours after parturition, and many breeding schemes utilize this estrus for increased production.

Representative breeding systems for mice include the colony, the monogamous, and the polygamous mating schemes. In colony mating, 1 male and 2 to 6 females are housed together continuously, and the young are removed at weaning. This system is the most efficient of all for utilization of space and labor, but record keeping is difficult.

The monogamous system involves the constant pairing of 1 male and 1 female. The young are removed prior to the next parturition. This system, which also utilizes the postpartum estrus, produces the maximum number of litters per female in the shortest time and provides for ease of record keeping and evaluation of individual production. Disadvantages of monogamous breeding include the need to maintain a larger population of males and the increased need for labor, space, and equipment to service the increased populations.

The polygamous or harem scheme combines 1 male and 2 to 6 females. Females are removed to separate cages prior to parturition; the postpartum estrus is not utilized. In this system each female provides more milk, larger young, and more young weaned per litter. Disadvantages include lowered total litter numbers and increased labor time per cage.

Pregnancy and Rearing

Detection of sperm by vaginal examination or the presence of a coagulation or postcopulatory plug in the vagina is evidence of mating within the past 24 hours. As gestation proceeds, the mammary structures develop, and the fetuses can be palpated. Daily weighing will reveal an increased rate of weight gain about 13 days into gestation. Mammary development is pronounced at 14 days.

In the nonsuckling dam the gestation period ranges from 19 to 21 days. Simultaneous lactation and pregnancy delays uterine implantation of conceptuses and prolongs the gestation period 3 to 10 days, depending on the size of the nursing litter. The gestation period in hybrid mice tends to be shorter than the period in inbred mice. If a mouse bred within the previous 24 hours is exposed to a strange male, the existing pregnancy may be aborted. This phenomenon is the Bruce effect. Pseudopregnancy in mice is rarely noticed but may follow male exhaustion and sterile matings. Pseudopregnancy may last 1 to 3 weeks.

Mice routinely prepare small sleeping nests in the bedding. The hollowed brood nest, which may contain one or more families, is prepared in late gestation. The female spends much time with the young and will retrieve scattered young to the nest. In harem breeding arrangements young may suckle several lactating dams.

Litter size varies considerably with strain and age. The first litter is usually smaller, with optimal production (10 to 12 young) occurring between the second and eighth litters. Inbred strains usually produce smaller litters than do hybrid, outbred, or random-bred animals.

Young mice are weaned at about 21 days (10 to 12 g). Small mice of slower-growing inbred strains may be weaned as late as 28 days of age. If the postpartum estrus is not utilized, the female resumes cycling 2 to 5 days postweaning. Mutilation and cannibalism of the young are uncommon among mice, but female mice whelping or recently whelped should remain undisturbed for at least 2 days postpartum. These aggressive behaviors may be strain related. Mice have a highly developed sense of smell and acute hearing. High-pitched sounds (ultrasonic) may cause audiogenic seizures in some strains (e.g., DBA/2) and associated destruction of litters.

Disease Prevention

In addition to the routine husbandry precautions listed in Chapter 1, a systematic diagnostic evaluation of colony animals should be conducted routinely to screen for subclinical infections. Filter cage covers prevent or reduce airborne transmission of microorganisms within a densely populated mouse room, although such covers increase intracage ammonia level, humidity, and cage temperature. Chlorination (10 to 12 ppm) or acidification (pH 2.5) of the drinking water will reduce the level of *Pseudomonas* contamination. A 10 ppm Cl_2 solution is prepared by mixing 2 ml of 5.25% sodium hypochlorite solution in 10 liters water. A solution with a pH of 2.5 is prepared by mixing 10 liters water with approximately 2.6 ml concentrated hydrochloric acid. Because of variations in the pH and hardness of local water supplies, the pH of the final mixture should be monitored.

Stress involved in shipment of mice (packing, unpacking, loading, and unloading) may cause immune reaction depression. A post-transport stabilization period of a minimum of 48 hours allows the immune system to return to normal.

Public Health Concerns

Salmonellosis and lymphocytic choriomeningitis are zoonotic diseases of rare occurrence in mouse colonies. Rabies vaccinations are not usually indicated for persons bitten by pet or laboratory rodents because these animals have little chance for exposure to the "rabies chain." The decision to use a killed rabies vaccine to protect a pet or research rodent or rabbit depends on the probability of exposure to the wild animal population.

Cutaneous and respiratory allergies to rodent dander or urinary proteins may develop in people. Such allergies are not always accompanied by a positive skin reaction to injected dander, and a physician's advice should be sought if itching, reddening, runny nose, sneezing, or difficult breathing follow exposure to rodents.

Uses in Research

Approximately 10.5 million mice are used annually in the United States in such research and testing areas as drugs and cosmetics, aging, virology, histocompatibility, hemolytic anemia, congenital defects, neoplasia, genetics, radiobiology, amyloidosis, giardiasis, autoimmune disease, congenital athymia, ameobiasis, obesity, dwarfism, monoclonal gammopathies, monoclonal antibody production, diabetes mellitus, renal disease, and behavior.

Sources of Information

Comprehensive works on the mouse are *The Mouse in Biomedical Research* (4 vols.), edited by H. Foster, et al., (1982, Academic Press, Inc., Orlando, FL 32887); *Biology of the Laboratory Mouse*, 2nd edition, edited by E. Green (1966, The Blakiston Division, McGraw-Hill Book Co., 1221 Avenue of the Americas, New York, NY 10020); *The UFAW Handbook on the Care and Management of Laboratory Animals*, 6th edition, pp. 275–308, edited by T.B. Poole (1987, Churchill Livingstone Inc., 1560 Broadway, New York, NY 10036); *The Guide to the Care and Use of Experimental Animals*, Vol. 2, Chapter 19, Mice, edited by H.C. Rowsell, et al. (1984, Canadian Council on Animal Care, 1105-151 Slater Street, Ottawa, ONTARIO K1P 5H3); and *Mouse News Letter* (M.R.C. Laboratories, Woodsmansterne Road, Carshalton, Surrey, SM5 4EF, UK or The Jackson Laboratory, Bar Harbor, ME 04609).

Other works on mice include:

Cook, M.: The Anatomy of the Laboratory Mouse. 1976. Academic Press, Inc., Orlando, FL 32887.

Gude, W.D., et al.: Histological Atlas of the Laboratory Mouse. 1982. Plenum Publishing, 233 Spring Street, New York, NY 10013.

Jones, T.: Encyclopedia of Pet Mice. 1979. T.F.H. Publications, 211 West Sylvania Avenue, Neptune, NJ 07753.

Paget, G.E.: *Clinical Testing of New Drugs*, pg. 33. Revere Publishing Co., NY, 1965.

Rafferty, K.A., Jr.: Methods in Experimental Embryology of the Mouse. 1970. Johns Hopkins University Press, Baltimore, MD 21218.

Rugh, R.: The Mouse: Its Reproduction and Development. 1968. Burgess Publishing Co., 7108 Ohms Lane, Minneapolis, MN 55435.

Sidman, R.L., Angevine, J.B., Jr., and Pierce, E.T.: Atlas of the Mouse Brain and Spinal Cord. 1971. Harvard University Press, 79 Garden Street, Cambridge, MA 02138.

Silverstein, A., and Silverstein, V.: Mice: All About Them. 1980. Harper and Row, 2350 Virginia Avenue, Hagerstown, MD 21740.

Theiler, K.: The House Mouse: Development and Normal Stages from Fertilization to Four Weeks of Age. 1972. Springer-Verlag, New York, Inc., 175 Fifth Avenue, New York, NY 10010.

Wirtschafter, Z.T.: The Genesis of the Mouse Skeleton: A Laboratory Atlas. 1960. Charles C Thomas, 2600 South First Street, Springfield, IL 62717.

THE RAT

The domestic variety of the brown or Norway rat, usually represented by the albino or piebald animal in research colonies, is an Old World import that occupies a longstanding and important place in biomedical and behavioral research. Rats are well-defined, easily maintained, inexpensive, relatively disease resistant, and suited for a wide range of research endeavors.

Origin and Description

The laboratory rat, *Rattus norvegicus*, is a rodent of the family Muridae. Wild rats apparently originated, in recent times at least, in the temperate regions of central Asia, from southern U.S.S.R. through northern China. Through migration along trade and military routes, the cosmopolitan rat has spread around the world. The rat has no special connection with Norway, as might otherwise be assumed from the species name.

Domesticated rats were raised by fanciers in the seventeenth century and for combat with terriers in subsequent centuries. By the mid-1800s, rats were being used in scientific experimentation. Rats, like mice, are now available in various ecologic (germfree, gnotobiotic, specific pathogen-free, and conventional) and genetic varieties; however, most research rats are barrier-raised in specific pathogen-free colonies. Commonly available strains or varieties of rats include the Sprague-Dawley, the Wistar, and the Long-Evans rats, although numerous other types are available.

The Sprague-Dawley rat, an albino, has a narrow head and tail longer than the body, whereas the Wistar rat has a wide head and shorter tail. The Long-Evans and other "hooded" varieties are smaller than the albino strains and have darker hair over portions of the head and anterior body. There is substantial evidence that strain-related aggressiveness and disease susceptibilities exist among the several varieties of domestic rats.

Examples of outbred and mutant stocks and inbred strains of rats of biomedical interest include: N:OM (Osborne-Mendel), highly susceptible to murine mycoplasmosis; SDN:SD (Sprague-Dawley), high incidence of spontaneous mammary tumors; BUF/N (Buffalo rats), thyroiditis and resistance to nephrosis; F344/N (Fischer 344), interstitial cell tumors; SHR/N, spontaneously hypertensive rat; Brattleboro rat, diabetes inspidus; and the Gunn rat, hepatic jaundice.

The black or roof rat, *Rattus rattus,* is occasionally seen in research colonies. This rat is smaller than *R. norvegicus* and more successfully adapted to tropical climates. Rats native to North America, members of the family Cricetidae, include the cotton rat *Sigmondon* and the kangaroo rat *Dipodormys.*

Anatomic and Physiologic Characteristics

The characteristics listed and briefly described below were obtained from Bivin, Crawford, and Brewer's "Morphophysiology" in *The Laboratory Rat,* Vol. 1, edited by H. Baker, et al. (New York, Academic Press, 1979); and from Barnes' *Special Anatomy of Laboratory Mammals* (Department of Anatomy, School of Veterinary Medicine, University of California, Davis, 1971).

Rats, like other laboratory murine rodents, have a baculum or *os penis,* a dental formula of 1/1 incisors and 3/3 molars with continuously erupting incisors, cheeks that close into the diastema separating incisors from the oral cavity, an open inguinal canal, a divided stomach and large cecum, an articulated mandibular symphysis, both pectoral and inguinal mammary glands extending dorsally, and a diffuse pancreas. Rats have no gallbladder. They have a tapered head with pinnae relatively smaller than those of the mouse, a long tail, and an anus usually pressed on the ground.

The gnawing apparatus, shared with other rodents, is remarkable. The continuously erupting (hypsodontic), chisellike incisors, powerful jaw muscles, articulated symphysis, and long diastema with loose cheek skin all contribute to the rat's omnivorous habits and gnawing ability. The 12 molars, on the other hand, are permanently rooted (brachiodontic), located far back in the mouth, and are used for grinding.

The rat, like other rodents, has, especially during youth, masses of brown fat in various cervical locations and between the scapulae. This fat probably serves neonatal and cold-stressed rodents as a thermogenic material and metabolic regulator. The quantity and significance of brown fat diminish into puberty and adulthood.

Albino rats have poor eyesight and depend on facial vibrissae and olfaction for sensory input; a blinded rat may, in fact, appear essentially normal. Behind the rat's eyeball lies a pigmented lacrimal gland called the harderian gland, after the seventeenth century anatomist Jakob Harder. This red-brown gland, larger than the eyeball itself, secretes a lipid and red porphyrin-rich secretion that lubricates the eye and lids and may play a neonatal role in determining pineal diurnal rhythms. In some stressful situations (restraint, swimming) or in acute illnesses, these red tears overflow and stain the face and nose. The dried, red porphyrin crust, which fluoresces

Rat—Physiologic Values

The values listed below are approximations only and may not represent the normal range in a given population. Sources consulted are included among the comprehensive texts listed in Chapter 1 and the publications on specific species mentioned in the respective sections of Chapter 2.

Adult body weight: male	450–520 g
Adult body weight: female	250–300 g
Birth weight	5–6 g
Body surface area (cm²)	10.5 (wt. in grams)$^{2/3}$
Body temperature	35.9–37.5° C
Diploid number	42
Life span	2.5–3.5 yr
Food consumption	10 g/100 g/day
Water consumption	10–12 ml/100 g/day
GI transit time	12–24 hr
Breeding onset: male	65–110 days
Breeding onset: female	65–110 days
Cycle length	4–5 days
Gestation period	21–23 days
Postpartum estrus	fertile
Litter size	6–12
Weaning age	21 days
Breeding duration, commercial	350–440 days 7–10 litters
Young production	4–5/mo
Milk composition	13.0% fat, 9.7% protein, 3.2% lactose
Respiratory rate	70–115/min
Tidal volume	0.6–2.0 ml
Oxygen use	0.68–1.10 ml/g/hr
Heart rate	250–450/min
Blood volume	54–70 ml/kg
Blood pressure	84–134/60 mm Hg
Erythrocytes	7–10 × 10⁶/mm³
Hematocrit	36–48%
Hemoglobin	11–18 g/dl
Leukocytes	6–17 × 10³/mm³
Neutrophils	9–34%
Lymphocytes	65–85%
Eosinophils	0–6%
Monocytes	0–5%
Basophils	0–1.5%
Platelets	500–1300 × 10³/mm³
Serum protein	5.6–7.6 g/dl
Albumin	3.8–4.8 g/dl
Globulin	1.8–3.0 g/dl
Serum glucose	50–135 mg/dl
Blood urea nitrogen	15–21 mg/dl
Creatinine	0.2–0.8 mg/dl

Rat—Physiologic Values (*Continued*)

Total bilirubin	0.20–0.55 mg/dl
Serum lipids	70–415 mg/dl
Phospholipids	36–130 mg/dl
Triglycerides	26–145 mg/dl
Cholesterol	40–130 mg/dl
Serum calcium	5.3–13.0 mg/dl
Serum phosphate	5.3–8.3 mg/dl

bright red under ultraviolet light, contains little or no blood.

Male rats exhibit a prolonged period of growth, and ossification of the long bones is not complete until into the second year. Murine rodents, however, mature in the first several months of life.

Rats as Pets

Behavior.   Despite a well-known association with plagues, garbage, and sorcery, domestic rats, if gently handled, make quiet, clean, easily trained pets. Rats are communal and several males and females may be combined in a single, uncrowded cage. Young are communally raised with shared nursing responsibilities. Fighting rarely occurs among adults, although sires may bother the young, and postparturient dams may fight among themselves.

Rats are burrowers, and even domestic rats do so if given the opportunity. They are also omnivorous, usually nocturnal, and year-round breeders in captivity. Escaped rats will often return to their cages.

Life Span.   Rats may live longer than 3 years and have a productive breeding life of approximately 9 or more months or until they are 12 or more months of age, during which time they may bear 7 to 10 litters with 6 to 14 offspring per litter. After 12 months, litter size decreases and litter interval increases until sexual senescence at 450 to 500 days.

Restraint.   Rats gently handled become tame and will rarely bite unless startled or hurt. Because the tail skin may tear, the tail should be grasped only at the base and for short periods. Rats in wire bottomed cages, in particular, should not be pulled by the tail because the animal can grasp the wire and injure its feet or tear the tail skin.

Rats are picked up by placing the hand firmly over the back and rib cage and restraining the head with thumb and forefinger immediately behind the mandibles (Fig. 27). Rats held upside down are more concerned with righting themselves than with biting.

Housing

Rats are housed either in metal cages with mesh or in plastic cages with solid floors. "Rat mesh," whether used for research or pet animal caging, should have openings of 1.6 cm² or 2 wires per inch. A homemade, shoe-box shaped cage with a hardware cloth top is satisfactory for pet rats, as are the several rodent cages available in pet stores. Terraria are commonly used to house pet rodents, but care must be taken to assure accessibility to water and feed. Care should be taken to prevent escape, loss of neonates through wire floors, and fractured limbs from too close a mesh. Also, young rats housed on wire and deprived of the nest may become dehydrated, especially during months when ambient humidity is low.

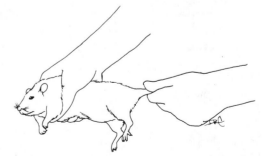

Fig. 27.   Forequarters and tail grip for restraining a rat.

Rat cages should provide at least 259 cm² (40 in.²) floor space per adult (300 g) rat. A female with litter requires 1000 cm² (155 in.²) floor space. Rat cages should be at least 18 cm (7 in.) high.

Bedding for rats, which may be paper, wood chips or shavings, ground corn cob, or sawdust, should be nonallergenic, dust-free, absorptive, nontoxic, and clean. Soft shavings are more easily formed into nests.

Temperatures in rat rooms should be maintained between 18° and 27° C (65° to 79° F), with the optimal ambient temperature at 22° C (72° F), and the relative humidity should be between 40% and 70%. The use of cage filter covers raises the ambient temperature, ammonia, and humidity levels in the cage above those of the room. Light intensity is reduced within translucent or filtered cages. A 12:12 dark:light cycle is usually used in rat colonies; continuous light will depress cycling. Rodents in general, especially nocturnal rodents, do very well in dimly lighted rooms. Apparently the bright light used in most animal facilities is more for man's benefit than for the rodent's well-being. To effect energy conservation and to promote both good animal health and good sanitation, lights in animal rooms can be controlled by a rheostat and timer.

Litter or bedding should be changed as often as necessary to keep odor minimal and the rats dry and clean; one to three bedding changes per week are usually sufficient. Cages, feeders, and water bottles are washed once or twice weekly. General husbandry concerns, described in Chapter 1, apply to facilities housing rats.

Feeding and Watering

Rats are fed a clean, wholesome, fresh, and nutritious pelleted rodent diet free choice. The commercially available diets are complete; that is, they do not require supplementation. Rats should be watered from water bottles with sipper tubes or through an automatic watering system. Several varieties of special-purpose rodent chows are available. Rats are cautious feeders (neophobic) and will avoid strange foods. Much

of the small-rodent feed found in pet stores, despite the attractive packaging, is inadequate in protein and energy. The boxed or bagged products claiming "high protein" (20% to 27% crude protein) are satisfactory, although similar products vary greatly in price.

An adult rat (300 g) will consume approximately 5 g feed and 10 ml water per 100 g body weight per day. Consumption varies with the ambient temperature, humidity, health status, breeding stage, diet, and the time of day. Rats are nocturnal and feed primarily at night.

Breeding

Sexing. The testicles are evident at an early age, especially if the male rat is held head up, and the testes drop from the inguinal canal into the scrotum. Males have a larger genital papilla and greater anogenital distance than those of females (5 mm in males and 2.5 mm in females at 7 days) (Figs. 28 and 29). Female nipples are visible when the young are between 8 and 15 days

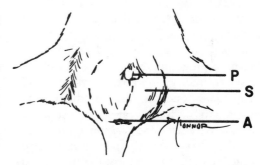

Fig. 28. External genitalia of male rat. P = tip of penis; S = scrotal sac; A = anus.

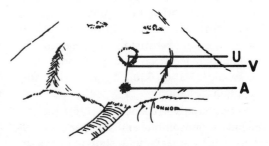

Fig. 29. External genitalia of female rat. U = urethral orifice; V = vaginal orifice; A = anus.

old. A rapid way to sex murine rodents is to lift tails and compare littermates.

Estrous Cycle. Rats reach puberty at 50 to 60 days, with the vagina opening at 35 to 90 days and the testes descending at 20 to 50 days. Copulation may occur early in puberty. Strong, healthy, vigorous offspring are produced if the rats are first mated at 65 to 110 days of age, when the females weigh approximately 250 g and the males 300 g. The age at first mating depends on the strain and growth rate of the rat.

The rat's estrous cycle lasts 4 to 5 days, with an estrous period of approximately 12 hours, which, as in mice, occurs in the evening. The Whitten effect, an induction and synchronization of estrus in some females following exposure to male odor, is less pronounced or certain in rats than in mice. Occasional induced ovulations may occur in rats.

The short estrous cycle reduces the necessity for cycle monitoring. During proestrus the vaginal smear contains epithelial cells with some cornified cells and leukocytes. In estrus, cornified cells become prominent. In late metestrus and during diestrus, cornification decreases and lymphocytes increase.

Breeding Programs. Considerations in rat breeding programs include space available, strain fecundity, inbreeding systems, epidemiologic concerns, and production requirements. Rats are continuously polyestrus, with minor seasonal variations, and they have a fertile, postpartum estrus occurring within 48 hours of parturition. Although about half the females bred postpartum have live births, most rat breeders do not utilize the postpartum estrus because the sire may bother the neonates. The sire is therefore removed from the cage prior to parturition and returned after weaning. Simultaneous lactation and pregnancy may delay implantation 3 to 7 days and therefore lengthen gestation.

Representative breeding systems for rats include monogamous and polygamous systems. The monogamous system involves the permanent pairing of one male and one female, with the young removed at weaning or prior to the next delivery. This system requires a large male population and increased labor and equipment. The postpartum estrus, however, would be utilized by approximately half the rats to maximize the number of litters per female per unit time.

The polygamous or harem system combines 1 male and 2 to 6 females. Females are removed to separate cages prior to parturition (usually about 16 days of gestation), and the postpartum estrus is not utilized. In this system the undisturbed females provide more milk and larger young and litters. The female is returned to the harem after the young have been weaned. If a colony system involves leaving the males and females together after delivery, removal of the young for 12 hours on the first postpartum day may facilitate postpartum mating and increase the survivability of the young.

Some commercial rat producers move the male from cage to cage; each cage contains a single female. The male remains 1 week in each cage and is reintroduced just after weaning. One male can be utilized for every 7 females. If the male is not removed before parturition, cannibalism, litter desertion, and agalactia may result. A similar system is used for hamsters, but the females are moved through the male's cage.

Pregnancy and Rearing

After rats mate, a white, waxy copulatory plug is present in the vagina for 12 to 24 hours postcoitum. A discharged plug may be found in the cage or litter pan. When one attempts timed matings, one finds that the rats' copulatory plugs tend to be dislodged more quickly than do those of mice. The placement of clean, dark paper under the breeding cage of newly mated rats allows one to search for the copulatory plug the morning following mating. If mating occurred, sperm will be present in the vaginal smear. Rats can be palpated, observed, or weighed to detect pregnancy. Mammary development is evident at 14 days.

Rats have a gestation period between 21 and 23 days long, unless there is concomitant lactation, in which case gestation is prolonged by several days. Exposure to a strange male within 24 hours of a fertile mating will not prevent implantation and luteotropic activity (Bruce effect), as may occur in mice. Pseudopregnancy is uncommon in rats, but when it occurs, it lasts approximately 13 days.

Rats build scant, shallow nests from bedding. Cotton, tissue paper, wood shavings, and shredded newspaper make good nesting materials. Pups born on ground corn cob bedding are often exposed to the cage floor, which may cool the young and cause the dam to abandon the litter.

A few hours before delivery, the dam licks her perineum, and a clear vaginal mucus appears. An average rat litter contains 6 to 12 young. Eyes open and hair is evident about 1 week postpartum. For the first few days following delivery, such disturbances as excessive handling, loud noises, and lack of nesting material may cause the female rat to destroy her young. It may be necessary, therefore, to isolate the mother before parturition. If the cage must be cleaned or changed, some of the nest and dirty bedding can be transferred to the new cage.

Young rats are weaned at 21 days (40 to 50 g). If the postpartum estrus is not utilized, the female rat resumes cycling 2 to 4 days postweaning.

Germ-free rats have been produced by hand raising under sterile conditions, and presumably young rats under 16 days of age could be fed a warmed formula from a pipette, but young rats are easily chilled, and the food may be aspirated. Simulated maternal stimulation of defecation and micturition must be provided. Precautions must be taken to assure that rodent diseases are not introduced *in utero* or via personnel, water, feed, cages, air, and equipment.

Disease Prevention

In addition to the routine husbandry precautions listed in Chapter 1, a systematic diagnostic evaluation of colony animals should routinely be conducted to screen for subclinical infections. Filter cage covers prevent or reduce airborne transmission of microorganisms within a densely populated rat room, although such covers alter the cage environment. Insects and wild rodents should be excluded from rodent colonies.

Public Health Concerns

Allergies to animal dander and urinary proteins occur commonly in humans, and cutaneous and upper and lower respiratory allergies to rats are very common. The urine of older rats may contain large amounts of serum protein, which may induce severe pulmonary allergies in people. Large, crowded rat populations, poor air circulation, and infrequent cage changing and cleaning that lead to accumulation of ammonia gas greatly compound the allergy problem in man.

Zoonotic diseases carried or transmitted by rats include leptospirosis, streptococcal infections, salmonellosis, cestodiasis, Korean hemorrhagic fever, and rat bite or Haverhill fevers. Sylvatic plague (*Yersinia pestis*) is carried by rat fleas. The mite *Liponyssus sylviarum* can transmit the St. Louis encephalitis virus, and *L. bacoti* attacks man directly. Rat-bite fever is caused by *Streptobacillus moniliformis,* which is carried in the nasopharynx of asymptomatic rats. People experiencing rat-bite fever may develop recurrent fever with petecchial hemorrhages, endocarditis, and polyarthritis. Although these diseases and infections may occur in wild rats brought into a laboratory, they are rare or unknown in domestic rats, as is rabies.

Uses in Research

Approximately 4.5 million rats are used annually in research and testing in the United States. Uses of laboratory rats include studies of aging, neoplasia, drug effects and toxicity, gnotobiology, dental caries, lipid metabolism, vitamin effects, behavior, alcoholism and cirrhosis, arthritis, phenylketonuria, jaundice, fructose intolerance, hypertension, embryology, teratology, diabetes insipidus, and infectious disease.

54 BIOLOGY AND HUSBANDRY

Sources of Information

Comprehensive works on the rat are *The Laboratory Rat* (2 vols.), edited by H.J. Baker, et al. (1979, Academic Press, Orlando, FL 32887); *The UFAW Handbook on the Care and Management of Laboratory Animals*, 6th edition, pp. 309–330, edited by T.B. Poole (1987, Churchill Livingstone Inc., 1560 Broadway, New York, NY 10036); and *The Guide to the Care and Use of Experimental Animals*, Vol. 2, Chapter 21, Rats, edited by H.C. Rowsell, et al. (1984, Canadian Council on Animal Care, 1105-151 Slater Street, Ottawa, ONTARIO K1P 5H3).

Other works on rats include:

Barnett, S.A.: The Rat: A Study in Behavior, 3rd Revised Edition. 1976. University of Chicago Press, 5801 Ellis Avenue, Chicago, IL 60637.

Brody, E.G.: Genetic Basis of Spontaneous Activity in the Albino Rat. 1942. Kraus Reprint, U.S. Division of Kraus-Thomson Organization, Ltd., Route 100, Millwood, NY 10546.

Burek, J.D.: Pathology of Aging Rats. 1978. CRC Press, 2000 Northwest 24th Street, Boca Raton, FL 33431.

Castaing, D., et al.: Hepatic and Portal Surgery in the Rat, 4th Edition. 1980. Masson Publishing U.S.A., Inc., 211 East 43rd Street, New York, NY 10017.

Chiasson, R.B.: Laboratory Anatomy of the White Rat, 4th Edition. 1980. William C. Brown Co., Publisher, 2460 Kerper Boulevard, Dubuque, IA 52001.

Cotchin, E., and Row, F.J.C.: Pathology of Laboratory Rats and Mice. 1967. Blackwell Scientific Publications, Inc., 52 Beacon Street, Boston, MA 02108.

Craigie, E.H.: Craigie's Neuroanatomy of the Rat, Revised Edition. W. Zeman and J.R. Innes, Editors. 1963. Academic Press, Inc., Orlando, FL 32887.

Geaves, P., and Faccini, J.M.: Rat Histopathology. Elsevier Science Publishing Co., Inc., 52 Vanderbilt Avenue, New York, NY 10017.

Greene, E.G.: Anatomy of the Rat. 1935, reprinted 1971. Hafner Press, Division of Macmillan Publishing Co., Inc., 866 Third Avenue, New York, NY 10022.

Hebel, R., and Stromberg, M.W.: Anatomy and Embryology of the Laboratory Rat. 1986. BioMed Verlag, Birgit, Hebel, D-8031 Worthsee, Fed. Rep. of Germany.

Herrick, C.J.: Brains of Rats and Men: A Survey of the Origin and Biological Significance of the Cerebral Cortex. 1983. Reprint of 1926 Edition. Box 148, Darby, PA 19023.

Konig, J., and Klippell, R.A.: The Rat Brain: A Stereotaxic Atlas of the Forebrain and the Lower Parts of the Brain Stem. 1974. Reprint of 1963 Edition. Robert E. Krieger Publishing Co., Inc., P.O. Box 9542, Melbourne, FL 32902.

Lambert, R.: Surgery of the Digestive System in the Rat. 1965. Charles C Thomas, 2600 South First Street, Springfield, IL 62717.

LeRoi, D.: Fancy Mice, Rats and Gerbils. Sportshelf & Soccer Associates, P.O. Box 634, New Rochelle, NY 10802.

Newberne, P.M., and Butler, W.H., Editors: Rat Hepatic Neoplasia. 1978. MIT Press, 28 Carleton Street, Cambridge, MA 02142.

Olds, R.J.: A Colour Atlas of the Rat: Dissection Guide. 1979. Halsted Press, Division of John Wiley & Sons, Inc., 605 Third Avenue, New York, NY 10158.

Paget, G.E.: *Clinical Testing of New Drugs*, pg. 33. Revere Publishing Co., NY, 1965.

Pellegrino, L.J., et al.: A Stereotaxic Atlas of the Rat Brain, 2nd Edition. 1979. Plenum Publishing Corp., 223 Spring Street, New York, NY 10013.

Richter, C.P.: Behavioristic Study of the Activity of the Rat. 1922. Kraus Reprint, U.S. Division of Kraus-Thomson Organization, Ltd., Route 100, Millwood, NY 10546.

Sherwood, N., and Timiras, P.: A Stereotaxic Atlas of the Developing Rat Brain. 1970. University of California Press, 2120 Berkeley Way, Berkeley, CA 94720.

Smith, E.M., and Calhoun, M.L.: The Microscopic Anatomy of the White Rat: A Photographic Atlas. 1968. Iowa State University Press, 2121 South State Avenue, 112C Press Office, Ames, IA 50010.

Thompson, S., et al.: The Adrenal Medulla of Rats: Comparative Physiology, Histology and Pathology. 1981. Charles C Thomas, 2600 South First Street, Springfield, IL 62717.

Thorpe, D.R.: Rat: A Dissection Guide in Color. 1968. Mayfield Publishing Co., 285 Hamilton Avenue, Palo Alto, CA 94301.

Walker, D.G., and Wirtschafter, Z.T.: The Genesis of the Rat Skeleton: A Laboratory Atlas. 1957. Charles C Thomas, 2600 South First Street, Springfield, IL 62717 (out of print).

Wayneforth, H.B.: Experimental and Surgical Techniques in the Rat. 1980. Academic Press, Inc., Orlando, FL 32887.

Wells, J.: The Rat. 1966. James H. Heineman, Inc., Publishing, 475 Park Avenue, New York, NY 10022.

Zinsser, H.: Rats, Lice and History. 1935. Little, Brown & Company, 34 Beacon Street, Boston, MA 02106.

Chapter 3
Clinical Procedures

The clinical principles underlying treatment of disease in rabbits and rodents include drug dosing and dosages, anesthetics and anesthesia, surgical procedures, analgesia, radiographic and serologic techniques, blood collection, and euthanasia. Specific treatments are discussed with the diseases described in Chapter 5.

Treatments for rabbits and rodents are often based on extrapolations from other species, sporadic and limited clinical trials, and hearsay. References should be reviewed before using the listed drugs. In some cases, treatments with drugs and regimens known to be efficacious in larger animals are seemingly ineffective in small mammals; nevertheless, for the benefit of the pet owner and the researcher with the valuable subject, a treatment should at least be considered, if not attempted.

DRUGS

Few drugs used in clinical practice are specifically approved by the US FDA for use in rabbits and rodents, a consideration that presents both legal and therapeutic complications.

Several products have been approved by the FDA for specific uses in rabbits, guinea pigs and mice. FDA approvals are old and specific products may no longer be on the market (Personal Communication—Dr. Emilio Viera, USDA Project IR-4, Animal Drugs of Minor Species, the State University of New Jersey, Cook College and Rutgers University, New Brunswick, NJ 08903), but other products containing the same drugs may be available. These products include:

Combiotic (Pfizer Inc., New York, NY 10017) contains 0.25 gm of dihydrostreptomycin sulphate and 200,000 units procaine penicillin G per ml and is approved for use in rabbits and is on the market. It must be discontinued 30 days before treated animals are slaughtered for food.

Distrycillin A.S. (Solvay Veterinary Inc., Box 788, Arlington Heights, IL 60006) is on the market and approved for rabbits and is similar to Combiotic.

Sulfaquinoxaline (40% Premix Merck & Co., Inc., Rahway, NJ 07065) has a 10 day withdrawal before treated animals are slaughtered for food. It was approved for use as a coccidiostat in rabbits, but the product has been discontinued.

TM-10 Premix (Pfizer, Inc., New York, NY 10017) contains 10 grams of oxytetracycline per pound of premix. It is approved as a growth stimulant and feed efficiency promoter in rabbits. There is no withdrawal time. It is on the market and is mostly sold directly to feed distributors for mixing in feed.

T-61 (Hoechst-Roussel Agri-Vet Company, Somerville, NJ 08876) was FDA-approved as an euthanasia solution for guinea pigs, mice, rats and rabbits but not to be used in animals intended for food. The specifically approved product is no longer on the market; however, a T-61 product is approved and available for IV injection in dogs.

Dyrex (Fort Dodge Labs., Inc., Fort Dodge, IA 50501) containing trichlorfon and atropine sulphate can be dissolved in drinking water to be used as the sole source of drinking water for 7 to 14 days. Although approved as an anthelmintic for mice, it is no longer on the market, but Dyrex T.F., a different formulation, is marketed for horses.

Yomesan 33% premix (Mobay, Shawnee Mission, KS 66201) is recommended for use at 6.6 pounds of premix per ton of feed, to be fed daily for 7 consecutive days. If infection recurs, treatment is to be repeated after 30 days. The specific product approved for treating tapeworms of mice is no longer on the market.

Apart from shortcomings resulting from interspecies extrapolations and limited clinical trials, several factors influence the effects of a drug. Sex, age, diet, health status, time of day, breeding status, metabolic rate, and nutritional level of the subjects, as well as composition of the bedding, experimental protocol, concomitant administration of other drugs, type of caging, and the ambient temperature, can affect a therapeutic outcome. In addition to these concerns, some drugs relatively safe and efficacious in other animals either are toxic in rodents or will suppress an active, clinical infection to a chronic, subclinical, carrier state.

Antibiotic therapy should include considerations of bacterial sensitivity, cost of treatment, toxicity of the antibiotic, public health significance of the pathogen involved, compatibility with other drugs or fluids, and the experimental protocol of the research study.

Fatal reactions to antibiotics are a major concern in the therapy of rabbit, guinea pig, and hamster diseases. The toxic effect may be direct, as with streptomycin and dihydrostreptomycin given in very large doses, or indirect through an alteration in the intestinal flora with a subsequent enterotoxemia. Following an antibiotic-induced reduction in anaerobes and gram-positive streptococci and lactobacilli, deaths from clostridial or coliform enterotoxemias begin within 3 days and continue for a week or more.

The penicillins have caused enterotoxemias in guinea pigs and hamsters. The morphologic effect, caused by doses of 10,000 U and more, is a hemorrhagic cecitis and colitis. Not all guinea pigs die, however, following penicillin injection, and rats, mice, gerbils, and rabbits are relatively resistant. Ampicillin given orally for 6 days at 10 mg/kg has killed rabbits; and procaine, a component of some penicillin preparations, may be fatal at 0.40 mg/kg in mice, guinea pigs, and rabbits.

Large doses of streptomycin are toxic in mice at 3 mg and dihydrostreptomycin is toxic in gerbils at 50 mg. Streptomycins cause an ascending flaccid paralysis, respiratory arrest, coma, and death. The tetracyclines may cause a fatal enterotoxemia when given at 50 mg/kg subcutaneously to hamsters, unless sulfaguanidine is given simultaneously.

According to the package insert, lincomycin is contraindicated for hamsters,

Adult Weights: Approximate Ranges

	Male	Female
Rabbit (medium)	4–5 kg	4.5–6 kg
Guinea pig	800–1000 g	750–900 g
Hamster	85–110 g	95–150 g
Gerbil	80–120 g	70–95 g
Mouse	20–40 g	25–40 g
Rat	300–400 g	250–300 g

guinea pigs, and rabbits. Low doses (5 to 30 mg) in rabbits may cause severe enterocolitis, anorexia, diarrhea, and death within 1 to 2 days. Gentamicin may counteract this effect to some extent. Erythromycin has caused enteric disease in hamsters, guinea pigs, and rabbits. Bacitracin orally at 2000 U has proven lethal in guinea pigs within 12 hours, but a combination of 5 mg neomycin and 3 mg polymyxin B given orally twice a day for 5 days counteracted this effect.

Perhaps the most popular therapeutic regimen in an animal colony is the administration of various concentrations of tetracycline or oxytetracycline in the drinking water. Dosages in some regimens are as high as 5 mg/ml drinking water, but the higher the concentration, especially with oxytetracycline, the greater the possibility of precipitation with some ions of tap water. This scale lines the bottles, blocks the sipper tubes, and reduces the antibiotic in solution. Thus, to retain a therapeutic level of the drug, fresh solutions should be prepared daily with deionized water. Another potential problem with long-term administration

of tetracycline is the inevitable development of a tetracycline-resistant microflora with abnormally high numbers of *Proteus* spp. Moreover, the treatment of a large room of rodents with a high concentration of tetracycline can cost hundreds of dollars. For the most part, however, drugs can be used in rabbits and rodents as they are in other species. The same concerns for dosage, duration of treatment, and compatibilities apply. Special precautions must be exercised to preclude drug residues in rabbits intended for human consumption.

The dosages recommended in the following tables are approximate for the average population for the indications stated. Individual responses may vary, and animals receiving these levels may be unaffected, respond as desired, or be fatally overdosed. The drugs included represent only a small sample of the drugs that are efficacious in small animals, and the intention is not to promote one product over another. The data in this section were assembled from personal observations and communications, from numerous scientific publications, from C.D. Barnes' and L.G. Eltherington's *Drug*

Table of Drug Dosages

Drug	Indication	Dosage	Route	Remarks
Anesthetics				
Ketamine HCl	Dissociative anesthetic	20 mg/kg for restraint 25–40 mg/kg in rabbits and guinea pigs 60–90 mg/kg in rodents	IM	Requires addition of acepromazine, xylazine, or diazepam; irritates IM injection site; IP given diluted
Pentobarbital sodium	Barbiturate anesthetic	10–20 mg/kg in young rodents 35–45 mg/kg in rats, rabbits, and guinea pigs 60–90 mg/kg in mice, gerbils, and hamsters	IV,IP	Dose can vary greatly; generally a high-risk anesthetic with several metabolic effects; a poor analgesic
Thiamylal sodium	Ultrashort-acting barbiturate	25–50 mg/kg or to effect	IV,IP	Causes tissue irritation if extravascular

Table of Drug Dosages (*Continued*)

Drug	Indication	Dosage	Route	Remarks
Antagonists				
Atropine sulfate	Organophosphate overdose	10 mg/kg every 20 min	SC	Large doses cause cardiovascular irregularities in guinea pigs
Vitamin K_1	Warfarin poisoning	1–10 mg/kg as needed	IM	Menadiols not for acute cases
Nalorphine	Narcotic overdose	2 mg/kg	IV	Antagonizes fentanyl but not barbiturates
Anthelmintics				
Dichlorvos	Endoparasitism	500 mg/kg feed for 24 hours	PO	Cholinesterase inhibitor
Ivermectin	Endoparasitism	200 to 400 mcg/kg	SQ	*Trichosomoides crassicauda* of rats and other parasitisms
Piperazine salts	Anthelmintic	200 mg/kg in rabbits 3 mg/ml in drinking water for rodents	PO	Piperazine flavor may be disguised by molasses
Thiabendazole	Anthelmintic	100–200 mg/kg 0.1% in feed	PO	Antipyretic and anti-inflammatory effects
Antimicrobials				
Chloramphenicol palmitate	Broad-spectrum antibiotic	50 mg/kg daily for 5–7 days	PO	May prolong barbiturate anesthesia
Chloramphenicol succinate	Broad-spectrum antibiotic	30 mg/kg daily for 5–7 days	IM	
Griseofulvin	Antifungal agent	25 mg/kg daily for 4 wk in rabbits 75 mg/kg daily for 2 wk in guinea pigs	PO	Cure is prolonged and difficult

Table of Drug Dosages (*Continued*)

Drug	Indication	Dosage	Route	Remarks
Procaine penicillin	Antibiotic	40,000 units/kg daily for 5–7 days	IM,SC	May cause fatal floral alteration in guinea pig and hamster gut
Sulfamethazine	Antimicrobial for *Citrobacter* and *Bordetella*	1 g/L drinking water for 5 days 1 mg/ml water for 20–60 days	PO	Rapid gut absorption with lower nephrotoxicity for sulfonamides
Sulfaquinoxaline	Coccidiostat and antimicrobial for rabbit pasteurellosis	0.025% in water for 30 days; treatment level 0.1% in water for 2 wk	PO	10 day withdrawal before treated animals are slaughtered for food
Tetracyclines	Antibiotics for mycoplasmosis, bacterial disease, Tyzzer's disease, or prevention of pasteurellosis	0.3–2 mg/ml water 5–50 mg/kg BW Recommendations in literature vary widely	PO, IM	5% sucrose in water may help to ensure consumption; mixed fresh daily in deionized water
Tylosin	Mycoplasmosis of rodents	66 mg/L water for 21 days	PO	Tylosin and the tetracyclines at best suppress *Mycoplasma*
Insecticides				
Carbaryl powder	Ectoparasitism	5% dust or diluted 1:1 with talc		Young rodents are susceptible to poisoning
Dichlorvos	Ectoparasitism	Resin strip sections left on cage for 48 hours 1 wk intervals or pellets in cage		Cholinesterase inhibitor, may inhibit breeding
Ivermectin	Parasitism	200–400 μg/kg	SC, IM	Eliminates ear mites on rabbits
Oxytocin				
Oxytocin	Delayed parturition or dystocia	0.2–3.0 units/kg	SC, IM	Also give manual assistance during delivery

Table of Drug Dosages (*Continued*)

Drug	Indication	Dosage	Route	Remarks
Preanesthetics				
Acepromazine maleate	Tranquilizer and antiemetic	1–2 mg/kg	IM	May precipitate seizures in gerbils; used with ketamine
Atropine sulfate	Anticholinergic; respiratory stimulant	0.1–3.0 mg/kg	SC	Many rabbits and rats possess a serum atropinesterase
Chlorpromazine	Tranquilizer, antiemetic	3–5 mg/kg 3–35 mg/kg	IV IM	Potentiates barbiturates; causes myositis
Diazepam	Tranquilizer, muscle relaxant	5–10 mg/kg in rabbits	IM	Used with ketamine and other anesthetics
Xylazine	Sedative, analgesic	3–5 mg/kg in rabbits and guinea pigs 4–8 mg/kg in rodents	IM	Widely used in combination with ketamine

Dosing and Injection Procedures

	Oral	SC	IM	IV
Rabbit	Feed or flexible tube	Flank, back	Thigh, lumbar muscles	Ear vein, forelimb vein
Guinea Pig	Feed, flexible tube, bulbed needle	Back	Thigh muscles	Carotid artery, cranial vena cava, caudal auricular (ear), cephalic, femoral, lateral saphenous, lateral metatarsal, and dorsolateral, penile veins
Hamster	Feed, bulbed needle	Back, abdomen	Thigh muscles	Femoral or jugular vein
Gerbil Mouse Rat }	Feed, bulbed needle	Back, abdomen	Thigh muscles	Tail vein, sublingual, or subclavian vein

Food and Water Needs of Adult Laboratory Animals

	Daily Feed Intake per 100 g Body Weight	Daily Water Intake per 100 g Body Weight
Rabbit	5 g	10 ml
Guinea pig	6 g	10 ml
Hamster	10–12 g	8–10 ml
Gerbil	5–8 g	4–7 ml
Mouse	15 g	15 ml
Rat	10 g	10–12 ml

Dosage in Laboratory Animals: A Handbook (Berkeley, University of California Press, 1973), and from I.S. Rossoff's *Handbook of Veterinary Drugs: A Compendium for Research and Clinical Use* (New York, Springer Publishing Company, 1974).

Specific drugs, dosages, and indications are listed, and information regarding average body weights, dosing and injection procedures, and daily water and feed consumption is included.

A soft rubber tube 2 to 3.5 mm in diameter or a No. 8 to No. 10 French catheter is a satisfactory stomach tube for a rabbit. The tube is passed through an open-ended syringe cover or through a hole in the sides of a 7.5-cm (3-in.) section of plastic hose or wooden dowel placed as a bit in the rabbit's diastema. The guide will prevent chewing on the tube. Bulbed needles for dosing rodents are available from several research animal equipment companies.

The food and water consumptions listed below are approximations for dosing purposes only; actual amounts consumed depend on the diet (dry, powder, pellets, supplements), ambient temperature, ventilation, breeding and health status, wastage, and competition for feed and water. Amounts may triple or be more during lactation.

ANESTHESIA

In this section anesthetics commonly available in a veterinary or animal facility clinic are described. Intraperitoneal and in-tramuscular routes for injectable drugs and chambers or nose cones for inhalant anesthetics are emphasized over intravenous and endotracheal routes. Information on the spectrum of drugs and techniques available is contained in the comprehensive reference list at the end of this chapter. A table of doses can be found in this chapter.

Many factors, such as interspecies and intraspecies variations, affect the response of rabbits and rodents to anesthetics. Specific factors, including sex, age, strain, weight, percent body fat, health and nutritional status, ingesta content, time of day, genetic background, type of bedding, environmental temperature during recovery, cardiovascular and respiratory system effects, and respiratory and metabolic rates also influence the response. The large, ingesta-filled ceca of rabbits and guinea pigs cause misleading body weights. Sleeping times are longer in the afternoon, when hepatic metabolic activity in the nocturnal species is at low ebb. Cedar and pine bedding emit substances that activate enzyme systems that degrade anesthetics; as a result, animals on these beddings have shorter sleeping times. Other chemicals may inhibit the same enzyme systems.

Preoperative Procedures

Candidates for anesthesia and surgical procedures should, if possible, be free from respiratory disease. Food, but not water, should be withheld from rabbits and guinea pigs for 8 to 12 hours preceding induction of anesthesia to reduce quantity of ingesta, which may cause variations in dose effects.

Atropine sulfate may be administered subcutaneously as a preanesthetic with ether, halothane, ketamine, droperidol-fentanyl, thiamylal, or thiopental anesthesia, according to species. Atropine, administered 30 minutes prior to anesthesia, reduces vagal tone and copious secretions caused by ether, halothane, ketamine, and thiobarbiturates. Atropine may be injected concurrently with droperidol-fentanyl to control salivation and bradycardia. Atropine is used as a preanesthetic for thiobarbiturates in rabbits only.

Some rabbits have atropinesterase, which hydrolyzes and inactivates atropine, circulating in the bloodstream. Therefore, the dosage of atropine for rabbits varies greatly and it may be necessary to increase the dose or administer it in intervals. The level of atropinesterase activity further depends on season of year and the sex, breed, age, and weight of rabbit. Rabbits can be tested qualitatively for atropinesterase activity prior to atropine administration by the following methods: 1) *in vivo* pupillary reflex method, 2) agar plate test, and 3) microhematocrit tube test. For *in vivo* testing, the rabbit is injected with 0.2 mg atropine/kg body weight and observed for pupillary dilation and loss of pupillary reflex (lacks atropinesterase) 45 minutes later.

Respiratory secretions accumulating during anesthesia are removed by suction or cotton swab; however, rabbits and rodents rarely regurgitate and therefore generally need not be intubated.

Acepromazine or diazepam is used as a tranquilizer or preanesthetic for ketamine or methoxyflurane. Acepromazine or the sedative-analgesic xylazine is often included in the syringe with ketamine. The area where surgery will be done should be warmed (30° to 35° C) by a circulating hot water pad, hot water bottle, or incandescent lamp, and the animal is positioned to ensure an open air passage. A bland ophthalmic ointment is placed on the cornea.

Administration of Anesthetics

Inhalant anesthetics are commonly delivered by wetted gauze in a covered chamber, face mask, or nose cone. The chamber may be plastic or glass. Paper cups and plastic syringe covers often make satisfactory nose cones, although the anesthetic may damage some plastics. The mask or cone is manipulated on and off the animal to supply effective combinations of air and anesthetic. Injectable anesthetics are usually given into the muscle mass of the rear limbs, into the skin folds on the back of the neck, or into the peritoneal cavity. Intraperitoneal injections are given by sharp puncture in the lower left abdominal quadrant while the animal's foreparts are tilted down.

Aspirating the syringe prior to injection helps to ensure that the anesthetic is administered into the peritoneal cavity and not into organs or blood vessels. An estimated 20% of intraperitoneal injections in guinea pigs inadvertently penetrate abdominal organs. Prolonged induction and recovery times may be indicative of injection of the anesthetic into an organ. Dosages and body weights should be accurately determined to avoid fatalities because entire doses are usually administered at once.

Tabletop anesthetic machines with gas and air pumps are available commercially and are described in several references.

Endotracheal intubation is difficult in rabbits and rodents because of the long distance from the incisors to the epiglottis, the sharp teeth, the small mouth, and occurrence of laryngospasms.

A Cole pediatric endotracheal tube or a Shiley nasal oral tube can be used for intubation. Recommended inside diameters of the tubes are 1.5 mm for rats and hamsters, 1.5 to 2.5 mm for guinea pigs, and 2.0 to 4.0 mm for rabbits. The tube should extend from the mouth to near the thoracic inlet. Long tubes must be cut. If inserted too far, the tube may enter a bronchus, thereby causing tissue trauma or occlusion.

A medium-sized laryngoscope with a curved blade facilitates passage of the tube over the extended tongue and into the rabbit's glottis. Rodents require a Miller O neonatal laryngoscope or a canine otoscope with a 4-mm backcone.

Animals are anesthetized prior to intubation. Relaxation of the jaw musculature signifies that the animal is sufficiently anesthetized to begin intubation. Topical anesthetic should not be sprayed into the oral cavity because of the danger of asphyxiation or drowning.

A topical anesthetic can be applied to the soft palate and pharynx with cotton swabs. Topical anesthesia to prevent laryngospasms in rabbits is not indicated for intubation because the anesthetic suppresses the convenient forward motion of the glottis during swallowing.

The animal to be intubated is placed in sternal recumbency with its neck extended to afford a straight line of vision from the mouth to larynx. The laryngoscope or otoscope is placed over the tongue, which is pulled forward, and the scope is guided into the oral cavity. A stylet is placed into the oral cavity and the epiglottis is moved forward to expose the larynx.

A lubricated endotracheal tube is rotated slowly while passing through the stylet and oral passages. Resistance is met when the endotracheal tube passes through the arytenoid cartilages. The endotracheal tube should be retracted at the point of resistance and reinserted gently while rotating. This step may need to be repeated until the tube successfully enters the larynx. If the tube advances without resistance, it is likely to be in the esophagus. The patency of the airway is verified by observing water condensation from air exhaled onto a smooth glossy surface.

The stylet and otoscope or laryngoscope are removed after the tube is in the larynx, and the tube is secured to the animal's head. Anesthetic gases (halothane, enflurane, methoxyflurane) are combined with a 1:1 oxygen:nitrous oxide mixture at 100 ml/mix for rodents and 2 L/mix for rabbits. Because small animals have a minute respiratory tidal volume, one must be certain not to use large equipment because of the potential for creating dead air spaces. The dead air spaces create dangers from repeated rebreathing of expired air.

Intravenous injections in rabbits are given into the marginal ear (pinna) vein. Visibility of the vein is improved by shaving the hair over the vein and rubbing the vein or applying alcohol or xylene. Rats and mice have small but accessible tail veins. Vasodilation is achieved by warming the tail with a heat lamp or water. Aspiration of blood from tail veins in mice is difficult because of the veins' small size and tendency to collapse. Limb, ear, and penile veins are superficial but small in hamsters and guinea pigs.

When small amounts of drugs are to be administered, doses can be more accurately calibrated by making diluted solutions. Anesthetics can be combined with distilled water or physiologic saline to achieve a 1:10 dilution ratio.

Anesthetics

Anesthetics are listed in the Table of Drug Dosages. Their pharmacologic actions are described in texts on anesthesia and anesthetics. Drugs used in anesthetic regimens and considered relatively safe, effective, available, and easy to administer are ketamine (100 mg/ml), xylazine (20 or 100 mg/ml), acepromazine (10 mg/ml), methoxyflurane, and atropine. In most cases the proper use of these drugs, in various combinations, will produce satisfactory anesthesia in rabbits or rodents, albeit experimental requirements or technical experience may require alternatives. Other useful drugs include ether, halothane, isoflurane, enflurane, fentanyl-droperidol, fentanyl-fluanisone, sodium pentobarbital, chlorpromazine, diazepam, and paraldehyde.

Cautions are to be observed. Ketamine alone provides poor muscle relaxation and minimal analgesia regardless of dosage level. It is a cardiovascular stimulant and gives complex cardiovascular responses. Its use is inappropriate for studies involving autonomic control. Some rodents may continue to perambulate despite ketamine doses of 100 mg/kg. Ketamine is an acidic solution and irritates muscle. Anesthetics given intraperitoneally should be diluted. For this reason, ketamine is inappropriate to use as the sole anesthetic in operations that require skeletal muscle relaxation and analgesia. Xylazine, acepromazine, and volatile anesthetics may be used in conjunction with ketamine to improve depth of anesthesia. Yohimbine hydrochloride given intravenously can be used to decrease anesthetic duration following ketamine/xylazine anesthesia in rabbits. Ether, halothane, and sodium pentobarbital pose risks and, in inexperienced hands, often cause fatalities. Fentanyl-droperidol, chlorpromazine, paraldehyde, and ketamine administered parenterally may cause pain, swelling, and tissue damage at the injection site and result in self mutilation. Acepromazine may induce convulsions in gerbils. Ether is highly flammable, irritates respiratory and oral mucous membranes, and causes laryngeal spasms or profuse salivation, which may block air passageways. Atropine may cause hypertension in guinea pigs and may increase susceptibility of hemorrhaging during surgery. Because of its explosive nature, the use of ether may be restricted to chemical fume hoods or be prohibited for insurance reasons. Halothane produces mild mucous membrane irritation and acts as a respiratory and cardiovascular depressant. Halothane and methoxyflurane may be hazardous to human health and should be used with a scavenger system or in a fume hood. Frequent use of halothane in guinea pigs may cause liver damage. The lethal and anesthetic dosages of pentobarbital are very close in guinea pigs, rabbits, and rats; therefore, great care must be taken in administration to avoid fatalities. Response to pentobarbital dosages varies widely among rabbits. Pentobarbital produces irritation if injected perivascularly. When methoxyflurane is used with F344 rats, a syndrome resembling diabetes mellitus may occur. Use of these agents should be avoided when possible.

Topical Anesthesia

Ethyl chloride spray (Chloroethane, Gebauer Chemical Co., Cleveland, OH) is used as a topical anesthetic on rabbits and guinea pigs during the tattooing process. A single application to intact skin provides anesthesia for up to 30 seconds. Re-application may produce frostbite and is not recommended.

Depth of Anesthesia

Indicators of depth of anesthesia in laboratory animals are similar to those in other species. Increased depth of respiration and slowed respiratory rate, relaxation of the abdominal and jaw musculature, color of the mucous membranes and albino iris, and certain reflex responses are used to determine the level. Signs of surgical anesthesia in the rabbit include the movement of the nictitating membrane over approximately one third of the cornea, a stable respiratory rate of not lower than 18 to 24 respirations per minute, relaxation of the abdominal musculature, and loss of the mouth, palpebral, and ear and toe pinch reflexes. Protrusion of the eyeball, loss of corneal reflex, dilated pupils, and cyanosis indicate an anesthetic overdose. A guinea pig in stage 3 of anesthesia may wave its rear legs and create the false impression of a return to consciousness, but additional anesthetic may kill the animal. The pedal reflex is not a reliable indicator of surgical anesthesia in the guinea pig but is useful for small rodents.

Anesthesia in the Rabbit

The rabbit has a respiratory rate between 35 and 60 respirations per minute and a tidal volume of 4 to 6 ml/kg body weight. Ether and sodium pentobarbital (35 to 55 mg/kg) are "high risk" anesthetics in rabbits because of the sensitivity of the rabbit's respiratory center to volatile anesthetics and injected barbiturates. Ketamine alone at 20 to 35 mg/kg will quiet a rabbit for examination or intubation, but ketamine is an acidic solution and causes tissue irritation.

Xylazine is a sedative and hypnotic. It has cardiovascular depressive and hypotensive actions. It is reversed by α-2 adrenergic (CNS) autagonists. Analgesia is primarily in the trunk and less so or even hyperalgesic in the appendages.

Acepromazine has a wide variety of receptor and membrane effects. These effects vary with dose and species. It has an adrenergic receptor blocking effect that affects physiologic and pharmacologic studies. It is a hypotensive and is contraindicated in cardiovascular research.

Ketamine in combination with acepromazine (1 mg/kg) is given intramuscularly at 25 mg/kg for tranquilization and 20 to 40 mg/kg for anesthesia. This injection may be followed as needed by a second ketamine injection (20 mg/kg) or by methoxyflurane or halothane to the desired level of anesthesia. The inhalational anesthetics may be administered by nose cone or endotracheal tube. Other combinations include ketamine (25 to 35 mg/kg) with xylazine (3 mg/kg) or diazepam (5 to 10 mg/kg). Diazepam enhances inhibitory CNS pathways. Induction is 4 to 13 minutes and anesthesia lasts 15 to 60 minutes. Ketamine alone is insufficient for anesthesia because it provides poor analgesia and skeletal muscle relaxation.

Ketamine (50 mg/kg) and paraldehyde (0.5 mg/kg) can be administered intramuscularly. After 2 hours ketamine may be given again at 20 mg/kg to maintain anesthesia. The combination of ketamine and paraldehyde produces long-lasting anesthesia. Paraldehyde produces pain on injection.

Methoxyflurane or halothane (induction at 3% and maintained at 0.5 to 1%) may be administered alone or with 1:1 oxygen:nitrous oxide until surgical anesthesia is reached, at which time the gas concentration is reduced for proper maintenance. Rabbits exposed to halothane often struggle and cough. The rabbits may be preanesthetized with acepromazine (1 mg/kg) or ketamine (20 mg/kg) intramuscularly.

Fentanyl-droperidol is administered intramuscularly at 0.13 to 0.22 ml/kg body weight. Sodium pentobarbital is given slowly via the marginal ear vein or intraperitoneally until the desired level of anesthesia is achieved; the dose is between 35 and 55 mg/kg, although respiratory depression may be present at doses of 50 mg/kg. The higher doses can be reduced to 20 or 30 mg/kg if the rabbit is given chlorpromazine at 7.5 mg/kg intravenously or 25 mg/kg intramuscularly.

Hypnosis has been used as a restraining technique for rabbits undergoing injections, radiographic procedures, and minor operations but the response varies greatly. The rabbit is firmly restrained on its back, with the eyes covered or uncovered, in a V-shaped rack. The ventrum is gently stroked until the rabbit relaxes and breathing slows from the normal average of 51 to approximately 21 respirations per minute. If the rabbit awakens during the procedure, it is restrained and again stroked to reinduce the trance. Some investigators claim that tranquilizers, soft music, and gentle words facilitate hypnosis, whereas loud noises may arouse the rabbit.

Anesthesia in the Guinea Pig

Highly volatile inhalational anesthetics (ether, halothane) are "high risk" anesthetics in guinea pigs because the guinea pig exposed to an irritating gas will first hold its breath and then breathe deeply. Halothane (1%) causes as a 35% to 40% decrease in blood pressure in the guinea pig. Fentanyl-droperidol injected intramuscularly in guinea pigs has produced necrosis of nerves, vessels, and muscles with subsequent self mutilation and sloughing of extremities. An alternate procedure is the administration of fentanyl-fluanisone at 1 ml/kg intramuscularly with diazepam at 2.5 mg/kg intraperitoneally. A guinea pig has a normal respiratory rate between 70 and 130 respirations per minute and an approximate tidal volume of 3.8 ml/kg body weight. Two cautions to observe with guinea pigs are the misleading body weights caused by the large ceca and the muscular movements occurring during surgical anesthesia.

Xylazine (5 mg/kg) and ketamine are used

to anesthetize or preanesthetize guinea pigs. This xylazine-ketamine mixture is injected intramuscularly at a ketamine dose of 20 to 40 mg/kg body weight. The injected drugs are followed by methoxyflurane or halothane via nose cone. An adequate oxygen supply must be provided. At 22 mg/kg, ketamine administered intramuscularly will quiet a guinea pig for approximately 15 minutes, although analgesia is poor. At 33 mg/kg, the anesthesia lasts approximately 26 minutes, and at 44 mg/kg, it lasts 35 minutes; however, surgical anesthesia is often not attained at the higher dose. Recovery requires 5 to 30 minutes. Ketamine stimulates salivation in some guinea pigs.

Methoxyflurane is a good anesthetic for use in guinea pigs, although it stimulates salivation. Induction is smooth and muscle relaxation is excellent, but control of the depth of anesthesia is difficult if the open drop or nose cone method is used. If methoxyflurane does not produce satisfactory anesthesia, the careful use of halothane is recommended. Atropine should be used as a preanesthetic to prevent excessive salivation. If a closed or semiclosed anesthetic machine is used, 1% methoxyflurane is used for induction and 0.3% for maintenance. Induction requires 15 to 18 minutes and recovery requires 10 minutes or more.

Sodium pentobarbital (1% solution) administered intraperitoneally is difficult to dose because of the indefinite injection site, differential absorption, and varying dispersion of the drug in the peritoneal cavity. Sodium pentobarbital may be given intraperitoneally at 28 to 35 mg/kg body weight. This dose will provide approximately 30 to 100 minutes of surgical anesthesia. A combination with chlorpromazine at 25 mg/kg will extend the anesthesia to 50 or more minutes. Diazepam (8 mg/kg) and pentobarbital (20 mg/kg) intraperitoneally produce surgical anesthesia after 13 minutes. Hypnosis may be used as a restraining technique in guinea pigs as well as in rabbits. As a result of its tranquilizing effects, administration of reserpine may help to induce hypnosis.

Anesthesia in the Hamster

Hamsters have a respiratory rate of approximately 40 to 120 respirations per minute and a tidal volume of 0.8 ml. The effect of ketamine in hamsters is variable; some investigators report anesthesia at 20 to 30 mg/kg, but others note only catalepsy and mild analgesia at 200 mg/kg body weight. Ketamine (80 mg/kg) plus xylazine intramuscularly or intraperitoneally produces anesthesia. Droperidol-fentanyl is not recommended for use in the hamster because central nervous system stimulation may occur.

Sodium pentobarbital administered intraperitoneally to hamsters at 60 to 90 mg/kg produces anesthesia lasting 30 to 45 minutes. The injection is given in the lower abdomen off the midline.

Volatile anesthetics (ether, methoxyflurane, halothane) placed on a gauze sponge in a covered container will anesthetize hamsters for a short period.

Anesthesia in the Gerbil

Sodium pentobarbital administered intraperitoneally at 60 mg/kg body weight to a maximum dose of 0.10 ml (6 mg) will anesthetize a gerbil for 30 to 45 minutes.

Ketamine (40 mg/kg) intramuscularly or intraperitoneally plus methoxyflurane and air via a nose cone (paper cup or plastic syringe cover) will produce anesthesia in 90 seconds and deep anesthesia in approximately 12 minutes. Diazepam at 10 mg/kg intraperitoneally can be given with ketamine.

Volatile anesthetics may be used in a covered container to anesthetize gerbils for short periods.

Anesthesia in the Mouse

The mouse has an average respiratory rate of 60 to 220 respirations per minute and an approximate tidal volume of 0.15 ml. Intravenous injections are made into the lateral tail vein, after the tail has been pulled through a hole in the cage or in a piece of cardboard. Warming the tail dilates the vessels and facilitates needle entry. Care should be taken not to aspirate blood or inject air emboli.

Sodium pentobarbital injected intraperitoneally at 80 to 90 mg/kg body weight is a satisfactory anesthetic in the mouse. The pentobarbital solution should be diluted 1:9 with physiological saline solution. Sleep time is 20 to 40 minutes. A combination of chlorpromazine intramuscularly (25 to 50 mg/kg) and sodium pentobarbital intraperitoneally (40 to 60 mg/kg) produces anesthesia in mice. Males and starved mice are more susceptible to pentobarbital effects.

Ether, halothane, and methoxyflurane may be applied to gauze and placed in a closed container. Ether causes profuse salivation in mice. Methoxyflurane, an excellent anesthetic for mice, has an induction time of 4 to 6 minutes in a chamber.

Although ketamine has been reported to produce anesthesia at 44 mg/kg body weight, other trials have reported that doses to 200 mg/kg are ineffective. A combination of ketamine (50 mg/kg) and xylazine (50 mg/kg) given intramuscularly gave anesthesia lasting 60 to 100 minutes. Ketamine (60 to 80 mg/kg) plus half that volume of xylazine (20 mg/ml stock solution) caused anesthesia lasting 60 to 80 minutes, whereas ketamine at 90 to 120 mg/kg plus xylazine (10 mg/kg) produced a 90-minute sleeping time.

Anesthesia in the Rat

Practical and safe methods for anesthetizing rats include methoxyflurane administered via a nose cone or various combinations of ketamine and sedatives or tranquilizers. Rats placed into a 2-L chamber with methoxyflurane-soaked gauze may not be anesthetized even after 40 minutes, but the same agent confined in a small nose cone provides rapid and effective anesthesia.

Ketamine alone is a poor anesthetic for rats, but at 75 to 95 mg/kg intramuscularly or intraperitoneally in combination with xylazine at 5 to 8 mg/kg, ketamine gives variable anesthesia for 15 to 60 minutes.

Recovery requires over 2 hours. If ketamine is given at 60 mg/kg intramuscularly and sodium pentobarbital at 20 mg/kg intraperitoneally, anesthesia will last approximately 1 hour.

Other regimens deemed useful involve fentanyl-droperidol at 0.20 ml/kg intramuscularly (anesthesia for 30 to 60 minutes) and fentanyl-fluanisone at 0.3 ml/kg intramuscularly with diazepam at 2.5 mg/kg intraperitoneally (anesthesia lasts 2 hours). Several other injectable anesthetics are mentioned in the references (Wixson, et al.: 1987).

Enflurane has been little used for rodent anesthesia, but its action resembles that of halothane. Halothane in rats causes a rapid induction (1 to 2 minutes) and frequent deaths; great care is needed to maintain a satisfactory level of halothane anesthesia in rats.

Much has been written about sodium pentobarbital anesthesia in rats, but with the anesthetic dose (30 to 50 mg/kg) near the LD_{50} (60 mg/kg) and with rats varying greatly in their response, the drug poses considerable risk. The young, the females, cooled animals, and possibly the albinos are more susceptible to the drug, whereas males, animals on low caloric diets, and animals on cedar bedding are more resistant.

Sodium pentobarbital administered intraperitoneally at 10 to 20 mg/kg in young rats and 30 to 50 mg/kg in older rats (males require doses higher by approximately 5 mg/kg) produces anesthesia in 5 to 10 minutes. Sodium pentobarbital has also been given orally (bulbed needle) and intraperitoneally at 25 mg/kg by each route.

Anesthetic Overdose

Anesthetic overdose is indicated by diaphragmatic breathing, gasping, loss of pupillary reflex, and cyanosis. Artificial respiration or respiratory stimulants can be used for barbiturate overdose and nalorphine (2 mg/kg subcutaneously) can be used for fentanyl overdose.

Gentle manual compression of the sternum 50 to 100 times per minute or blowing by tube or tightly fitting face mask into the animal's mouth may stimulate respiration, as may gently swinging or tipping the animal head down and horizontally. Oxygen may be supplied.

Postoperative Care

Because little research has been done and little has been written about long-lasting postoperative analgesia, doses are not well defined for use of the various products in rabbits and rodents (see pp. 69–70). Anesthetized animals should be kept warm (30° to 35° C) and hydrated (0.9% saline or lactated Ringer's solution subcutaneously or intraperitoneally). Heating pads, water bottles, or lamps, however, can overheat and dehydrate or burn these small species, and great care must be taken. Isothermic pads are safer for warming animals and maintaining their body temperatures. Guinea pigs recovering from anesthesia should be turned every 30 minutes to reduce hypostatic pulmonary congestion. Antibiotics may be given to rats and mice when asepsis is not possible or is compromised. Antibiotics should be used carefully, if at all, however, in guinea pigs, hamsters, and rabbits. Wounds should be cleaned and bedding changed often. Postsurgical deaths frequently result from anesthetic overdose, hypothermia, fluid loss, or acidosis.

SURGICAL PROCEDURES

Descriptions of surgical procedures on rabbits and rodents are frequently encountered in the biomedical literature, but usually the procedures are a means to an end in an experiment rather than a method used to correct an abnormality. Commonly performed corrective surgical procedures in small mammalian pets are nail trimming, trimming of incisors in rabbits and premolars in guinea pigs, repair of fractures and lacerations, removal of tumors, drainage of abscesses, removal of uroliths, amputation of a necrotic tail, castration, and removal of a blood sample.

Preoperative and postoperative concerns

described in the section on anesthesia should be observed. Animals are weighed before injectable anesthetics are used, and during and after the procedure patients are kept warm and hydrated. Hair should be shaved carefully to prevent skin injury. Hypothermia, hemorrhage, and fluid loss are major concerns. Surgery performed on small animals should involve, insofar as possible, the same concerns for asepsis and analgesia as are practiced with larger animals.

Toenail trimming is particularly important with rabbits and rats, because rabbits scratch people and may catch their long nails in caging, whereas rats often self-traumatize a pruritic infection. Nails can be clipped with nail clippers designed for use on animals or with some kinds of forceps or wire cutters.

Rabbit incisors that are maloccluded and overgrown are trimmed without splitting or shattering with a sharp clipper or a dental saw. Malocclusion in rabbits is an inherited condition, and overgrowth will reoccur. Overgrown guinea pig premolars are more difficult to see and trim. A spatula and a small light or an otoscope and assistance opening the anesthetized animal's mouth are prerequisites. Clipped bits of teeth should be removed from the oral cavity.

Fracture reduction and fixation present no unique problems in rodents except for the small size and low density of the long bones (see Rickards, et al., Surgical Procedures reference). The animals' movements and chewing behaviors may cause external fixation devices to break down, but a promazine tranquilizer may be used to reduce splint chewing and suture removal. Radiographic evaluation is performed using a nonscreen film. As the most common limb fractures of rabbits and rodents involve the tibia and femur, intramedullary pins (hypodermic needles) should be considered for midshaft breaks. Splints of rolled film or a small stick can be used for tibial fractures, but because the femur is enclosed within the abdominal skin, a padded tubular-traction cast made from a plastic syringe cover or barrel should be used for femoral fractures. Cage rest may suffice for some fractures.

Benign mammary tumors in rats are easily removed, although larger tumors become necrotic and highly vascularized. Rats are anesthetized with ketamine-xylazine, ether, or methoxyflurane (via cone), and the encapsulated tumor is removed through a cutaneous incision. Hemostasis will be necessary. Mammary tumors in mice, on the other hand, are usually invasive and difficult to remove.

Opening and draining subcutaneous abscesses poses the problem of possible septicemia. An additional concern exists when abscesses are multiple and interconnected. Large subcutaneous abscesses in rabbits or guinea pigs are opened, drained, cleaned, and left to heal.

If the tail of a rodent is grasped near the end, the tail skin may be pulled off. This is a common problem with gerbils. The skinless tail may be amputated or allowed to slough.

The most common indications for castration of rabbits are birth control, urine spraying, territoriality, and aggressiveness. Castration of rabbits and rodents is complicated by the open inguinal canal, which predisposes to intestinal herniation. Elevating the forequarters during surgery causes the testes to fall into the sacs, and during castration the testes may have to be manipulated from the inguinal canal into the scrotal pouches. Cutaneous incisions may be ample, but the tunica vaginalis should be opened distally only enough to permit removal of the testes. Simple ligation of vessels and ducts leading to or from the testes will effectively castrate a guinea pig. Castrated rabbits maintain mounting activity for a month or more, but other aggressive and sexual activities diminish rapidly.

For orchidectomy in the mouse, the peritoneum is entered via a small, transverse, suprapubic incision. The vas deferens and vessels are transected, and the testes, epididymis, fat, and gubernaculum are removed. Hemostasis is rarely necessary. The cutaneous wound is closed with fine silk sutures.

Ovariectomy in the mouse requires a dorsal, transverse, cutaneous incision and muscle separation at the second lumbar level. The skin is closed with silk sutures. Ovariectomy of guinea pigs involves bilateral, retroperitoneal incisions made just ventral to the erector spinae muscles and 1 cm posterior to the last rib. As guinea pigs may have ovarian rests on the oviduct, cycling may continue despite the ovariectomy.

Hysterectomy in the rabbit involves tying the animal on a board tipped "head down" to displace the viscera craniad. A 5-cm midline incision beginning just cranial to the pubic symphysis exposes the peritoneum. The uterine horns, lying on either side of the bladder, are retracted and removed with the ovaries. Ligations are performed as in larger species.

ANALGESIA

Pain is assumed to be perceived to the same extent in rabbits and rodents as in humans. Administration of postoperative analgesia in rabbits and rodents is essential in some cases.

Most available analgesics, i.e., morphine and meperidine, are considered narcotics and are controlled substances. Their use requires special state and federal registration. Registration forms are available through state agencies and the U.S. Department of Justice, Drug Enforcement Agency, Washington, DC 20537.

Morphine (Duramorph, morphine sulphate injection, Elkins-Sinn Inc., Cherry Hill, NJ 08034) is a narcotic agonist analgesic that may cause severe central nervous system, cardiovascular, respiratory, and thermoregulatory center depression, as well as nausea, emesis, and diarrhea. The side effects vary greatly among species. Subcutaneous injection is twice as potent as intraperitoneal administration. Use of morphine is contraindicated in animals with gastrointestinal obstruction or in animals recovering from ophthalmic surgery.

Meperidine (Demerol, Winthrop-Breou, New York, NY 10016) is a synthetic narcotic analgesic one-fifth as potent as morphine with fewer side effects. Meperidine has a more rapid onset and shorter duration than morphine. As with morphine, meperidine depresses the central nervous, respiratory, and cardiovascular systems, but unlike morphine, meperidine does not induce vomiting, nausea, or diarrhea. Rapid intravenous administration may cause convulsions.

Pentazocine lactate (Talwin V, Winthrop Veterinary, New York, NY 10016) is approximately one-third as potent as morphine and may be appropriate for the control of mild or chronic pain.

Subcutaneous injection of the synthetic narcotic nalbuphine hydrochloride (Nubain, Dupont, Wilmington, DE 19898) provides relief from discomfort for 3 to 8 hours (in dogs) with minimal cardiovascular and respiratory depression. Nalbuphine hydrochloride is not a controlled substance.

Oxymorphone hydrochloride (Numorphan, DuPont, Wilmington, DE 19898) is a narcotic analgesic that may be administered to rodents via any parenteral route. It is 10 times more potent than morphine and is effective for 4 to 6 hours. Possible side effects include bradycardia, slight respiratory depression, and increased sensitivity to auditory stimuli.

Butorphanol tartrate (Torbutrol, Bristol-Myers, Evansville, IN 47721) is an injectable synthetic antagonist analgesic three to five times more potent than morphine. Use of butorphanol tartrate in pregnant or lactating rats is contraindicated because it increases nervousness and decreases newborn caretaking behavior.

Hydroxyzine hydrochloride (Vistaril, Pfipharmecs Div., Pfizer, New York, NY 10017) has antihistaminic effects and analgesic properties. As with butorphanol tartrate, use of hydroxyzine hydrochloride on pregnant rats is contraindicated.

Innovar-Vet injection (Pitman-Moore, Washington Crossing, NJ 08560) is an analgesic-tranquilizer that is given IM and contains 0.4 mg fentanyl and 20 mg droperidol per ml. It is counterindicated in guinea pigs because muscle and tissue necrosis results at the injection site when effective doses are given.

Dimethyl sulfoxide (DMSO, RIMSO-50, Research Ind., Salt Lake City, UT 84119) may provide relief from inflammation when applied topically to the affected area.

Per os dosages of non-narcotic acetylsalicylic acid (aspirin) at 400 mg/kg given subcutaneously or per os, phenacetin (Empirin), acetaminophen (Tylenol), and ibuprofen (Motrin, Upjohn, Kalamazoo, MI 49001) may help to relieve minor or chronic pain in laboratory rodents.

The effects of analgesics in mice may be assessed through application of the formalin test. Injection of dilute formalin into the dorsal surface of the hindpaw induces licking responses. Licking of the site of the noxious stimuli is indicative of the mouse's ability to perceive localized pain. Licking behavior secondary to formalin injection is observed less consistently in rats than in mice.

RADIOGRAPHY

Radiography for laboratory animals contributes to both experimental and diagnostic efforts. Abnormal conditions that may require radiography include luxation or fracture of a limb or the spine, arthritis, scorbutus, malocclusion, congenital abnormalities, otitis media, pyometra, neoplasia, abscessation, radiopaque cystic calculi, gastric hairballs, and intestinal impactions. Visualization of a gastric hairball may be facilitated if barium sulfate and a few milliliters of air are passed into the stomach, but normal controls often show a similar picture.

Diagnostic ultrasound may be useful in examining soft-tissue lesions, such as neoplasms, pregnancy abnormalities, and abscesses. Animals may be restrained for radiographic examination by mechanical means alone or in combination with tranquilizers or anesthetics. Mechanical restraint alone generally results in malpositioning of the animal and, hence, radiographs of poorer quality. Whole body studies are often practical because of the small size of rabbits and rodents. The choice of kilovoltage depends on the tissue to be examined, the films and screens used, and the techniques employed. Higher kilovolt ranges permit reduction of exposure time and allow for more exposure latitude (more shades of grey) and organ observation; however, fine bone detail is lost. Lower kilovolt levels, which require longer exposure times, will result in finer bone detail, but will increase patient motion. The increased contrast also reduces definition of soft tissues. Nonscreen films or detailed film and screens combinations should be used. Exposure times of $1/60$ of a second or shorter most effectively reduce patient motion caused by rapid respiratory rate. It is also best to use a small focal spot.

Approximate exposure values that might serve as an exposure guide for radiographing small rodents are 50 to 60 kvp and 70 to 150 mas at 40 inches with nonscreen film. Film used with a screen requires about 50 to 60 kvp and 1.5 to 5 mas.

Comprehensive discussions of radiologic techniques in small mammals are contained in *Textbook of Veterinary Diagnostic Radiology*, by D.E. Thrall (Philadelphia, W.B. Saunders Company, 1986) and in *Techniques of Veterinary Radiology*, by J.P. Morgan and S. Silverman (Davis, CA, Veterinary Radiology Associates, 1984).

EUTHANASIA

Euthanasia (from the Greek words for "easy death") is the humane killing of an animal. Euthanatize is the preferred verb form but euthanize is also acceptable (*Webster's Ninth New Collegiate Dictionary*, 1985). Implied in humane killing are a transition from consciousness to unconsciousness and a death that is painless, quick, quiet, and free of fear and distress. Euthanasia should be carried out by trained personnel using methods, substances, and facilities permitted under institutional policy, governmental regulations, and social, ethical, or religious convention.

Animals should not be killed while crowded or in the presence of animals not being killed. The method of euthanasia varies with the species, number, and intended postmortem use of the animal. Products and procedures specifically prohibited in the American Veterinary Medical Association's report on euthanasia (Smith, et al., 1986) should not be used for euthanasia.

Inhalant anesthetics and carbon dioxide are commonly used for euthanasia. Gas flow noise should not frighten animals. To reduce anxiety during induction of CO_2 anesthesia and to rapidly effect euthanasia, the chamber should be precharged prior to the introduction of animals. Animals should not be wetted with liquid forms of gas anesthetics. Chloroform, once widely used, has almost vanished from animal laboratories because of its hepatotoxic and nephrotoxic effects on animals and possible carcinogenic effects on humans. Halothane and methoxyflurane, popular anesthetics, are expensive for euthanasia and animals may struggle to avoid inhaling vapors.

Volatile agents, such as ether, irritate the mucous membranes and induce excessive salivation. Risks associated with the flammability and combustibility of ether must be considered.

Of the inhalant methods of euthanasia, carbon dioxide in a closed chamber is frequently the procedure of choice, particularly when many animals are involved. CO_2 is inexpensive, nonflammable, nonexplosive, and relatively safe to use with proper equipment. Because CO_2 is heavier than air, the chamber should open from the top. If the euthanizing chamber is incompletely filled with gas, animals may avoid the agent by climbing to the top. The time required to euthanatize immature animals with CO_2 is longer than for adults. When possible, compressed CO_2 gas in cylinders should be used in lieu of dry ice, as direct contact with ice causes pain. The addition of oxygen (e.g., 30% O_2:70% CO_2 in cylinders) or air reduces the discomfort of hypoxia during the induction phase of inhalant-induced euthanasia. If dead rodents or rabbits are to be used to feed other animals (snakes, carnivores, and raptors), euthanasia with CO_2 is recommended to increase palatability and to avoid secondary poisoning with euthanasia agents.

Carbon monoxide also produces relatively painless and rapid death. As with CO_2, the chamber should be prefilled. The system must be carefully controlled to prevent animal discomfort and to reduce CO exposure to personnel. Exhaust from a combustion engine is an inappropriate source of CO.

Nitrogen gas (N_2) introduced into a closed chamber displaces oxygen and produces death by hypoxia, but its use may be aesthetically objectionable. N_2 should not be used with animals under 4 months of age because of the excessive amount of time required to produce euthanasia. Prior to N_2-induced euthanasia, animals may appear panicked, distressed, and sensitive to pain. Animals may convulse, tremble, or yelp after loss of consciousness. For these reasons, N_2 is generally not recommended for euthanasia and may be prohibited by law in some areas.

A common method of euthanasia for larger animal species and neonates is the intravenous or, less desirably, intraperitoneal injection of an overdose of a noninhalant pharmacologic agent. Intravenous administration incurs less trauma to the animal, i.e., tissue damage or pain, and induces unconsciousness faster than does intraperitoneal, oral, or rectal administration. Aural veins are used in rabbits and tail veins are used in mice, rats, and gerbils. For small rodents, intravenous injections may be difficult; thus, the intraperitoneal approach is commonly used. Concentrated barbiturate solutions specifically intended for euthanasia are available. The recommended dosage for euthanasia with sodium pentobarbital is 150 to 200 mg/kg, but care must be taken to ensure that the heart beat is stopped. Barbiturates are controlled substances and are not available to all animal care personnel.

If barbiturates cannot be used, T-61® (Hoechst-Roussel Agri-Vet Co, Somerville, NJ 08876), a non-controlled, non-barbiturate mixture of drugs, is an acceptable euthanatizing agent. T-61® causes pain when injected extravascularly, and therefore, intravenous administration by a skilled person is recommended. The first two thirds of the intravenous dose should be administered slowly and the last third rapidly. T-61® and sodium pentobarbital, given intracardially, may cause pulmonary edema. T-61® in large doses may also cause hemorrhage and tissue necrosis. Intracardiac, intraperitoneal, and intrathoracic administration of T-61® are not recommended.

Although aesthetically unpleasant, physical methods of euthanasia, such as cervical dislocation, are satisfactory for mice, immature rats (< 200 grams) and rabbits, and poultry if conducted by experienced personnel. Decapitation of rodents and small rabbits with a well-designed guillotine may be necessary to recover chemically uncontaminated tissues or body fluids. Devices for cervical dislocation (Cervical Dislocator, Box 157, Schofield, WI 54476; Harvard, 22 Pleasant St., So. Natick, MA 01760) and decapitation (Braintree, Box 361, Braintree, MA 02184; Edco, Box 2305, Chapel Hill, NC 27515; Stoelting, 1350 So. Kostner, Chicago, IL 60623) are commercially available. When possible, animals should be sedated or lightly anesthetized prior to cervical dislocation or decapitation. Other methods, including decompression, electrocution, rapid freezing, and microwave irradiation, are discussed in the material listed in the references.

Rabbits and rodents are sometimes exsanguinated to obtain hyperimmune antisera. Because of anxiety associated with hypovolemia, exsanguination should be done only on sedated or anesthetized animals.

Muscle paralytics, such as curare, quaifenesin, pancuronium, and succinylcholine, should not be used as euthanatizing agents because of their ineffectiveness in rendering the thalamus or cerebral cortex nonfunctional. Use of magnesium and potassium salts, chloral hydrate, nicotine, strychnine, and hydrocyanic acid is also contraindicated. These drugs produce undesirable side effects, such as emesis, defecation, seizures, gasping, salivation, and vocalization, or they fail to induce loss of consciousness before death.

BLOOD COLLECTION

The choice of procedure for blood collection will depend on the animal species, volume of blood required, frequency of bleeding, and kind of anesthetic that can be used.

In most cases orbital sinus bleeding is the method of choice for small rodents. Many people routinely anesthetize rodents for orbital bleeding. The anesthetized rodent is held on a flat surface in lateral recumbency. A microhematocrit tube or small-bore Pasteur pipette is rotated and directed caudally and medially at the angle or medial canthus of the eye into the orbital sinus. Blood from the ruptured sinus flows through the tube. When the blood ceases to flow or the required amount is obtained, the tube is withdrawn. Eye trauma, harderian gland lesions, and nasal hemorrhage may occur with this technique; however, the extent of trauma is inversely proportional to the skill of the bleeder. With care and experience the technique is minimally stressful and the ruptured vessels repair quickly. Lesions of the harderian gland caused by the bleeding procedure must be differentiated from those caused by SDAV infections in the rat.

Blood is frequently removed from the rodent tail. The animal is placed in a restraint chamber or is otherwise restrained, and the ventral artery is partially lacerated or the tip of the tail is amputated. Tail bleeding can be painful for unanesthetized animals, and blood may clot or not flow freely. Warming the entire animal for 5 to 10 minutes, immersing the tail in warm water, applying xylene to the warmed tail, or applying heparin or citrate solution to the wound all increase blood flow and help to prevent clot formation. Xylene should be washed off the tail with alcohol following bleeding. The collected sample often contains hemolyzed blood and extravascular tissue fluid. Pre-cooling of test tubes used for collection and refrigeration of samples

awaiting centrifugation may reduce hemolysis.

In rabbits and guinea pigs, the marginal ear vein or the central artery of the ear can be used for bleeding or injections. It may be useful to shave the area, disinfect with alcohol, apply petroleum jelly, distend the vein with digital pressure at the base of the ear, nick the vein (not the entire ear) with a scalpel, and then collect blood into a pipette or tube. Application of heat or a small amount of xylene will increase blood flow. Xylene is irritating and should be removed later with 70% alcohol after the procedure is completed.

Also, an implanted jugular catheter can be used to collect blood. Depending on species, other sites to be considered for blood collection are the saphenous, jugular, femoral, penile, cephalic, and sublingual veins. Additionally toe nail clips may be used with varying degrees of success in several species.

Large volumes of blood may be collected rapidly by cardiac puncture; however, this is a difficult technique to master and should be performed only on animals under general anesthesia. In some laboratories cardiac puncture is permitted only as a terminal procedure. Cardiac puncture is useful in smaller species without otherwise accessible sites for collecting large volumes of blood. The needle is inserted toward the heart (and the nose) at a 10° to 30° angle above the ventral abdomen lateral to the xiphoid process. In larger animals, the needle can be inserted into the lateral thoracic region toward the area of maximal heart palpitation. In rats, for example, a 24-gauge ½-inch needle inserted between the 5th and 6th ribs is directed at the left ventricle. Care must be taken to avoid injury to thoracic organs and tissues, which frequently leads to hemothorax, pulmonary hemorrhage, blood in the pericardial sac and death. Large volumes of blood also can be collected by severing the brachial vessels.

Blood volume limits need to be considered in blood collection because overbleeding can leave the animal hypovolemic, anemic, and weak. Total blood volume is

SUMMARY OF RABBIT AND RODENT DIAGNOSTIC TESTS

Virus	Species Tested	Test	Secondary Test	Comments
PVM	M, R, H, Gp, (Rb)	E	IFA, HI, Path	Very common—minimal lesions
Reo-3	M, R, H, Gp	E	IFA	Uncommon
TMEV	M	E	IFA	Uncommon—also called TO or GDVII
KRV	R	HI	IFA, Path	Common—abortions and resorptions
H-1	R	E	IFA	Rare—no reports of clinical diseases
K	M	(HI)	E	Rare—need for testing in question
MVM	M	E	IFA	Uncommon
Polyoma	M, (H)	E	HI, IFA	Unlikely to occur naturally
Sendai	M, R, H, Gp	E	IFA, Path	Very common—lung lesions
MHV	M	E	IFA, Path	Very common—many strains of virus, gut syncytia
SDAV—RCV	R	E	IFA, Path	Common—many strains of virus, salivary gland lesions
SV-5	H, Gp	HI	IFA	Significance unknown
EDIM	M	(E)	(IFA, Path)	Incidence not well defined

SUMMARY OF RABBIT AND RODENT DIAGNOSTIC TESTS (*Continued*)

Virus	Species Tested	Test	Secondary Test	Comments
Adenoviruses	M, H, (R)	E	IFA, Path	Common in hamsters and conventional mice
Ectromelia	M	E	Path, IFA	Uncommon—may see high mortality
LCM	M, R, H, Gp	IFA	(E)	Uncommon—zoonotic disease
LDHV	M	(LDH plasma enzyme assay)		Incidence uncertain—colorimetric assay—no clinical signs
Mouse Thymic Virus	M	(IFA)	(Path)	Incidence uncertain—thymic necrosis in neonates
Cytomegaloviruses	M, R, Gp	Path	(E)	Common in guinea pigs—lesions in salivary glands
Mycoplasma pulmonis	M, R	E	Culture, Path	Common in conventional colonies—readily cultured
CAR Bacillus	M, R, Rb	Path	(E)	Incidence uncertain—silver stain of respiratory epithelium
Tyzzer's Disease (*Bacillus piliformis*)	M, R, H, Gp, Gb, Rb	Path	(E)	Incidence uncertain—gerbil highly susceptible

M = mouse; R = rat; H = hamster; Gp = guinea pig; Rb = rabbit; Gb = gerbil; IFA = immunofluorescence; HI = hemagglutination inhibition; E = ELISA = enzyme-linked immunosorbent assay; Path = pathology; In parenthesis—test not commonly performed or beyond the expertise of many diagnostic laboratories at this time (1988).

about 6% of body weight. As a guideline, up to 25% of total blood volume can be collected in a 2-week period. Replacement of blood with fluids (sterile 0.9% saline) should be considered. With rodents, microanalytic methods should be used to maximally reduce the blood volume that must be withdrawn.

SEROLOGIC TESTING

Viral infections in rodents, although often latent, are common and may influence experimental results or predispose the infected animal to clinical viral or bacterial disease. Ectromelia virus (mousepox) can cause devastating epizootics in mouse colonies, and Sendai and mouse hepatitis virus infections are the most common and serious diseases among mice in the United States. The detection of rodents carrying antiviral antibodies, and possibly the virus itself, is therefore an important consideration in disease prevention in research animal colonies. The table summarizes the more common serologic tests suggested for screening sera from various species of rodents.

The collection, processing, dilution, and storage of rodent sera intended for serologic testing must follow an established protocol if results are to be meaningful. Collection tubes must be clean. Hemolysis, excessive tissue products, and bacterial growth must be avoided.

For small rodents, 0.5 ml of blood is collected into 1.0 ml of isotonic phosphate buffered saline. For larger animals (over 100 g), 1.0 ml blood is collected into 2.0 ml saline. Assuming a 50% packed cell volume, the final serum concentration will be 1:5. The serum may also be diluted after centrifugation. The saline-blood mixture is allowed to clot for 30 minutes and is then ringed, covered, and refrigerated overnight. The following day, sera are separated and heat-inactivated in a water bath at 56° C for 30 minutes. A minimum of 0.2 ml of diluted serum is required for virus screens. Several commercial and service institutions perform serologic testing on rodent samples.

REFERENCES

Drugs

Baer, H.: Long-term isolation stress and its effects on drug response in rodents. Lab. Anim. Sci., 21:341–349, 1971.

Barnett, S.A., and Munro, K.M.H.: 'Terramycin' and breeding mice. Lab. Anims., 2:45–47, 1968.

Beaucage, C.M., Fox, J.G., and Whitney, K.M.: Effect of long-term tetracycline exposure (drinking water additive) on antibiotic-resistance of aerobic gram-negative intestinal flora of rats. Am. J. Vet. Res., 40:1454–1457, 1979.

Bolton, G.R.: Aerosol therapy. In Current Veterinary Therapy VI. Edited by R.W. Kirk. Philadelphia, W. B. Saunders Co., 1977.

Brown, A.M.: Pharmacogenetics of the mouse. Lab. Anim. Care, 15:111–118, 1965.

Escorela, L., et al.: Sensibility of rabbits to treatment with ampicillin and gentamycin. Ann. Rech. Vet., 12:11–17, 1981.

Fesce, A., et al.: Ecophylaxis: Preventive treatment with gentamicin of rabbit lincomycin-associated diarrhea. Folia Vet. Lat., 7:225–242, 1977.

Finco, D.R.: Fluid therapy. In Current Veterinary Therapy VI. Edited by R.W. Kirk. Philadelphia, W. B. Saunders Co., 1977.

Findon, G., and Miller, T.E.: Treatment of Trichosomoides crassicauda in laboratory rats using ivermectin. Lab. Anim. Sci., 37:496–499, 1987.

Flecknell, P.A.: The relief of pain in laboratory animals. Lab. Anim., 18:147–160, 1984.

Freireich, E.J., et al.: Quantitative comparison of toxicity of anticancer agents in the mouse, rat, hamster, dog, monkey, and man. Cancer Chemother. Rep., 50:219–244, 1966.

Galloway, J.H.: Antibiotic toxicity in white mice. Lab. Anim. Care, 18:421–425, 1968.

Gardner, M.C., Owens, D.R., and Wagner, J.E.: In vitro activity of select antibiotics against Mycoplasma pulmonis from rats and mice. Lab. Anim. Sci., 31:143–145, 1981.

Goldblatt, D., et al.: Effect of anticonvulsants on seizures in gerbils. Neurology, 21:433–434, 1971.

Gray, J.E., and Lewis, C.: Enigma of antibiotic-induced diarrhea in the laboratory rabbit. Toxicol. Appl. Pharmacol., 8:342, 1966.

Griffin, H.C.: Lincomycin-associated enterocolitis in rabbits. J. Am. Vet. Med. Assoc., 185:670–671, 1984.

Hagen, K.W.: Effect of antibiotic-sulfonamide therapy on certain microorganisms in the nasal turbinates of domestic rabbits. Lab. Anim. Care, 17:77–80, 1967.

Hamre, D.M., et al.: The toxicity of penicillin as prepared for clinical use. Am. J. Med. Sci., 206:642–652, 1943.

Hankinson, G., White, W.J., and Lang, C.M.: The effects of selected antimicrobials on glucose transport in the rat intestine. Lab. Anim. Sci., 29:35–39, 1979.

Harrison, J.B., Sussman, H.H., and Pickering, D.E.: Fluid and electrolyte therapy in small animals. J. Am. Vet. Med. Assoc., 137:637–645, 1960.

Henness, A.M., et al.: Use of drugs based on square meters of body surface area. J. Am. Vet. Med. Assoc., 171:1076–1077, 1977.

Hooper, D.G., and Hirsh, D.C.: Changes of resistance of enteric bacteria in mice given tetracycline in the drinking water. Am. J. Vet. Res., 38:565–567, 1977.

Jondorf, W.R., Maickel, R.P., and Brodie, R.B.: Inability of newborn mice and guinea pigs to metabolize drugs. Biochem. Pharmacol., 1:352–354, 1958.

Kaipanien, W.J., and Faine, S.: Toxicity of erythromycin. Nature, 174:969–970, 1954.

Katz, L., et al.: Experimental clindamycin-associated colitis in rabbits: Evidence of toxin-mediated mucosal damage. Gastroenterol., 74:246–252, 1978.

Killby, V.A.A., and Silverman, P.H.: Toxicity of antibiotics in laboratory animals. Science, 156:264, 1967.

King, J.O.L.: Effects of feeding antibiotics to growing and breeding mice. Lab. Anims., 3:1–6, 1969.

King, J.O.L.: Inclusion of intestinal antiseptics in rabbit diets. Lab. Anims., 15:205–206, 1981.

King, J.O.L.: The continuous feeding of two antibiotics to growing rabbits. Br. Vet. J., 123:453–458, 1967.

King, J.O.L.: The feeding of penicillin to rabbits. Br. Vet. J., 122:112–116, 1966.

King, J.O.L.: The feeding of zinc bacitracin to growing rabbits. Vet. Rec., 99:507–508, 1976.

King, J.O.L.: The response of growing rabbits to the feeding of an antibiotic. Vet. Rec., 74:1411–1414, 1962.

Kruckenberg, S.M., Cook, J.E., and Feldman, B.F.: Clinical toxicities of pet and caged rodents and rabbits. Vet. Clin. North Am., 5:675–684, 1975.

Liebenberg, S.P., and Linn, J.M.: Seasonal and sexual influences on rabbit atropinesterase. Lab. Anims., 14:297–300, 1980.

Maiers, J.D., and Mason, S.J.: Lincomycin-associated enterocolitis in rabbits. J. Am. Vet. Med. Assoc., 185:670–671, 1984.

Miller, W.F.: Aerosol therapy in acute and chronic respiratory disease. Arch. Intern. Med., 131:148–155, 1973.

Milner, N.A.: Letter: Penicillin toxicity in guinea pigs. Vet. Rec., 96:554, 1975.

Morgan, D.R.: Routine birth induction in rabbits using oxytocin. Lab. Anims., 8:127–130, 1974.

Muller, G.H.: Topical dermatologic therapy. In Current Veterinary Therapy V. Edited by R.W. Kirk. Philadelphia, W. B. Saunders Co., 1974.

Olfert, E.D.: Ampicillin toxicity in rabbits. Can. Vet. J., 22:218–220, 1986.

Owens, D.R., Wagner, J.E., and Addison, J.B.: Antibiograms of pathogenic bacteria isolated from laboratory animals. J. Am. Vet. Med. Assoc., 167:605–609, 1975.

Rehg, J.E., and Pakes, S.P.: Clostridium difficile antitoxin neutralization of cecal toxin(s) from guinea pigs with penicillin-associated colitis. Lab. Anim. Sci., 31:156–160, 1981.

Schneierson, S.S., and Perlman, E.: Toxicity of penicillin for the Syrian hamster. Proc. Soc. Exp. Biol. Med., 91:229–230, 1956.

Shah, D.V., and Suttie, J.W.: Vitamin K requirement and warfarin tolerance in the hamster. Proc. Soc. Exp. Biol. Med., 150:126–128, 1975.

Small, J.D.: Fatal enterocolitis in hamsters given lincomycin hydrochloride. Lab. Anim. Care, 18:411–420, 1968.

Stunkard, J.A., Schmidt, J.P., and Cordaro, J.T.: Consumption of oxytetracycline in drinking water by healthy mice. Lab. Anim. Care, 21:121–122, 1971.

Thilsted, J.P., et al.: Fatal diarrhea in rabbits resulting from the feeding of antibiotic-contaminated feed. J. Am. Vet. Med. Assoc., 179:360–362, 1981.

Wade, A.E., Wu, B., and Smith, P.B.: Effects of ascorbic acid deficiency on kinetics of drug hydroxylation in male guinea pigs. J. Pharm. Sci., 61:1205–1208, 1972.

Wightman, S.R., Mann, P.C., and Wagner, J.E.: Dihydrostreptomycin toxicity in the Mongolian gerbil, Meriones unguiculatus. Lab. Anim. Sci., 30:71–75, 1980.

Woodard, G.: Principles in drug administration. In Methods of Animal Experimentation. Vol. 1. Edited by W.I. Gay. New York, Academic Press, 1965.

Wright, F.C., and Riner, J.C.: Comparative efficacy of injection routes and doses of ivermectin against Psoroptes in rabbits. A.J.V.R., 46(3):752–754, 1985.

Young, J.D., et al.: An evaluation of ampicillin pharmacokinetics and toxicity in guinea pigs. Lab. Anim. Sci., 37:652–656, 1987.

Anesthesia

General Anesthesia

Adams, H.R.: Prolongation of barbiturate anesthesia by chloramphenicol in laboratory animals. J. Am. Vet. Med. Assoc., 157:1908–1913, 1970.

Clifford, D.: Preanesthesia, anesthesia, analgesia, and euthanasia. In Laboratory Animal Medicine. Edited by J.G. Fox, B.J. Cohen, and F.M. Loew. Orlando, Academic Press, 1984.

Clifford, D.: Restraint and anesthesia of small laboratory animals. In Textbook of Veterinary Anesthesia. Edited by L.R. Soma. Baltimore, Williams & Wilkins, 1971.

Cook, R., and Dorman, R.G.: Anaesthesia of germ-free rabbits and rats with halothane. Lab. Anim., 3:101–106, 1969.

Dobson, C., and Tschirky, H.: Development of an anesthetic apparatus for experimental surgery on rats. Pharmacology, 5:307–313, 1971.

Dudley, W.R., et al.: An apparatus for anesthetizing small laboratory animals. Lab. Anim. Sci., 25:481–482, 1975.

Duke, D., Clark, B.F., and Askill, S.: A small-animal operating table for use with halothane anaesthetic administered by intubation or inhalation. Lab. Anims., 5:233–237, 1971.

Flecknell, P.A., and Mitchell, M.: Midazolam and fentanyl-fluanisone: Assessment of anesthetic effects in laboratory rodents and rabbits. Lab. Anim., 18:143–146, 1984.

Green, C.J.: Anaesthesia in experimental animals. Proc. R. Soc. Med., 69:366–367, 1976.

Green, C.J.: Animal anaesthesia. In Laboratory Animal Handbook 8. London, Laboratory Animals Ltd., 131–162, 1979.

Green, C.J.: Neuroleptanalgesic drug combination in the anesthetic management of small laboratory animals. Lab. Anims., 9:161–178, 1975.

Green, C.J., Halsey, M.J., Precious, S., and Wardley-Smith, B.: Alphaxolone-alphadolone anesthesia in laboratory animals. Lab. Anim., 12:85–89, 1978.

Green, C.J., et al.: Ketamine alone and combined with diazepam or xylazine in laboratory animals: A 10-year experience. Lab. Anims., 15:163–170, 1981.

Greenberg, S.R.: The use of combined inhalation anesthesia in laboratory animal surgery. J. Am. Vet. Med. Assoc., 149:935–937, 1966.

Heidt, G.A.: A portable anesthesia chamber for intractable small animals. Lab. Anim. Sci., 28:212–213, 1978.

Holland, A.J.: Laboratory animal anesthesia. Can. Anaesth. Soc. J., 20:693–705, 1973.

Hughes, H.C., Jr., White, W.J., and Lang, C.M.: Guidelines for the use of tranquilizers, anesthetics, and analgesics in laboratory animals. Vet. Anesth., 2:19–24, 1975.

Janssen, P.A., Niemergeers, C.J., and Marsboom, R.P.: Etomidate, a potent non-barbiturate hypnotic. Intravenous etomidate in mice, rats, guinea pigs, rabbits, and dogs. Arch. Int. Pharmacodyn. Ther., 214:92–132, 1975.

Keenaghan, J.B., and Boyes, R.N.: The tissue distribution, metabolism, and excretion of lidocaine in rats, guinea pigs, dogs and man. J. Pharmacol. Exp. Ther., 180:454–463, 1972.

Levy, D.E., Zwies, A., and Duffy, T.E.: A mask for delivery of inhalation gases to small laboratory animals. Lab. Anim. Sci., 30:868–870, 1980.

Lewis, G.E., and Jennings, P.B.: Effective sedation of laboratory animals using Innovar-Vet. Lab. Anim. Sci., 22:430–432, 1972.

Lipman, N.S., Phillips, P.A., and Newcomer, C.E.: Reversal of ketamine/xylazine anesthesia in the rabbit with yohimbine. Lab. Anim. Sci., 37:474–477, 1987.

Mauderly, J.L.: An anesthetic system for small laboratory animals. Lab. Anim. Sci., 25:331–333, 1975.

McIntyre, J.W.R.: An introduction to general anesthesia of experimental animals. Lab. Anims., 5:99–114, 1971.

McKay, D.H., and Clement, J.G.: A heated operating board for maintenance of normal body temperature during anesthesia in small laboratory animals. Lab. Anim. Sci., 27:1036–1037, 1977.

Mulder, J.B.: A unit for inhalation anesthesia of small laboratory animals. Anesth. Analg. (Cleve.), 52:369–372, 1973.

Mulder, J.B., and Brown, R.V.: An anesthetic unit for small laboratory animals. Lab. Anim. Sci., 22:422–423, 1972.

Parbrook, G.D.: Exposure of experimental animals to nitrous-oxide-containing atmospheres. Br. J. Anaesth., 39:114–118, 1967.

Rassaert, C.C.: A respiratory mucus extractor and resuscitator for use in rodents. Lab. Anim. Care, 21:420–421, 1971.

Schaffer, A.: Anesthesia and sedation. In Methods of Animal Experimentation. Vol. 1. Edited by W.I. Gay. New York, Academic Press, 1965.

Sebesteny, A.: Fire-risk-free anaesthesia of rodents with halothane. Lab. Anims., 5:225–231, 1971.

Silverman, J., et al.: Evaluation of a combination tiletamine and zolazepam as an anesthetic for laboratory rodents. Lab. Anim. Sci., 33:457–460, 1983.

Simmons, M.L., and Smith, L.H.: An anesthetic unit for small laboratory animals. J. Appl. Physiol., 25:324–325, 1968.

Smith, D.M., et al.: An apparatus for anesthetizing small laboratory rodents. Lab. Anim. Sci., 23:869–871, 1973.

Soma, L.R. (ed.): Textbook of Veterinary Anesthesia. Baltimore, Williams & Wilkins, 1971.

Stevens, W.C., et al.: Comparative toxicities of enflurane, fluroxene, and nitrous oxide at subanaesthetic concentrations in laboratory animals. Can. Anaesth. Soc. J., 24:479–490, 1977.

Stevens, W.C., et al.: Comparative toxicities of halothane, isoflurane, and diethyl ether at subanesthetic concentration in laboratory animals. Anesth., 42:408–419, 1975.

Stunkard, J.A., and Miller, J.C.: An outline guide to general anesthesia in exotic species. Vet. Med. Small Anim. Clin., 6:1181–1186, 1974.

Torkelson, T.R., Oyen, F., and Rowe, V.K.: The toxicity of chloroform as determined by single and repeated exposure of laboratory animals. Am. Ind. Hyg. Assoc. J., 37:697–705, 1976.

Ward, G.S., Johnsen, D.O., and Roberts, C.R.: The use of CI 744 as an anesthetic for laboratory animals. Lab. Anim. Sci., 24:737–742, 1974.

Weisbroth, S.H., and Fudens, J.H.: Use of ketamine HCl as an anesthetic in laboratory rabbits, rats, mice, and guinea pigs. Lab. Anim. Sci., 22:904–906, 1972.

White, W.J., and Field, K.J.: Anesthesia and surgery of laboratory animals. In: The Veterinary Clinics of North America. Vol. 17(5). Edited by J.E. Harkness. Philadelphia, W.B. Saunders Co., 1987.

Wright, E.M., Marcella, K.L., and Woodson, J.F.: Animal pain: Evaluation and control. Lab. Anim., 14:20–36, 1985.

Anesthesia in the Rabbit

Alexander, D.J., and Clark, G.C.: A simple method of oral endotracheal intubation in rabbits (*Oryctolagus cuniculus*). Lab. Anim. Sci., 30:871–873, 1980.

Bertolet, R.P., and Hughes, H.C.: Endotracheal intubation: An easy way to establish a patent airway in rabbits. Lab. Anim. Sci., 30:227–230, 1980.

Bree, M.M., and Cohen, B.J.: Effects of urethane anesthesia on blood and blood vessels in rabbits. Lab. Anim. Care, 15:254–259, 1965.

Bree, M.M., Cohen, B.J., and Abrams, G.D.: Injection lesions following intramuscular administration of chlorpromazine in rabbits. J. Am. Vet. Med. Assoc., 159:1598–1602, 1971.

Chaffee, V., and Parkash, V.: A satisfactory method of anesthetizing rabbits for major or minor surgery. J. Am. Vet. Med. Assoc., 163:664, 1973.

Danneman, P.J., et al.: An evaluation of analgesia associated with the immobility response in laboratory rabbits. Lab. Anim. Sci., 38:51–57, 1988.

Davis, N.L., and Malinin, T.I.: Rabbit intubation and halothane anesthesia. Lab. Anim. Sci., 24:617–621, 1974.

Fick, T.E., and Schalm, S.W.: A simple technique for endotracheal intubation in rabbits. Lab. Anims., 21:265–266, 1987.

Flecknell, P.A., et al.: Neuroleptanalgesia in the rabbit. Lab. Anims., 17:104–109, 1983.

Freeman, M.J., Bailey, S.P., and Hodesson, S.: Premedication, tracheal intubation, and methoxyflurane anesthesia in the rabbit. Lab. Anim. Sci., 22:576–580, 1972.

Gardner, A.F.: The development of general anesthesia in the albino rabbit for surgical procedures. Lab. Anim. Care, 14:214–225, 1964.

Guerreiro, D., and Page, C.P.: The effect of neuroleptanalgesia on some cardiorespiratory variables in the rabbit. Lab. Anims., 28:205–209, 1987.

Hodesson, S., et al.: Anesthesia of the rabbit with Equi-Thesin following the administration of preanesthetics. Lab. Anim. Care, 15:336–344, 1965.

Hoge, R.S., et al.: Intubation technique and methoxyflurane administration in rabbits. Lab. Anim. Care, 19:593–595, 1969.

Kent, G.M.: General anesthesia in rabbits using methoxyflurane, nitrous oxide, and oxygen. Lab. Anim. Sci., 21:256–257, 1971.

Kisloff, B.: Ketamine-paraldehyde anesthesia for rabbits. Am. J. Vet. Res., 36:1033–1034, 1975.

Lacy, M.J., Kent, C.R., and Voss, E.W., Jr.: d-Limonene: An effective vasodilator for use in collecting rabbit blood. Lab. Anim. Sci., 37:485–487, 1987.

Lindquist, P.A.: Induction of methoxyflurane anesthesia in the rabbit after ketamine hydrochloride and endotracheal intubation. Lab. Anim. Sci., 22:898–899, 1972.

Linn, J.M., and Leibenberg, S.P.: In vivo detection of rabbit atropinesterase. Lab. Anim. Sci., 29:335–337, 1979.

Ludders, J.W., et al.: An anesthetic technique for repeated collection of blood from New Zealand White rabbits. Lab. Anim. Sci., 37:803–805, 1987.

McCormick, M.J., and Ashworth, M.A.: Acepromazine and methoxyflurane anesthesia of immature New Zealand White rabbits. Lab. Anim. Sci., 21:220–223, 1971.

Mersereau, W.A.: Endotracheal ether anesthesia in the rabbit. J. Surg. Res., 21:63–66, 1976.

Morgan, W.W., et al.: Pentobarbital anesthesia in the rabbit. Am. J. Vet. Res., 27:1133–1134, 1966.

Mulder, J.B.: Anesthesia in the rabbit using a combination of ketamine and promazine. Lab. Anim. Sci., 28:321–322, 1978.

Murdock, H.R.: Anesthesia in the rabbit. Fed. Proc., 28:1510–1516, 1969.

Pandeya, N.K., and Lemon, H.M.: Paraldehyde: An anesthesia for recovery experiments in albino rabbits. Lab. Anim. Care, 15:304–306, 1965.

Sanford, T.D., and Colby, E.D.: Effect of xylazine and ketamine on blood pressure, heart rate and respiratory rate in rabbits. Lab. Anim. Sci., 30:519–523, 1980.

Sartick, M., et al.: Recovery rate of the cardiovascular system in rabbits following short-term halothane anesthesia. Lab. Anim. Sci., 29:186–190, 1979.

Skartvedt, S.M., and Lyon, N.C.: A simple apparatus for inducing and maintaining halothane anesthesia of the rabbit. Lab. Anim. Sci., 22:922–924, 1972.

Strack, L.E., and Kaplan, H.M.: Fentanyl and droperidol for surgical anesthesia of rabbits. J. Am. Vet. Med. Assoc., 153:822–825, 1968.

Tucker, F.S., and Beattie, R.J.: Qualitative microtest for atropine esterase. Lab. Anim. Sci., 33:268–269, 1983.

Walden, N.B.: Effective sedation of rabbits, guinea pigs, rats and mice with a mixture of fentanyl and droperidol. Aust. Vet. J., 54:538–540, 1978.

Wass, J.A., Keene, J.R., and Kaplan, H.M.: Ketamine-methoxyflurane anesthesia for rabbits. Am. J. Vet. Res., 35:317–318, 1974.

Watson, S.C., and Cowie, A.T.: A simple closed-circuit apparatus for cyclopropane and halothane anesthesia of the rabbit. Lab. Anim. Care, 16:515–519, 1966.

White, G.L., and Holmes, D.D.: A comparison of ketamine and the combination of ketamine-xylazine for effective surgical anesthesia in the rabbit. Lab. Anim. Sci., 26:804–806, 1976.

Zahavi, I., et al.: A simple and safe method of anesthetizing infant rabbits for abdominal surgery. J. Surg. Res., 34:94–95, 1983.

Anesthesia in the Guinea Pig

Blouin, A., and Cormier, Y.: Endotracheal intubation in guinea pigs by direct laryngoscopy. Lab. Anim. Sci., 37:244–245, 1987.

Chang, P.: Letter: The effects of ketamine on guinea pig heart. Br. J. Anaesth., 45:929–930, 1973.

Cannell, H.: Pentobarbitone sodium anesthesia for oral and immunological procedures in the guinea pig. Lab. Anims., 6:55–60, 1972.

Clifford, D.R.: What the practicing veterinarian should know about guinea pigs. Vet. Med. Small Anim. Clin., 68:678–685, 1973.

Del Pozo, R., and Armas, M.C.: The use of diazepam as premedication in pentobarbital anesthesia in guinea pigs. Experientia, 29:635–636, 1973.

Douglas, J.S., et al.: Airway dilatation and constriction in spontaneously breathing guinea pigs. J. Pharmacol. Exp. Ther., 180:98–109, 1972.

Evans, E.F.: Neuroleptanesthesia for the guinea pig. An ideal anesthetic procedure for long-term physiological studies of the cochlea. Arch. Otolaryngol., 105:185–186, 1979.

Frisk, C.S., Herman, M.D., and Senta, K.E.: Guinea pig anesthesia using various combinations and concentrations of ketamine, xylazine, and/or acepromazine. Lab. Anim. Sci., 32:434, 1982.

Gilroy, B.A., and Varga, J.S.: Use of ketamine-diazepam and ketamine-xylazine combinations in guinea pigs. Vet. Med. Small Anim. Clin., 75:508–509, 1980.

Hart, M.V., et al.: Hemodynamics in the guinea pig after anesthetization with ketamine/xylazine. Am. J. Vet. Res., 45:2328–2335, 1984.

Hegab, E.-S., and Miller, J.A., Jr.: Ether anesthesia, hypothermia, and carbohydrate metabolism in adult guinea pigs. J. Appl. Physiol., 25:130–133, 1968.

Hoar, R.M.: Anesthesia in the guinea pig. Fed. Proc., 28:1517–1521, 1969.

Hoar, R.M.: The use of metofane (methoxyflurane) anesthesia in guinea pigs. Allied Vet., 36:131–134, 1964.

Kinsey, V.E.: The use of sodium pentobarbital for repeated anesthesia in the guinea pig. J. Am. Pharmacol. Assoc., 29:342–346, 1940.

Lamkin, R.H., and McPherson, D.L.: Inhalation anesthesia for the short-term guinea pig experiment. Arch. Otolaryngol., 101:138–139, 1975.

Latt, R.H., and Ecobichon, D.J.: Self-mutilation in guinea pigs following intramuscular injection of ketamine-acepromazine. Lab. Anim. Sci., 34:516, 1984.

Leash, A.M., Beyer, R.D., and Wilber, R.G.: Self-mutilation following Innovar-Vet injections in the guinea pig. Lab. Anim. Sci., 23:720–721, 1973.

Maykut, M.O.: The combined action of pentobarbital and meperidine and of procaine and meperidine in guinea pigs. Can. Anaesth. Soc. J., 5:161, 1958.

Moll, W., Künzel, W., and Ross, H.G.: Gas exchange of the pregnant uterus of anaesthetized and unanaesthetized guinea pigs. Respir. Physiol., 8:303–310, 1970.

Moore, C.J.: The use of CT 1341 (Saffan) as an anesthetic for guinea pigs. Guinea Pig Newsletter, *11*:37–39, 1977.

Newton, W.M., Cusick, P.K., and Raffe, M.R.: Innovar-Vet induced pathologic changes in the guinea pig. Lab. Anim. Sci., *25*:597–601, 1975.

Reuter, R.E.: Venipuncture in the guinea pig. Lab. Anim. Sci., *37*:245–246, 1987.

Reves, J.G., and McCracken, L.E., Jr.: Halothane in the guinea pig. Anesthesiol., *42*:230–231, 1975.

Rubright, W.C., and Thayer, C.B.: The use of Innovar-Vet as a surgical anesthetic for the guinea pig. Lab. Anim. Care, *20*:989–991, 1970.

Shucard, D.W., Andrew, M., and Beauford, C.: A safe and fast-acting surgical anesthetic for use in the guinea pig. J. Appl. Physiol., *38*:538–539, 1975.

Thomasson, B., Ruuskanen, O., and Merikanto, J.: Spinal anaesthesia in the guinea pig. Lab. Anim., *8*:241–244, 1974.

Ward, P.H., and Honrubia, V.: The effects of local anesthetics on the cochlea of the guinea pig. Laryngoscope, *79*:1605–1617, 1969.

Watson, R.T., and McLeod, K.: Inhalation anesthesia with methoxyflurane for guinea pig ear surgery. Arch. Otolaryngol., *104*:179–180, 1978.

Anesthesia in the Hamster

Carvell, J.E., and Stoward, P.J.: Halothane anaesthesia of normal and dystrophic hamsters. Lab. Anims., *9*:345–352, 1975.

Curl, J.L., and Peters, L.L.: Ketamine hydrochloride and xylazine hydrochloride anaesthesia in the golden hamster (*Mesoiricetus auratus*). Lab. Anims., *17*:290–293, 1983.

Ferguson, J.W.: Anaesthesia in the hamster using a combination of methohexitone and diazepam. Lab. Anims., *13*:305–308, 1979.

Gaertner, D.J., Boschert, K.R., and Schoeb, T.R.: Muscle necrosis in Syrian hamsters resulting from intramuscular injections of ketamine and xylazine. Lab. Anim. Sci., *37*:80–83, 1987.

Mulder, J.B., et al.: Anesthesia with Ketaset-Plus in guinea pigs and hamsters. Vet. Med. Small Anim. Clin., *74*:1807–1808, 1979.

Turner, T.T., and Howards, S.S.: Hyperglycemia in the hamster anesthetized with Inactin (5-ethyl-5-(1-methylpropyl)-2-thiobarbituate). Lab. Anim. Sci., *27*:380–382, 1977.

Willoughby, C.R.: Anesthetizing hamsters and small rodents. Vet. Rec., *92*:572–573, 1973.

Anesthesia in the Gerbil

Flecknell, P.A., et al.: Injectable anesthetic techniques in two species of gerbil (*Meriones libycus* and *Meriones unguiculatus*). Lab. Anim., *17*:118–122, 1983.

Lightfoote, W.E., and Molinari, G.F.: A comparison of ketamine and pentobarbital anesthesia in the Mongolian gerbil. Am. J. Vet. Res., *39*:1061–1063, 1978.

Norris, M.L.: Portable anaesthetic apparatus designed to induce and maintain surgical anaesthesia by methoxyflurane inhalation in the Mongolian gerbil (*Meriones unguiculatus*). Lab. Anims., *15*:153–155, 1981.

Norris, M.L., and Turner, W.D.: An evaluation of tribromo-ethanal (TBE) as an anesthetic agent in the Mongolian gerbil (*Meriones unguiculatus*). Lab. Anims., *17*:324–329, 1983.

Smith, S.M., and Kaplan, H.M.: Ketamine-methoxy-flurane anesthesia for the Mongolian gerbil, *Meriones unguiculatus*. Lab. Anims., *8*:213–216, 1974.

Anesthesia in the Mouse

Dolowy, W.C., Mombelloni, P., and Hesse, A.L.: Chlorpromazine premedication with pentobarbital anesthesia in the mouse. Am. J. Vet. Res., *21*:156–157, 1960.

Gomwalk, N.E., and Healing, T.D.: Etomidate: a valuable anaesthetic for mice. Lab. Anims., *15*:151–152, 1981.

Green, C.J., et al.: Metomidate, etomidate and fentanyl as injectable anaesthetic agents in mice. Lab. Anims., *15*:171–175, 1981.

Mulder, J.B.: Anesthesia in the mouse using a combination of ketamine and promazine. Lab. Anim. Sci., *28*:70–74, 1978.

Mulder, K.J., and Mulder, J.B.: Ketamine and xylazine anesthesia in the mouse. Vet. Med. Small Anim. Clin., *74*:569–570, 1979.

Sawyer, D.C.: Anesthetic techniques of rabbits and mice. *In* Experimental Animal Anesthesiology. Edited by D.C. Sawyer. Brooks Air Force Base, Texas, 1965.

Taber, R., and Irwin, S.: Anesthesia in the mouse. Fed. Proc., *28*:1528–1532, 1969.

Tarin, D., and Sturdee, A.: Surgical anaesthesia of mice: evaluation of tribromo-ethanol, ether, halothane, and methoxyflurane and development of a reliable technique. Lab. Anims., *6*:79–84, 1972.

Anesthesia in the Rat

Ben, M., Dixon, R.L., and Adamson, R.H.: Anesthesia in the rat. Fed. Proc., *28*:1522–1527, 1969.

Collins, T.B., and Lott, D.F.: Stock and sex specificity in the response of rats to pentobarbital sodium. Lab. Anim. Care, *18*:192–194, 1968.

Cooke, W.J.: Small inexpensive anesthetic apparatus for rats. J. Appl. Physiol., *41*:429–430, 1976.

Dahloef, L.G., van Dis, H., and Larsson, K.: A simple device for inhalational anesthesia in restrained rats. Physiol. Behav., *5*:1211–1212, 1970.

Dolowy, W.C., Thompson, I.D., and Hesse, A.L.: Chlorpromazine premedication with pentobarbital anesthesia in the rat. Proc. Anim. Care Panel, *9*:93–96, 1959.

Fleischman, R.N., McCracken, D., and Forbes, W.: Adynamic ileus in the rat induced by chloral hydrate. Lab. Anim. Sci., *27*:238–243, 1977.

Fowler, J.S.L., Brown, J.S., and Flower, E.W.: Comparison between ether and carbon dioxide anaesthesia for removal of small blood samples from rats. Lab. Anims., *14*:275–278, 1980.

Garcia, D.A., et al.: Deep anesthesia in the rat with the combined action of droperiodol-fentanyl and pentobarbital. Lab. Anim. Sci., *25*:585–587, 1975.

Glen, J.B., Cliff, G.S., and Jamieson, A.: Evaluation of a scavenging system for use with inhalation anaesthesia techniques in rats. Lab. Anims. *14*:207–211, 1980.

Hoar, R.M.: Anesthetic techniques for the rat and guinea pigs. *In* Experimental Animal Anesthesiology. Edited by D.C. Sawyer. Brooks Air Force Base, Texas, 1965.

Ingall, J.R.F., and Hasenpusch, P.H.: A rat resuscitator. Lab. Anim. Care, *16*:82–83, 1966.

Jones, J.B., and Simmons, M.C.: Innovar-Vet as an intramuscular anesthetic for rats. Lab. Anim. Care, 18:642–643, 1968.

Matthews, H.B.: A simple method for extending the period of surgical anesthesia in rats. Lab. Anims. Sci., 28:720–722, 1978.

Medd, R.K., and Heywood, R.: A technique for intubation and repeated short-duration anaesthesia in the rat. Lab. Anims., 4:75–78, 1970.

Molello, J.A., and Hawkins, K.: Methoxyflurane anesthesia of laboratory rats. Lab. Anim. Care, 18:581–583, 1968.

Mulder, J.B., and Johnson, H.B.: Ketamine and promazine for anesthesia in the rat. J. Am. Vet. Med. Assoc., 173:1252–1253, 1978.

Munson, E.S.: Effect of hypothermia on anesthetic requirement in rats. Lab. Anim. Care, 20:1109–1113, 1970.

Paterson, R.C., and Rowe, A.H.R.: Surgical anesthesia in conventional and gnotobiotic rats. Lab. Anims., 6:147–154, 1972.

Pena, H., and Crescencio, C.: Improved endotracheal intubation technique in the rat. Lab. Anim. Sci., 30:712–713, 1980.

Secord, D.C., Taylor, A.W., and Fielding, W.: Effect of anesthetic agents on exercised, atropinized rats. Lab. Anim. Sci., 23:397–400, 1973.

Shearer, D., Creel, D., and Wilson, C.E.: Strain differences in the response of rats to repeated injections of pentobarbital sodium. Lab. Anim. Sci., 23:662–664, 1973.

Stickrod, G.: Ketamine/xylazine anesthesia in the pregnant rat. J. Am. Vet. Med. Assoc., 175:952–953, 1979.

Tran, D.Q., and Lawson, D.: Endotracheal intubation and manual ventilation of the rat. Lab. Anim. Sci., 36:540–541, 1986.

Valenstein, E.S.: A note on anesthetizing rats and guinea pigs. J. Exp. Anal. Behav., 4:6, 1961.

Van Pelt, L.F.: Ketamine and xylazine for surgical anesthesia in rats. J. Am. Vet. Med. Assoc., 171:842–847, 1977.

Wilson, P., and Wheatley, A.M.: Ketamine hydrochloride in the laboratory. Nat. Lab. Anims., 15:349.

Wixson, S.K., et al.: A comparison of pentobarbital, fentanyl-droperidol, ketamine-xylazine and ketamine-diazepam anesthesia in adult male rats. Lab. Anim. Sci., 37:726–730, 1987.

Wixson, S.K., et al.: The effects of pentobarbital, fentanyl-droperidol, ketamine-xylazine and ketamine-diazepam on noxious stimulus perception in adult male rats. Lab. Anim. Sci., 37:731–735, 1987.

Wixson, S.K., et al.: The effects of pentobarbital, fentanyl-droperidol, ketamine-xylazine and ketamine-diazepam on arterial blood pH, blood gases, mean arterial blood pressure and heart rate in adult male rats. Lab. Anim. Sci., 37:736–742, 1987.

Wixson, S.K., et al.: The effects of pentobarbital, fentanyl-droperidol, ketamine-xylazine and ketamine-diazepam on core and surface body temperature regulation in adult male rats. Lab. Anim. Sci., 37:743–749, 1987.

Youth, R.A., et al.: Ketamine anesthesia for rats. Physiol. Behav., 10:633–636, 1973.

Hypnosis

Gruber, R.P., and Amato, J.J.: Hypnosis for rabbit surgery. Lab. Anim. Care, 20:741–742, 1970.

Klemm, W.R.: Drug potentiation of hypnotic restraint of rabbits as indicated by behavior and brain electrical activity. Lab. Anim. Care, 15:163–167, 1965.

Rapson, W.S., and Jones, T.C.: Restraint of rabbits by hypnosis. Lab. Anim. Care, 14:131–133, 1964.

Surgical Procedures

Agmo, A.: Sexual behaviour following castration in experienced and inexperienced male rabbits. Z. Tierpsychol., 40:390–395, 1976.

Alpert, M., Goldstein, D., and Triner, L.: Technique of endotracheal intubation in rats. Lab. Anim. Sci., 32:78–79, 1982.

Anderson, M., and Froimovitch, M.: Simplified method of guinea pig castration. Can. Vet. J., 15:126–127, 1974.

Archer, R.K., and Riley, J.: Standardized method for bleeding rats. Lab. Anims., 15:25–28, 1981.

Barrow, M.V.: Modified intravenous injection technique in rats. Lab. Anim. Care, 18:570–571, 1968.

Bergstein, I.L., and Agee, J.: Successful lobectomy in the guinea pig. Lab. Anim. Care, 14:519–523, 1964.

Bergström, S.: A simple device for intravenous injections in the mouse. Lab. Anim. Sci., 21:600–601, 1971.

Bett, N.J., Hynd, J.W., and Green, C.J.: Successful anaesthesia and small bowel anastomosis in the guinea pig. Lab. Anims., 14:225–228, 1980.

Bivin, W.S., and Smith, G.D.: Techniques of Experimentation. In Laboratory Animal Medicine. Edited by J.G. Fox, B.J. Cohen, and F.M. Loew. Orlando, Academic Press, 1984.

Bivin, W.S., and Timmons, E.H.: Basic biomethodology. In The Biology of the Laboratory Rabbit. Edited by S.H. Weisbroth, R.E. Flatt, and A.L. Kraus. Orlando, Academic Press, 1974.

Brocklehurst, W.E., and Cashin, C.H.: An injection block for small animals. Lab. Pract., 21:731, 1972.

Brouwer, E.A., Dailey, R., and Brouwer, J.B.: Ovariectomy of newborn rats: A descriptive surgical procedure. Lab. Anim. Sci., 30:546–548, 1980.

Brown, P.M.: A laryngoscope for use in rabbits. Lab. Anims., 17:208–209, 1983.

Castro, J.E.: Surgical procedures in small laboratory animals. J. Immunol. Methods, 4:213–216, 1974.

Chaffee, V.W.: Surgery of laboratory animals. In Handbook of Laboratory Animal Science, Vol. I. Edited by E.C. Melby and N.H. Altman. Cleveland, Chemical Rubber Company Press, 1974.

Cockerell, G.L., et al.: Necessity for end colostomy with transection of the descending colon in guinea pigs. J. Am. Vet. Med. Assoc., 175:954–956, 1979.

Cooley, R.K., and Vanderwolf, C.H.: Stereotaxic surgery in the rat: A photographic series. 2nd Ed. London, Ontario, A.J. Kirby, 1978.

DiPasquale, G., and Campbell, W.A.: A gag for gastric intubation of rabbits. Lab. Anim. Care, 16:294–295, 1966.

Duggan, K.A., Macdonald, G.J., and Rose, M.A.: Chronic renal artery catheterization for infusion studies in the conscious rabbit. Lab. Anim. Sci., 37:499–501, 1987.

Fleischman, R.W.: A technique for performing an ovariectomy on a hamster. VM/SAC, 76:1006–1007, 1981.

Frankenberg, L.: Cardiac puncture in the mouse

through the anterior thoracic aperture. Lab. Anims., 13:311–312, 1979.

Fraser, T., and Ascoli, R.C.: The castration of guinea pigs. J. Inst. Anim. Tech., 21:21–24, 1970.

Furner, R.L., and Mellett, L.B.: Mouse restraining chamber for tail vein injection. Lab. Anim. Sci., 25:648–649, 1975.

Gilroy, B.A.: Endotracheal intubation of rabbits and rodents. J. Am. Vet. Med. Assoc., 179:1295, 1981.

Gordon, H.A.: Comments on surgery and terminal procedures in gnotobiotic experiments. Lab. Anim. Care, 13:588–590, 1963.

Grice, H.C.: Methods for obtaining blood and for intravenous injections in laboratory animals. Lab. Anim. Care, 14:483–493, 1964.

Hall, L.L., et al.: A procedure for chronic intravenous catheterization in the rabbit. Lab. Anim. Sci., 24:79–83, 1974.

Hard, G.C.: Thymectomy in the neonatal rat. Lab. Anims., 9:105–110, 1975.

Hoar, R.M.: A technique for bilateral adrenalectomy in guinea pigs. Lab. Anim. Care, 16:410–416, 1966.

Hoar, R.M.: Methodology. In The Biology of the Guinea Pig. Edited by J.E. Wagner and P.J. Manning. Orlando, Academic Press, 1976.

Hopcroft, S.C.: A technique for the simultaneous bilateral removal of the adrenal glands in guinea pigs using a new type of safe anesthetic. Exp. Med. Surg., 24:12–19, 1966.

Hurwitz, A.J.: A simple method for obtaining blood samples from rats. J. Lab. Clin. Med., 78:172–174, 1971.

Kaplan, H., and Timmons, E.: The rabbit: A model for the principles of mammalian physiology and surgery. Orlando, Academic Press, 1979.

Kromka, M.C., and Hoar, R.M.: An improved technic for thyroidectomy in guinea pigs. Lab. Anim. Sci., 25:82–84, 1975.

Kujime, K., and Natelson, B.H.: A method for endotracheal intubation of guinea pigs (Cavia porcellus). Lab. Anim. Sci., 31:715–716, 1981.

Lambert, R.: Surgery of the digestive system in the rat. Springfield, IL, Charles C. Thomas, 1965.

Lang, C.M.: Animal Physiologic Surgery. 2nd Ed. New York, Springer-Verlag, 1982.

Mandel, M.: Castration of the domestic rabbit. Vet. Med. Small Anim. Clin., 71:365, 1976.

McGlinn, S.M., Shepherd, B.A., and Martan, J.: A new castration technic in the guinea pig. Lab. Anim. Sci., 26:203–205, 1976.

Moreland, A.F.: Collection and withdrawal of body fluids and infusion techniques. In Methods of Animal Experimentation. Vol. 1. Orlando, Academic Press, 1965.

Rickards, D.A., Hinko, P.J., and Morse, E.M.: Orthopedic procedures for laboratory animals and exotic pets. J. Am. Vet. Med. Assoc., 161:728–732, 1972.

Rosenhaft, M.E., Bing, D.H., and Knudson, K.C.: A vacuum-assisted method for repetitive blood sampling in guinea pigs. Lab. Anim. Sci., 21:598–599, 1971.

Sedgwick, C.J.: Spaying the rabbit. Mod. Vet. Pract., 63:401–403, 1982.

Talseth, T.: Vasoconstriction limits to the use of the central ear artery pressure in conscious rabbits. Lab. Anims., 15:1–3, 1981.

Toofanian, F., and Targowski, S.: Small intestinal anastomosis and preparation of intestinal loops in the rabbit. Lab. Anim. Sci., 32:80–82, 1982.

Watson, R.T., Leslie, W.G., and Jennings, E.H.: Operating table for guinea pig ear surgery. Arch. Otolaryngol., 104:177–178, 1978.

Waynforth, H.B.: Experimental and surgical technique in the rat. London, Academic Press, 1980.

Waynforth, H.B., and Parkin, R.: Sublingual vein injection in rodents. Lab. Anims., 3:35–37, 1969.

Wilson, P.: Bone marrow biopsy in the rabbit. Lab. Anims., 5:203–206, 1971.

Wixson, S.K., Murray, K.A., and Hughes, H.C., Jr.: A technique for chronic arterial catheterization in the rat. Lab. Anim. Sci., 37:108–111, 1987.

Wright, B.A.: A new device for collecting blood from rats. Lab. Anim. Care, 20:274–275, 1970.

Analgesia

Aranson, C.E.: Veterinary Pharmaceuticals and Biologicals. 4th Ed. Edwardsville, KS, Vet. Med. Pub. Co., 1984.

Flecknell, P.A.: The relief of pain in laboratory animals. Lab. Anims., 18:147–160, 1984.

Hunskaar, S., Fasmer, O.B., and Hole, K.: Formalin test in mice, a useful technique for evaluating mild analgesics. J. Neurosci. Meth., 14:69–76, 1985.

Morton, D.B., and Griffiths, P.H.M.: Guidelines on the recognition of pain, distress and discomfort in experimental animals and an hypothesis for assessment. Vet. Rec., 116:431–436, 1985.

Physician's Desk Reference, 40th Ed. Oradell, NJ, Medical Economics Company, Inc., 1986.

Sawyer, D.C.: Use of narcotics and analgesics for pain control. Scientific Proceedings AAHA 52nd Annual Meeting, 1985.

Physiologic Data

Extensive reviews of physiologic data are available in the general references listed in Chapter 1 and in the major texts on specific species listed in Chapter 2.

Archer, R.K., and Jeffcott, L.B. (eds.): Comparative Clinical Haematology. Oxford, Blackwell Scientific Publications, 1977.

Bannon, P.D., and Friedell, G.H.: Values for plasma constituents in normal and tumor bearing golden hamsters. Lab. Anim. Care, 16:417–420, 1966.

Bruckner-Kardoss, E., and Wostmann, B.S.: Oxygen consumption of germfree and conventional mice. Lab. Anim. Sci., 28:282–286, 1978.

Cubitt, J.G.K., and Barrett, C.P.: A comparison of serum calcium levels obtained by two methods of cardiac puncture in mice. Lab. Anim. Sci., 28:347, 1978.

Dillon, W.G., and Glomski, C.A.: The Mongolian gerbil: Qualitative and quantitative aspects of the cellular blood picture. Lab. Anims., 9:283–287, 1975.

Federation of American Societies of Experimental Biology: Biology Data Book, Vol. II. 2nd Ed. Bethesda, FASEB, 1973.

Firth, C.H., Suber, R.L., and Umholtz, R.: Hematologic and clinical chemistry findings in control BALB/c and C57BL/6 mice. Lab. Anim. Sci., 30:835–840, 1980.

Hong, C.C., et al.: Measurement of guinea pig body surface area. Lab. Anim. Sci., 27:474–476, 1977.

Jilge, B.: Soft feces excretion and passage time in the laboratory rabbit. Lab. Anims., 8:337–346, 1974.

Jilge, B.: The gastrointestinal transit time in the guinea pig. Z. Versuchstierkd., 22:204–210, 1980.

Kozma, C.K., et al.: Normal biological values for Long-Evans rats. Lab. Anim. Sci., 19:746–755, 1969.

Mauderly, J.L., et al.: Respiratory measurements of unsedated small laboratory mammals using nonrebreathing valves. Lab. Anim. Sci., 29:323–329, 1979.

Mays, A., Jr.: Baseline hematological and blood biochemical parameters of the Mongolian gerbil (Meriones unguiculatus). Lab. Anim. Sci., 19:838–842, 1969.

Mitruka, B.M., and Rawnsley, H.M.: Clinical, Biochemical and Hematological Reference Values in Normal Experimental Animals. New York, Masson Pub. USA, Inc., 1977.

Nachbaur, J., et al.: Variations of sodium, potassium, and chloride plasma levels in the rat with age and sex. Lab. Anim. Sci., 27:972–975, 1977.

Paget, G.E.: Clinical Testing of New Drugs. Edited by A.D. Herrick and M. Cattell. New York, Revere Publishing Co., 1965, p. 33.

Pettersson, G., Ahlman, H., and Kewenter, J.: A comparison of small intestinal transit time between the rat and the guinea pig. Acta Chir. Scand., 142:537–540, 1976.

Pickard, D.W., and Stevens, C.E.: Digesta flow through the rabbit large intestine. Am. J. Physiol., 222:1161–1166, 1972.

Quillec, M., Debout, C., and Izard, J.: Red cell counts in adult female guinea-pigs. Pathol. Biol., 25:443–446, 1977.

Schalm, O.W., Jain, N.C., and Carroll, E.J.: Veterinary Hematology. 3rd Ed. Philadelphia, Lea & Febiger, 1975.

Schermer, S.: The Blood Morphology of Laboratory Animals. 3rd Ed. Philadelphia, F.A. Davis Co., 1967.

Streett, R.P., and Highman, B.: Blood chemistry values in normal Mystromys albicaudatus and Osborne-Mendel rats. Lab. Anim. Care, 21:394–398, 1971.

Radiography

Carlson, W.D.: Radiography. In Methods of Animal Experimentation. Vol. 1. Edited by W.I. Gay. Orlando, Academic Press, 1965.

Engeset, A.: Intralymphatic injections in the rat. Cancer Res., 19:277–278, 1959.

Felson, B. (ed.): Roentgen Techniques in Laboratory Animals. Philadelphia, W.B. Saunders Co., 1968.

Freyschmidt, J., et al.: X-ray enlargement modified for use in experimental animal science. Lab. Anims., 9:305–311, 1975.

Gibbs, C.: Radiological examination of the rabbits. The head, thorax and vertebral column. J. Small Anim. Pract., 22:687–703, 1981.

James, A.E., et al.: Magnification in veterinary radiology. J.A.V.R.S., 16:52–64, 1975.

Margulis, A.R., Carlsson, E., and McAlister, W.H.: Angiography of malignant tumors in mice. Acta Radiol., 56:179–192, 1961.

Matthew, H.G., and Barnard, H.J.: Radiographic techniques. In Roentgen Techniques in Laboratory Animals. Edited by B. Felson. Philadelphia, W.B. Saunders Co., 1968.

Pyke, R.E., and Stover, B.J.: A lead shielded cabinet for use in X-raying rats. Lab. Anim. Care, 16:292–293, 1966.

Serota, K.S., Jeffcoat, M.K., and Kaplan, M.L.: Intraoral radiography of molar teeth in rats. Lab. Anim. Sci., 31:507–509, 1981.

Shively, M.J.: Xeroradiographic anatomy of the domesticated rabbit (Oryctolagus cuniculus). Part I: Head, thorax, and thoracic limb. Southwestern Veterinarian, 32:219–233, 1979.

Shively, M.J.: Xeroradiographic anatomy of the domesticated rabbit (Oryctolagus cuniculus). Part II: Abdomen, pelvis, and pelvic limb. Southwestern Veterinarian, 33:57–67, 1980.

Tirman, W.S., et al.: Microradiography: its application to the study of the vascular anatomy of certain organs of the rabbit. Radiology, 57:70–80, 1951.

Wagner, J.E., et al.: Otitis media of guinea pigs. Lab. Anim. Sci., 26:902–907, 1976.

Euthanasia

See Lab Animal, 11(4):17–47, 1982, for several articles on euthanasia of small animals.

Blackshaw, J.N., et al.: The behaviour of chickens, mice and rats during euthanasia with chloroform, carbon dioxide and ether. Lab. Anims., 22:67–75, 1988.

Booth, N.H. Inhalant anesthetics. In Veterinary Pharmacology and Therapeutics. 5th Ed. Edited by N.H. Booth, and L.E. McDonald. Ames, IA, Iowa State University Press, 1982, pp. 175–202.

Clifford, D.H. Preanesthesia, anesthesia, analgesia, and euthanasia. In Laboratory Animal Medicine. Edited by J.G. Fox, B.J. Cohen, and F.M. Loew. Orlando, Academic Press, 1984, pp. 528–563.

Fawell, J.K., Thomson, C., and Cooke, L.: Respiratory artefact produced by carbon dioxide and pentobarbitone sodium euthanasia in rats. Lab. Anims., 6:321–326, 1972.

Feldman, D.B., and Gupta, B.N.: Histopathologic changes in laboratory animals resulting from various methods of euthanasia. Lab. Anim. Sci., 26:218–221, 1976.

Hatch, R.C.: Euthanatizing agents. In Veterinary Pharmacology and Therapeutics. Edited by N.H. Booth, and L.E. McDonald. Ames, IA, Iowa State University Press, 1982, pp. 1059–1064.

Hernett, T.D., and Haynes, A.P.: Comparison of carbon dioxide/air mixture and nitrogen/air mixture for the euthanasia of rodents. Design of a system for inhalation euthanasia. Anim. Technol., 35:93–99, 1984.

Hughes, H.C.: Euthanasia of laboratory animals. In Handbook of Laboratory Animal Science, Vol. III. Edited by E.C. Melby, Jr., and N.H. Altman. Cleveland, CRC Press, 1976, pp. 553–559.

Kitchell, R.L.: Animal Pain: Perception and Alleviation. Edited by E. Carsten, et al. Bethesda, MD, American Physiological Society, 1983.

Kitchell, R.L., and Johnson, R.D.: Assessment of pain in animals. In Animal Stress. Edited by G.P. Moberg. Bethesda, MD, American Physiological Society, 1983, pp. 113–140.

Lumb, W.V., and Jones, E.: Veterinary Anesthesia. Philadelphia, Lea & Febiger, 1985.

Mikeska, J.A., and Klemm, W.R.: EEG evaluation of humaneness of asphyxia and decapitation euthanasia of the laboratory rat. Lab. Anim. Sci., 25:175–179, 1975.

Rudolph, H.S.: A small animal euthanasia chamber. Lab. Anim. Care, 13:91–95, 1963.

Smith, A.W., et al.: Report of the AVMA panel on euthanasia. J. Am. Vet. Med. Assoc., 188:252–268, 1986.

Chapter 4

Clinical Signs and Differential Diagnoses

The clinical signs and disease syndromes commonly encountered in rabbits, guinea pigs, hamsters, gerbils, mice, and rats are caused by a variety of cutaneous, gastrointestinal, respiratory, reproductive, neuromuscular, and miscellaneous conditions. The content of this chapter, divided by species, provides several differential diagnoses for each sign. These diagnostic alternatives, based on literature references and personal observations and communications, are not exhaustive; additional information may be extrapolated from other species or gleaned from the bibliographies.

Detailed information concerning specific diagnoses may be found through the Index (those marked by an asterisk); in the species-categorized bibliographies following this chapter; in Chapter 5; or in the general reference works listed in Chapters 1 and 2.

Diagnosis based on clinical signs alone is presumptuous, and confirmation by necropsy, histopathologic, cultural, serologic, and epidemiologic means is necessary.

HISTORY PROTOCOL

The discussion of factors predisposing to disease (Chapter 1) and the topics in this listing provide a basis for establishing an information base before proceeding toward a diagnosis.

Animal Involved
  Species and strain
  Sex
  Age
  Source
  Breeding status
Environment of the Animal
  Barrier or conventional housing
  Cage type and location
  Feeder and waterer type
  Bedding type
  Sanitation level
  Waste disposal
  Other species or vermin in the colony

Disturbances in the colony
  Recent changes or new animals
  Light cycles
  Ventilation
  Room temperature
  Humidity
  Experimental procedure
Diet of the Animal
  Source and composition
  Storage facilities
  Milling date
  Dietary supplementation
  Recent changes in diet
Health History
  Previous diseases
  Enzootic diseases
  Treatments
  Vaccinations
  Breeding record
  Quarantine procedure

Complaint
  Onset and progression
  Number exposed

Morbidity and mortality
  Ages and sex affected
  Clinical signs

THE RABBIT

CUTANEOUS SIGNS

Alopecia

Hair loss or thinning in rabbits may be due to hair pulling for nests (Sawin et al. 1960), a behavioral vice, malnutrition (usually a fiber deficiency), or an ectoparasite infection.* Rubbing on cage or feeder, dermatophyte* or bacterial (O'Donoghue and Whatley 1971) dermatopathies, and hair pulling by predators or humans also cause hair loss. Seasonal molting may produce hair coat irregularities and thinning.

Cutaneous or Subcutaneous Swelling

The common causes of single or multiple subcutaneous masses in a rabbit are abscesses caused by *Pasteurella multocida** or *Staphylococcus aureus,** although other bacteria, such as *Fusobacterium* (Beattie, Yates and Donaldson 1913; Ward, Crumrine and Mattloch, 1981) may be involved. Neoplasia* (fibromas, myxomas, lymphomas, papillomas), hematomas, cuterebriasis (Baird 1983; Jacobson, McGinnes and Catts 1978; Weisbroth, Wang and Scher 1973; Lopushinsky, 1970), and diseased mammary glands (Burrows 1940) are other causes of swelling. *Cuterebra* fly larvae may be seen in rabbits housed outdoors in the summer months. The larvae are usually located subcutaneously on the throat or neck and are indicated by a 2- to 3-cm, cone-shaped swelling with an apical pore. The larva is killed with chloroform or other anesthetic, the pore is opened, and the larva is removed intact. Crushing the larva in the subcutis may result in a fatal shock reaction. The cleaned wounds heal rapidly.

Dermatitis

Inflammation of the skin, indicated variably by reddening, alopecia, scaliness, crust or scab formation, seborrhea, or ulceration, can be caused by several types of infectious

agents. The ear mite *Psoroptes cuniculi** causes crust formation and reddening of the ear canal. The fur mite *Cheyletiella parasitivorax** may cause scaley dandruff accumulations deep in the hair coat and occasionally causes hair loss and an oily dermatitis on the face or back, whereas sarcoptid mites (Low 1911) are found rarely on the face and limbs. The rabbit louse *Haemodipsus ventricosis* (also rare) causes anemia, emaciation, and pruritus.

Traumatic dermatopathies include pododermatitis (sore hocks),* bites by flies, fleas, or other rabbits, and such moist dermatopathies as sore dewlap (Williams and Gibson 1975), hutch or urine burn, and extension of a conjunctival infection over the face.

Ulcerative dermatitis in rabbits, a common condition known as "sore hocks," occurs unilaterally or bilaterally on the plantar metatarsal (rear feet) or, less often, on the metacarpal (front feet) surfaces. The lesion consists of focal, scab-covered, often hemorrhagic, cutaneous ulcers. Factors predisposing to sore hocks include reduced plantar fur pad thickness (inherited or through wetting), thumping and bruising, pressure from excessive weight or restraint, lack of movement in a small cage, or abrasions from irregular cage flooring. Sore hocks can occur in rabbits on both solid and wire floors. The incidence of sore hocks may increase with age and weight.

Sore hocks, which can occur in any type of cage or confinement device, is prevented by housing rabbits in clean cages on soft, clean, dry bedding. Treatment includes cleansing, toenail trimming, and the application of a topical antiseptic. Healing is prolonged and complicated by secondary bacterial infection.

Bacterial dermatopathies can be caused by *Pseudomonas aeruginosa* (blue fur disease) (O'Donoghue and Whatley 1971), *Pasteurella multocida,** *Staphylococcus aureus,** *Fusobacterium*

---

*Asterisks refer to an Index listing.

necrophorum (Beattie, Yates and Donaldson 1913), *Corynebacterium pyogenes,* streptococci, and *Treponema cuniculi** infections on the genitalia and face.* Other causes of dermatopathies in rabbits include *Trichophyton mentagrophytes** and *Microsporum gypseum** on the face and forelegs, myxomatosis,* and, rarely, zinc deficiency (Shaw et al. 1974) and rabbit pox (Christensen, Bond and Matanic 1967; Greene 1934). Multifocal, moist, crusted areas over the head and trunk may be caused by *Staphylococcus* or *Aspergillus* from contaminated bedding.

GASTROINTESTINAL SIGNS

Diarrhea

The specific cause of a case or outbreak of diarrhea in rabbits is often difficult to determine. In some instances a definitive diagnosis would probably involve two or more organisms and associated toxins. Specific examples of such relationships are colibacillosis and coccidiosis and enterotoxemia preceded by Tyzzer's or rotavirus infection. Specific conditions that might be considered, if not conclusively diagnosed, are colibacillosis,* enterotoxemia from *Clostridium* infections,* Tyzzer's disease,* intestinal coccidiosis,* lack of dietary fiber, mucoid enteropathy,* antibiotic administration* (Thilsted et al. 1981), and such rarely reported conditions as salmonellosis,* pseudomoniasis (McDonald and Pinheiro 1967), aflatoxicosis, and rotavirus infection.* The first four differential conditions are usually seen in weanling rabbits. A less definitive diagnosis for diarrhea in the rabbit is simply enteropathy complex or nonspecific enteropathy.*

Ptyalism

Drooling over the chin and dewlap may accompany incisor malocclusion,* abrasion of the mouth on feeder or cage, heat stress, or abdominal pain.

Pendulous Abdomen

Intestinal disorders,* hepatic coccidiosis,* mastitis,* metritis, uterine adenocarcinomas, pregnancy, and obesity can cause an enlarged abdomen.

Polydipsia

Increased water intake by rabbits accompanies lactation, heat stress, febrile disease, dry feed, salty diets, diabetes mellitus, and enteritis. Potable water must be provided at all times.

Constipation

Gastric hairball,* pyloric stenosis in young (Cardy 1973; Weisbroth and Scher 1975), mucoid enteropathy,* water deprivation, and anorexia may all lead to constipation.

Anorexia

Malocclusion, gastric hairballs, dietary change, pain, systemic disease, environmental stress, insufficient water supply, unpalatable (especially moldy) feed (Cheeke 1974), changed feeding or watering devices, loss of olfaction, and acidotic or ketotic conditions may lead to anorexia.

RESPIRATORY SIGNS

Nasal Discharge

Mucopurulent rhinitis in most cases is caused by *Pasteurella multocida.** *Staphylococcus aureus* and, in rare circumstances, *Bordetella bronchiseptica** may also be upper respiratory pathogens. A nasal discharge may accompany heat stress, allergies to hay or feed dust, myxomatosis,* and rabbit pox (Christensen, Bond and Matanic, 1967; Greene 1934). Myxomatosis is a rare condition in most areas except California, Australia, and parts of Europe and Brazil.

Dyspnea

Labored breathing is usually associated with pneumonia caused by *Pasteurella multocida,** although *Staphylococcus aureus** and possibly *Bordetella bronchiseptica* and other agents (Patton 1975) may occasionally be involved. Pneumonia in rabbits is usually first noticed as a sudden death, but careful antemortem examination may reveal bluing of the iris and lips, weight loss, depression, and difficult breathing.

Ocular Discharge

Conjunctivitis caused by *Pasteurella multocida** is common. Other bacteria, chlamydia, dust (Buckley and Lowman 1979), trauma, myxomatosis,* entropion (Fox et al. 1979), and rabbit pox (Christensen, Bond and Matanic 1967; Greene 1934) may also cause an ocular discharge. Exudate and infection spreading from the conjunctiva may extend over the face and under the chin. If *Pseudomonas* infection is involved, fur adjacent to the lesion may be stained blue-green.

REPRODUCTIVE SIGNS

Infertility

Infertility in rabbits is difficult to diagnose. Factors that may contribute to a real or apparent infertility are immaturity or senescence (Maurer and Foote 1972), heat stress,* crowding (Mykytowycz and Fullagar 1973), sexual exhaustion (Adams and Singh 1981; McNitt 1981), autumnal depression (Doggett 1956; Doggett 1958; Farrell, Powers and Otani 1968; Sittmann 1964), an incompatible mate, estrogenic stimulation, uterine infection or neoplasia,* orchitis, venereal spirochetosis,* poisoning (Confer, Ward and Hines 1980; Kendall, Salisbury and Vandermark 1950), caloric or other nutritional deficiencies (Shaw et al. 1974), noise (Zondek and Tamari 1960), pseudopregnancy (Kaufmann, Quist and Broderson 1971), and oxyuriasis.*

Prenatal Mortality

Resorption, abortion, and stillbirths occur at different times of gestation, but the causes can be similar (Adams 1960, 315–316; Adams 1960, 36–44; Adams 1960, 325–344). Fetuses are particularly susceptible to dislodgment at 13 days and between 20 and 23 days. In the first case, a placental change is occurring; in the second, the fetal mass is round, turgid, and easily torn loose.

Other possible causes of fetal loss are excessively large or small litters (Adams 1970), pregnancy toxemia (Abitol, Driscoll and

Ober 1976), inadequate caloric intake or deficiency of the pregnant doe, especially in dwarf breeds in cool weather, vitamin A excess (Millen and Dickson 1957), arsanilic acid (Confer, Ward and Hines 1980), DDT (Hart et al. 1972), congenital abnormalities, heat (Hellmann 1979), pasteurellosis,* listeriosis (Gray, Singh and Thorp 1955; Vetesi and Kemenes 1967; Watson and Evans 1985), salmonellosis,* aspergillosis (Boro et al. 1978), chlamydial infection (Parker, Hawkings and Brenner 1966), systemic disease, uterine crowding, colony crowding, and nitrates (Dollahite and Rowe 1974; Kruckenberg 1974).

Vitamin A excess may occur in rabbits fed a diet high in alfalfa and vitamin supplement containing vitamin A.

Nitrates enter the drinking water from fertilizer, organic matter, or manure runoff. Nitrates are present in many feed crops, including alfalfa, clover, orchard grass, timothy hay, and blue grass. Drought, high temperatures, cloudy days, chemical contaminants, and immaturity at harvest contribute to nitrate accumulation in plants. Rabbits may abort if chronically exposed to 50 ppm nitrate in the drinking water. Nitrate analyses of feed and water should be obtained in suspect cases.

Litter Desertion or Death

Maternal inexperience, handling the young, disturbing the nest, cannibalism during placental consumption, insufficient nest material, outside disturbances, deformed or injured young, chilling of the young, thirst, aspiration pneumonia, caloric deficiency or vitamin A excess in the doe, young out of the nest, a divided nest, and agalactia can lead to neonatal neglect or death. Agalactia can result from a gastric hairball, malocclusion, anorexia, mastitis, and water deprivation.

Death of young rabbits can also be caused by enterotoxemia,* trampling, rat bites, enteropathies, *Pasteurella multocida* septicemia,* starvation or dehydration, pyloric stenosis (Cardy 1973; Weisbroth and Scher 1975), and ingestion of quaternary ammonium residues. Young may die after suckling a doe with staphylococcal mastitis.

Vaginal Discharge

Normal rabbit urine can be pale yellow to red-orange to brown and may vary from clear to opaque and resemble pus or blood. Uterine adenocarcinomas often bleed, and a *Pasteurella*- (Thigpen, Clements and Gupta 1978) or, rarely, *Listeria*-caused (Holmes 1961) metritis can produce a purulent vaginal discharge. A variety of gram-negative, enteric bacteria may also cause metritis in rabbits. Abortions may also result in a sanguineous discharge.

NEUROMUSCULAR SIGNS

Torticollis

Wry neck is due in almost all cases to an otitis interna established from migration of bacteria (usually *Pasteurella multocida*) via the eustachian tube and middle ear.* Extension of a mite infection from the external ear to the inner ear is rare, and in any case, such an invasion would probably be complicated by bacterial infection. Scoliosis (Sawin and Grary 1964) also occurs.

Encephalopathies from *Pasteurella multocida*,* listeriosis (Traub 1942), encephalitozoonosis (Kunstyr, Naumann and Kaup 1986),* and *Ascaris columnaris* (Dade et al. 1975; Nettles et al. 1975) migrations also occur.

Incoordination or Convulsions

Incoordination can be caused by spinal or limb injury, muscular dystrophy, ascariasis (Dade et al. 1975; Nettles et al. 1975), encephalitozoonosis,* otitis interna, bacterial encephalitis, poisoning, pregnancy toxemia,* enterotoxemia,* heat exhaustion,* congenital abnormalities, rabies (Rabbit Rabies, CDC Vet Notes 1981), magnesium deficiency (Kunkel and Pearson 1948), and ataxia (Sawin, Anders and Johnson 1942).

Paresis and Paralysis

Fracture or luxation of the lumbar spine is common in rabbits (Baxter 1975; Jones et al. 1982; Mendlowski 1975; Templeton

1946). If rabbits are dropped, allowed to thrash during restraint, or excited in a cage, they may fracture the lower spine and damage the spinal cord. The degree and duration of the resulting paresis or paralysis and loss of bladder and sphincter control depend on the severity and location of the cord lesion. If radiography or palpation reveals displacement of the vertebral canal and probable cord laceration, the prognosis is poor and recovery is unlikely. If the cord is only swollen or locally inflamed, recovery may follow cage rest for up to 1 week, although this situation seldom occurs.

Splay leg is a descriptive term applied to a variety of inherited or acquired abnormalities in which the rabbit is unable to adduct one or more limbs (Arendar and Milch 1966). These abnormalities include improper development of the spine, pelvis, coxofemoral junction, or long bones. The muscles of the affected limbs may remain functional or become partially or totally paralyzed. Unlike paralysis associated with vertebral fracture or luxation, splay leg usually has a familial inheritance pattern. Splay leg may be evident at a few days after birth or after several months. Other conditions leading to weakness of rear limbs are ataxia (Sawin, Anders, and Johnson 1942), malnutrition (Ringler and Abrams 1971), and severe *Sarcocystis* infection (Cosgrove, Wiggins and Rothenbacher 1982; Cerna et al. 1981; Dubey 1976).

MISCELLANEOUS SIGNS

Weight Loss

Rabbits lose weight from anorexia of various causes (Cheeke 1974), but medium-sized rabbits limit fed 90 to 120 g (3 to 5 oz) of pelleted feed daily rarely have this problem. Malocclusion,* gastric hairball,* nutritional deficiency (Cheeke and Patton 1978; Stevenson, Palmer and Findley 1976), water deprivation, food pellets that are too small, bacterial disease (Daniels 1961; Harkins and Saleeby 1928), consumption of moldy feed, encephalitozoonosis,* neoplasia,* pyloric stenosis (Cardy 1973; Weisbroth and Scher 1975), pain, oxyuriasis,* chronic renal disease, ectoparasitism, extreme cold, and arteriosclerosis (Haust and Geer 1970) also lead to weight loss.

Death

Death of mature rabbits with few or no preceding signs may occur with neoplasia,* acute bacterial disease (especially pasteurellosis), septicemia from a staphylococcal mastitis, colibacillosis,* enterotoxemia,* mucoid enteropathy,* and, rarely, salmonellosis,* toxoplasmosis (Harcourt 1967; Lainson 1955), myxomatosis,* cardiomyopathy (Weber and Van der Walt 1975), rabbit pox (Greene 1934), tularemia,* and urolithiasis.* The death of does when the litter is a few weeks old is most likely due to a toxemia caused by *Clostridium* spp (Patton and Cheeke 1980)* or retained placentas.

Anemia

The rabbit louse *Haemodipsus ventricosis* is rarely seen. It sucks blood and can cause anemia. Anemia may also accompany lymphosarcoma, certain genetic abnormalities, repeated blood sampling, bleeding from a uterine adenocarcinoma, tropical rat mite infestation, and marrow depression from a chronic infection.

Yellow Fat

The occasional presence of a distinctive yellow fat in rabbits is due to a homozygous recessive condition resulting in a failure to metabolize xanthophylls. The accumulation of xanthophylls has no known deleterious effect on rabbits (Wilson and Dudley 1946).

Eye Enlargement

Buphthalmia (infantile or congenital) is caused by an autosomal recessive defect with incomplete penetrance and a variable age of onset. Clinical signs include unilateral or bilateral progressive enlargement of the eye globe, corneal clouding, scarring, flattening and vascularization, and conjunctivitis.

Bloody Urine

Bloody urine is rare in rabbits and rodents. If the urine is indeed bloody, then cystitis, bladder polyps, pyelonephritis, renal infarcts, urolithiasis,* disseminated intravascular coagulation (DIC), and leptospirosis should be considered (Garibaldi et al. 1987). Many cases of "bloody" urine turn out to be porphyrin-pigmented basic urine or a sanguineous vaginal discharge associated with uterine adenocarcinoma, polyps, or abortion. Thick white urine containing reddish-orange pigment is indicative of an excess of dietary calcium.

Liver Spots

The most common cause of white, tan, or yellow foci on the liver are hepatic coccidiosis (numerous, irregular, fuzzy edges),* migrating tapeworm larvae (few, elongated, clear edges),* extension of an intestinal infection via the portal vein (Tyzzer's disease),* and colibacillosis.* Less common conditions affecting the liver include tularemia ("milky way" liver),* listeriosis (Vetesi and Kemenes 1967), salmonellosis,* toxoplasmosis (Harkins and Saleeby 1928), rabbit pox (Greene 1934), and tuberculosis (Harkins and Saleeby 1928).

THE GUINEA PIG

CUTANEOUS SIGNS

Alopecia

Hair loss or thinning may be seen in intensively bred sows, particularly in the last trimester of pregnancy when metabolic demands on the sow are considerable. Repeatedly bred sows lose increasing amounts of hair with each succeeding pregnancy and lactation. Hair loss is also seen in weakened young near weaning age. Guinea pigs often barber one another, and an animal low on the social hierarchy and young animals may lose considerable amounts of hair. Abrasion on rough surfaces and fungal (*Trichophyton mentagrophytes*) and bacterial dermatopathies may also result in hair loss.

Rough Hair Coat

Abyssinian guinea pigs normally have a rough hair coat characterized by several rosettes or whorls of hair. Generalized roughness of hair coat is associated with illness, a leaky water bottle, a moist dirty cage, or an aggressive posture (Grant and MacKintosh 1963).

Dermatitis

Bite wounds (Lee, Johnson and Lang 1978), usually on the ears or back, are associated with cutaneous inflammation, as are dermatophytoses,* staphylococcal dermatitis,* pediculosis,* and *Trixacarus** and perhaps *Chirodiscoides** mite infestations. Dermatophytes cause patchy lesions on the face and back; mites affect the trunk, as do bacterial infections, and lice are found primarily on the head and neck. Lesions of ectoparasitism are usually exacerbated by scratching, as occurs during the highly pruritic *Trixacarus* infection and louse infestations.

Cutaneous or Subcutaneous Swelling

Abscesses are relatively common in guinea pigs. *Streptococcus zooepidemicus** (cervical "lumps" and generalized abscessation) and *Staphylococcus aureus* (pododermatitis)* are commonly involved, whereas *Streptobacillus moniliformis* (Aldred, Hill and Young 1974) and *Yersinia pseudotuberculosis* (Bishop 1932) are occasionally seen. Mastitis (Gupta, Langham and Conner 1970), arthritis (Gupta, Conner and Meyer 1972), cystitis (Wood 1981), and joint swelling from scorbutus* also occur. Neoplasia, including cavian leukemia with lymphadenopathy,* is rarely seen.

GASTROINTESTINAL SIGNS

Diarrhea

When enteropathies do occur in guinea pigs, they are rapidly fatal with few clinical signs. Some conditions that might cause diarrhea include Tyzzer's disease,* colibacillosis,* salmonellosis,* coccidiosis,* *Arizona* spp. infection, and clostridial infection.* Rapidly fatal, acute enteropathies include acute typhlitis and the reaction to antibiotics (Farrar and Kent 1965; Rehg, Yarbrough and Pakes 1980), both of which may be related to clostridial proliferation and toxin release. Although guinea pigs may have nematode infections (Porter and Otto 1934), these infections rarely cause clinical problems. *Cryptosporidium* spp. (Tzipori et al. 1981) may cause diarrhea, poor weight gain, and possibly death in young guinea pigs.

Ptyalism

Guinea pigs slobber in cases of malocclusion (Olson 1971), hypovitaminosis C,* heat stress,* and fluorosis (Hard and Atkinson 1967). The teeth most often overgrown are the premolars or anterior molars; an otoscope is helpful in examining the cheek teeth.

RESPIRATORY SIGNS

Nasal Discharge and Dyspnea

A nasal and ocular discharge frequently accompanies respiratory disease in guinea pigs. Bacterial organisms commonly involved in respiratory disease in guinea pigs include *Bordetella bronchiseptica,** *Streptococcus pneumoniae,** and *Streptococcus zooepidemicus.** Organisms infrequently involved include *Klebsiella pneumoniae* (Dennig and Eidmann 1960), *Pasteurella multocida,** *Pseudomonas aeruginosa* (Bostrum et al. 1969), and *Staphylococcus aureus.** Noninfectious conditions causing apparent respiratory distress are heat stress, diaphragmatic hernia, pregnancy toxemia,* and gastric torsion (Lee, Johnson and Lang 1977).

Ocular Discharge

An ocular discharge, like a nasal discharge, rough hair coat, and reluctance to move, is a nonspecific sign of illness in guinea pigs. Specific conditions commonly causing conjunctival inflammation include *Bordetella bronchiseptica** and *Streptococcus pneumoniae** infection, salmonellosis,* and chlamydial neonatal conjunctivitis (Murray 1964).

REPRODUCTIVE SIGNS

Infertility

Real or apparent infertility or decreased production in guinea pigs may be related to age at the time of breeding, use of wire floors, estrogens in the feed (Wright and Seibold 1958), bedding adhering to the genitalia (Plank and Irwin 1966), elevated ambient temperature,* nutritional deficiency, metritis (Juhr and Obi 1970), or environmental stress (Haines 1931).

Prenatal Mortality

Abortions and stillbirths in guinea pigs occur with nutritional deficiencies, dystocias, *Bordetella, Salmonella, Streptococcus,* and cytomegalovirus infections, asphyxia at birth, and with pregnancy toxemia.* Dystocias occur if the fetuses are too large for the pelvic canal or if the uterine musculature is weakened by toxins or infection.

If the guinea pig is first bred after 8 months of age, or if the guinea pig is excessively obese, a subsequent dystocia is probable. Dystocia may be associated with uterine hemorrhage, exhaustion, toxemia, and death. Cesarean section or 0.2 to 3.0 units/kg oxytocin (Rossoff 1974) may be used to alleviate dystocia, if the pelvic canal is large enough. Small litters (1 or 2) carried less than 66 days or longer than 74 days usually die, as do large litters carried longer than 72 days. Stillbirths are most likely when parturition occurs at less than 62 days or more than 75 days; the optimal gestation is approximately 69 days (Goy, Hoar and Young 1957).

Litter Desertion or Death

As neonatal guinea pigs are remarkably precocious, maternal attention other than nursing and licking is not necessary for survival, although the mother should certainly remain with the young, if at all possible. Young weighing less than 50 g at birth usually die.

Factors that reduce maternal attention are mastitis, maternal inexperience, and environmental disturbances, such as noise or active cagemates.

NEUROMUSCULAR SIGNS

Torticollis

Otitis interna is the common cause of a head tilt. Encephalopathies leading to torticollis are rarely seen.

Incoordination or Convulsions

A common cause of incoordination or paralysis is nerve or bone injury from a fall. Pregnancy toxemia, enterotoxemias, and various muscle disorders are other causes.

Reluctance to Move or Paralysis

Hypovitaminosis C,* with swollen and painful joints, is a common cause of immobility, as are systemic diseases such as *Bordetella bronchiseptica* * infection, streptococcal disease,* salmonellosis,* and malnutrition, specifically vitamin E deficiency (Seidel and Harper 1960). Osteoarthritis (Taylor et al. 1971), spinal fracture, histoplasmosis (rare) (Correa and Pacheco 1967), and the various myopathies (Bender, Schottelius and Schottelius 1959; Saunders 1958; Ward et al. 1977; Webb 1970) are other possible causes of reluctance to move. Back strain with edema and possibly hemorrhage around the spinal cord may result in temporary paralysis.

MISCELLANEOUS SIGNS

Weight Loss

Premolar malocclusion (Olson 1971) and hypovitaminosis C* are the common causes of weight loss in guinea pigs. Less commonly, neophobia regarding feed and feeders, protein deficiency, metastatic calcification (Sparschu and Christie 1968), ectoparasitism (Griffiths 1971), urolithiasis (Spink 1978), chronic renal disease,* neoplasia,* and bacterial disease cause weight loss in guinea pigs. A relatively common bacterial problem is staphylococcal pododermatitis,* which causes local pain and, eventually, systemic disorders such as amyloidosis.

Death

Abrupt death with few or no preceding signs may be caused by chilling or overheating (90° F); a septicemia or toxemia (salmonellosis,* cecitis, enteritis, pregnancy toxemia,* antibiotic reaction) (Farrar and Kent 1965; Rehg, Yarbrough and Pakes 1980); pneumonia; volvulus or torsion of stomach or cecum (Lee, Johnson and Lang 1977); dystocia, or dehydration.

Stampeding or Circling

Guinea pigs, especially if housed in large round or square cages, may run wildly in circles when excited. Stampeding guinea pigs, circling or not, will often leap from an open cage regardless of the height above the floor. Rectangular cages and barriers discourage circling.

THE HAMSTER

CUTANEOUS SIGNS

Alopecia

Aged, immunologically compromised hamsters may develop demodectic mange* with alopecia, scaliness, and pustules over the rump and back. Hamsters may also, but rarely, have dermatophytosis,* hereditary hairlessness (Nixon 1972), or hair loss from an endocrine neoplasm, from effects of chronic renal disease, or from constant rubbing on a cage or feeder. Low protein diets (<16%) may contribute to alopecias. The dark, partially haired patches prominent on the flanks of adult male hamsters are scent glands and not foci of alopecia. Male hamsters housed together may cannibalize the flank glands of one another. Loss of facial hair occasionally occurs around the vibrissae.

Rough Hair Coat

Hamsters in cages with leaking water bottles and soiled bedding, hamsters with polyuria, endocrine dysfunction, or diarrhea, and those housed with incompatible cagemates may have rough hair coats. A rough hair coat is a nonspecific sign of fighting and a variety of diseases.

Dermatitis

Inflammation of the skin is most often seen in association with bite wounds (Frisk, Wagner and Kusewitt 1977), but demodectic mange,* Notoedres infection (Fulton 1943), dermatophytosis,* and bacterial dermatopathies also occur. Pasteurella pneumotropica,* Streptococcus spp., and Staphylococcus aureus* have caused isolated cases of cutaneous abscessation in hamsters. Abrasion from wood shavings has also been reported (Meshorer 1976).

Cutaneous or Subcutaneous Swelling

Normal structures in hamsters often erroneously thought to be abnormal swellings are the pendulous testes, distended cheek pouches, and flank glands. Pathologic processes causing palpable swellings include abscesses (McKenna, South and Musacchia 1970; Waterman, Myers and Flannagan 1958), hernias (Raymond, Chesterman and Sebesteny 1967), neoplasms (Pour et al. 1976), arthropathies (Alspaugh and Van Hoosier 1973; Silberberg et al. 1952), mastitis (Frisk, Wagner and Owens 1976), and granulomas.

GASTROINTESTINAL SIGNS

Diarrhea

The vague term "wet tail" is a popular synonym for the diarrheal complex proliferative ileitis or transmissible ileal hyperplasia.* Diarrhea in hamsters may also be a sign of colibacillosis,* clostridial infection (Davis and Jenkins 1986),* salmonellosis,* cestodiasis,* Bacillus piliformis infection,* cecal mucosal hyperplasia (Barthold, Jacoby and Pucak 1978), and the result of antibiotic administration (probably a clostridial enterotoxemia) (Bartlett et al. 1978; Schneienson and Perlman 1950; Small 1968).

Constipation

Constipation in hamsters may be related to *Hymenolepis* in the small intestine,* myiasis (Wantland and Lichtenstein 1954), bedding ingestion with obstruction of the intestine, and intussusceptions associated with an enteritis.

Rectal Prolapse

A dark, red, tubular protrusion from the anus is a prolapsed rectum (Pollock 1975). This problem usually results from hypermotility associated with acute enteritis and proliferative ileitis. Reduction or operation is usually not successful.

RESPIRATORY SIGNS

Dyspnea

Respiratory disease is uncommon in hamsters, but pneumonia caused by Sendai virus,* streptococci (Kummeneje, Nesbakken and Mikkelsen 1975; Weinstein et al. 1951), *Pasteurella pneumotropica,* * and pneumonia virus of mice (unlikely) (Pearson and Eaton 1940) is occasionally noted.

Ocular Discharge

Bacteria, dust, bite wounds, and lymphocytic choriomeningitis virus* can cause a rare conjunctival reaction or infection in hamsters.

REPRODUCTIVE SIGNS

Infertility

Actual or apparent infertility in hamsters may be due to immaturity (under 5 weeks), senescence (over 15 months) (Soderwall and Britenbaker 1955), starvation (Granados 1951; Printz and Greenwald 1970; Printz and Greenwald 1971), cold (7.5°C for 2 weeks) (Grindeland and Folk 1962), prolonged darkness with single, short light exposures every 24 hours (Rudeen and Reiter 1980), normal winter breeding quiescence, pair incompatibility, inadequate nest, a transparent cage, and using anestrus females just removed from an all-female group (Lisk, Langenber and Buntin 1980). Virgin hamsters older than 6 to 7 months exhibit a high degree of infertility.

Prenatal Mortality

Nutritional deficiency, systemic disease, large fetal loads, and environmental stress may lead to fetal loss (Purdy and Hillemann 1950).

Litter Desertion or Death

Litter destruction, usually accompanied by cannibalism, is common among hamsters, especially in group-housed hamsters with a strong sense of territory (Haley 1965; Rowell 1961). Cannibalism can be precipitated by handling the young or the nest, disturbing the mother, not providing adequate nesting material or privacy, agalactia and mastitis, noise, a break in husbandry routine, a wire cage floor, and reduced feed or water. The sipper tube should extend to the level of the young, and food pellets should be placed on the cage floor within easy reach of the female and her litter. There is probably little difference in maternal behaviors between primiparous and multiparous dams (Swanson and Campbell 1979).

NEUROMUSCULAR SIGNS

Torticollis

Encephalitis or otitis interna are possible causes of a rarely seen head tilt.

Incoordination or Convulsions

Lymphocytic choriomeningitis,* tick paralysis (Gregson 1959), and hereditary seizures (Yoon, Peterson and Corrow 1976) are rarely seen causes of incoordination or paralysis in hamsters.

MISCELLANEOUS SIGNS

Weight Loss

Malocclusion, usually from teeth broken on tubes or feeders, can lead not only to weight loss but to agalactia, neonatal deaths, and cannibalism (Harkness et al. 1977). Nutritional deficiencies, gastric hairball (Nelson 1975), and such diseases as proliferative ileitis,* Tyzzer's disease,* salmonellosis,* amyloidosis (Gleiser et al. 1971), myocardial degeneration (McMartin 1979), nephrosis, hepatic cirrhosis (Chesterman and Pomerance 1965), and neoplasia (Pour et al. 1976) also lead to weight loss.

Death

Death following nonspecific signs or without preceding signs may be due to such geriatric conditions (McMartin 1979) as atrial thrombosis, amyloidosis (Gleiser et al. 1971), myocardial degeneration, renal disease,* and neoplasia.* Other fatal conditions include trauma, warfarin poisoning, water deprivation, enteric disease, chilling or overheating, and antibiotic effects (Bartlett et al. 1978; Schneienson and Perlman 1950; Small 1968). Rare conditions include salmonellosis,* tularemia,* streptococcal infection (Weinstein et al. 1951), Sendai virus infection,* and pregnancy ketosis (Galton and Slater 1965).

Urogenital Exudate

Hamster urine is light yellow, opaque, and thick, as is the normal vaginal exudate on the day following estrus. Both products on first glance resemble pus.

Aggression

Female hamsters are generally aggressive toward males, especially young and inexperienced males, but this is not always the case, especially during estrus. Other factors contributing to dominance include body weight, darkness of pectoral patches, age, and territoriality (Floody and Pfaff 1977; Marques and Valenstein 1977; Vandenbergh 1971; Wise 1974). Care should be taken when pairing breeders. Males housed together may bite one another. Adult females are frequently housed separately, especially during pregnancy and lactation.

Hibernation

A "hibernating" hamster, perhaps even a sleeping hamster, appears curled in a ball and either comatose or dead, but some hamsters do "hibernate" for 1 to 3 days in response to cool temperatures (6° C) and other poorly defined stimuli. If the photoperiod is short (2 hours), some hamsters will "hibernate" at room temperature (Hoffman, Hester and Townes 1965). Hamsters have a natural breeding quiescence in late winter.

THE GERBIL

CUTANEOUS SIGNS

Sore nose

Gerbils are highly prone to acute nasal dermatitis, with or without conjunctivitis. Excessive burrowing activity in gerbils may aggravate ulcerations of the nose. *Staphylococcus* spp. can frequently be isolated from the affected area (Bresnahan et al. 1983). Stress induces increased secretion by the harderian glands behind the eyes. The secreted porphyrin tends to be highly irritating and is normally removed by grooming with the forepaws. The condition can also be created by applying Elizabethan collars (Farrar et al. 1988). It can be prevented in collared animals by harderian adenectomy. The condition may be alleviated by removing stressors, cleansing the face, using clay

bedding products, and applying ophthalmic ointments.

Alopecia

Although gerbils tolerate crowding, they may nevertheless chew and denude one another's hair coat, particularly on the dorsum of the tail. Hair may also be lost through rubbing on the cage and other equipment. Runted young often have scant hair coats (Marston and Chang 1965). Gerbils in the early stages of infectious dermatopathies (acariasis, bacterial infection) may show hair loss with minimal inflammation.

Rough Hair Coat

A rough hair coat is a nonspecific sign of fighting, disease, or malnutrition. Hair may be matted and damp if the water bottle leaks, if the bedding is soiled and damp, or if a gerbil has diarrhea or polyuria. Gerbils also have matted or seemingly damp, rough hair coats when the ambient humidity is high.

Dermatitis

Bite wounds on the rump and tail, nose abrasions from burrowing and staphylococcal infection, demodectic mange,* and bacterial infection* (usually *Staphylococcus*) produce dermatitis in gerbils.

Cutaneous or Subcutaneous Swelling

Abscesses and neoplasms are probable causes of swellings. Tail and perianal abscesses are common in rodents because of frequent biting at these sites. Neoplasia of the skin is relatively common among gerbils. The yellow-tan, midventral scent gland prominent in male gerbils is occasionally mistaken as abnormal.

GASTROINTESTINAL SIGNS

Diarrhea

Tyzzer's disease,* caused by *Bacillus piliformis,* and possibly colibacillosis* or enterotoxemia* are causes of diarrhea in gerbils. Salmonellosis,* dietary change, and *Hymenolepis* infection* are less common causes. Although unlikely to cause clinical signs, eggs of an oxyurid, *Dentostomella* spp., may be found in the feces (Wightman, Pilitt and Wagner 1978). Similarly, rat and mouse pinworms, *Syphacia muris* and *S. obvelata,* may occur in gerbils (Ross et al. 1980).

RESPIRATORY SIGNS

Nasal Discharge and Dyspnea

A nasal discharge is associated with heat stress, allergy, or, rarely, a respiratory infection. Pneumonia is infrequently seen in pet or laboratory gerbils. Fluids in the thoracic cavity or diaphragmatic hernias may also cause dyspnea.

Ocular Discharge

Rodents may have a primary conjunctivitis, a dacryoadenitis (usually associated with *Pasteurella pneumotropica* or *Staphylococcus* spp. infections or, in the case of rats, coronavirus infection), or an ocular infection secondary to a debilitating systemic disease. Dust can also cause increased lacrimation. Gerbils often have red (porphyrin-containing) tears associated with nonspecific stresses.

REPRODUCTIVE SIGNS

Infertility

Gerbils are generally considered monogamous animals, and the loss of a mate may abrogate mating interest in the remaining animal. Gerbils have a high incidence of reproductive disorders, especially ovarian or periovarian cysts or neoplasms (Norris and Adams 1972; Vincent, Rodrick and Sodeman 1979). Females with large ovarian cysts may be presumed pregnant because of abdominal enlargement caused by the cyst. Other causes of infertility in the gerbil include pair incompatibility, sexual immaturity or senescence (over 18 months) (Marston and Chang 1965), overcrowding, pesticide/toxin ingestion or absorption (Collins, Hansen and Keeler 1971), nutritional deficiencies (Yahr and Kessler 1975), environmental disturbances, very low temperatures, and various systemic diseases.

Prenatal Mortality

Abortions and stillbirths are difficult to detect because the remains are often buried or consumed. Prenatal deaths may be induced by nutritional deficiencies, systemic or genital disease, and trauma or stress.

Litter Desertion or Death

Approximately 20% of young gerbils fail to survive through weaning because of lack of maternal care, lactation failure, suffocation, crushing, old food that is too hard to gnaw, or inability to reach or operate feeders or waterers (Ahroon and Fidura 1976; Elwood 1975; Norris and Adams 1972).

Maternal neglect or lactation failure may also result from wire-floored or transparent caging, lack of nesting material, lack of water, small litters (1 or 2), environmental disturbances, presence of an aggressive male, and abnormal or injured young. Loss of an imperfect or threatened litter results in the dam's return to estrus. One of the most common causes of cannibalism among murine rodents is food or water deprivation.

NEUROMUSCULAR SIGNS

Torticollis

Otitis media, otitis interna, and encephalitis may cause a head tilt. Gerbils have an apparent resistance to otitis media, a common and debilitating condition of other rodents and of young humans. Such resistance may be associated with the gerbil's relatively high resistance to pathogens and with ear drainage that is superior to that in other rodents.

Incoordination and Convulsions

Epileptiform seizures in gerbils are hypnotic, cataleptic, or convulsive episodes with a variable threshold for onset. These seizures, initiated by handling, startling, or environmental change, occur in approximately 20% of the gerbil population, although the incidence of this genetically influenced trait ranges from near zero to very high (Kaplan and Miezejeski 1972; Loskota, Lomas and Rich 1974; Robbins 1976; Thiessen, Lindzey and Friend 1968).

Although the violent convulsive body jerking, erratic locomotion, and prostrate passive posture following epileptiform seizure, characteristic of gerbils, appears severe, there are no apparent lasting or long-term adverse effects, and normal activity is resumed within a few minutes. Although there are resistant strains, the seizure trait is influenced by cage, diet, and environmental influences. Seizures may have a protective effect in the wild, because a carnivore/predator would be unlikely to consume a seemingly sick convulsing animal. Once a seizure has occurred, a refractory period seems to follow, although this reaction may be an adaptation/habituation effect. Encephalitis, poisoning, toxemias, and trauma may also produce convulsions.

MISCELLANEOUS SIGNS

Aggression

Extreme crowding may lead to tail and rump biting. Newly arrived animals placed into established cages may be attacked and killed, especially in male versus male encounters. Pairing sexually mature males and females, especially in the female's cage, may lead to the injury or death of the male (Ginsburg and Braud 1971; Norris and Adams 1972, 447–450; Norris and Adams 1972, 295–299; Rieder and Reynierse 1971; Wallace, Owen and Thiessen 1973; Yahr et al. 1977).

Weight Loss

Incisor malocclusion (Loew 1967), neoplasia, and enteric disease are among several factors leading to progressive weight loss and death. Weight loss in breeding females is often accompanied by infertility, abortion, or lactation failure. Food or water deprivation (Dunstone, Krupski and Weiss 1971) are among the most common causes of precipitous weight loss and death among gerbils and other small, caged rodents. Although difficult to establish in a history, the usual cause is negligence in providing or making accessible feed and water.

Death

Chilling or overheating, septicemia, streptomycin, toxicity (Wightman, Mann and Wagner 1980), starvation or dehydration, neoplasia, and trauma can cause sudden death in gerbils.

Loss of Tail Skin

The skin of the tip of a gerbil's tail is easily pulled off. The skinless tail becomes necrotic and sloughs, but the stump usually heals without complication.

Foot Stomping

Gerbils rapidly stomp their rear feet to express territoriality in their own cages (Spatz and Granger 1970). They also thump during such aversive states as the interruption of coitus or the cessation of a rewarding stimulus.

THE MOUSE

CUTANEOUS SIGNS

Alopecia

Barbering (Litterst 1974; Thornburg, Stowe and Pick 1973) is a common vice among mice, particularly among adult breeding mice. The muzzle and other areas of the body are frequently "shaved" by cagemates. Removal of the barber mouse will allow hair (may be off-color) to regrow on remaining mice; however, another mouse may assume the barber's role. Microscopic examination of the barbered hair shaft reveals its blunt end, which looks as if it were cut with a scissors.

Hair thinning or loss in mice is often associated with acariasis* and the resulting pruritis with grooming and self-traumatization. Another frequent cause of hair loss is abrasion on feeders or cage tops. Uncommon causes are endocrine imbalances, hereditary characteristics, and reovirus infection.*

Rough Hair Coat

Rough hair coats result from ectoparasitism, dermatophytoses, systemic illness, cold, fighting, damp bedding, old age, leaking water bottles, and urine soiling, and are seen especially in old males housed on wire.

Dermatitis

Acariasis and bite wounds are common causes of dermatitis. Mite lesions, exacerbated by self-traumatization, are usually on the back, neck, head, and shoulders.* These lesions range from simple alopecia to small scabs to extensive cutaneous ulceration, particularly in deeply pigmented strains. A necrotic dermatitis of yet unknown origin commonly occurs in C57BL mice (Stowe, Wagner and Pick 1971). Bite wounds, common among group-caged male mice, are usually on the rump and tail (Les 1972). Because of the tendency of some group-housed male mice to mutilate one another, one should not group house male mice unless absolutely necessary. Tail biting among weanling mice is attributed to crowding,

weaning stresses, diet, and strain differences. Some or all cagemates may show tail, ear, and leg mutilation. C3H mice commonly tail bite and produce lesions resembling those of mouse pox.

Other causes of dermatitis in mice are mouse pox,* arthritis (Alspaugh and Van Hoosier 1973), dermatophytosis,* pediculosis,* bacterial infections (Maronpot and Chavannes 1977; Stewart et al. 1975; Taylor and Neal 1980; Weisbroth, Scher and Boman 1969), and, rarely, leprosy (Krakower and Gonzales 1940). A dry gangrene of the ear has been described (Bell et al. 1970).

Cutaneous or Subcutaneous Swelling

Swellings on or under the skin can be preputial gland abscesses caused by *Escherichia coli* or *Staphylococcus aureus;* * granulomas or pyogranulomas caused by *Mycobacterium avium-intracellulare* organisms; lymphomas;* mammary or other tumors;* bacterial abscesses caused by Group A streptococci (Nelson 1954), *Staphylococcus, * Pasteurella pneumotropica, * Actinobacillus* spp. (Simpson and Simmons 1980), *Corynebacterium, * or *Klebsiella* (Schneemilch 1976); fungal lesions (Mullink 1968); pneumothorax (Fisk 1976); or the follicular mite *Psorergates simplex.* * Such normal protuberances as fetuses and the testes are occasionally thought abnormal by persons unfamiliar with the mouse. The intestines of young adult male mice are free to move in and out of the inguinal canal prior to development of normal inguinal fat, which tends to fill and block the canal. Biting of the scrotum by other male mice may cause localized scrotal infections, peritonitis, and even spinal meningitis. Adhesions may develop between the intestines and the scrotal wall.

Appendage Inflammation or Amputation

Biting and cage trauma are common causes of limb or tail injury. Other less common limb conditions are those associated with mouse pox,* *Streptobacillus moniliformis* infection (Freudt 1959), and arthropathies caused by *Corynebacterium kutscheri.* *

GASTROINTESTINAL SIGNS

Diarrhea

Diarrhea in unweaned mice is usually caused by rotavirus (EDIM),* reovirus,* or corona virus (MHV).* The diarrhea is usually yellow and sticky and mats the neonatal hair. Tyzzer's disease,* salmonellosis,* the pathogenic biotype (4280) of *Citrobacter freundii*,* spironucleosis,* and dietary factors may also lead to diarrhea in mice.

Pendulous Abdomen

Pregnancy is the common cause of a distended abdomen in mice, although spironucleosis,* neoplasia, renal disease, ascites associated with monoclonal antibody production, giardiasis (Sebesteny 1969), cardiomyopathy, salmonellosis,* and cestodiasis* can also cause abdominal distention.

Prolapsed Rectum

Estrogenic stimulation, *Citrobacter* colitis,* severe oxyuriasis,* and a rectal perianal edema in AKR mice may cause a prolapsed rectum.

RESPIRATORY SIGNS

Nasal and Ocular Discharge and Dyspnea

Respiratory disease in mice can be caused by Sendai virus,* pneumonia virus of mice,* *Mycoplasma pulmonis*,* and *Pasteurella pneumotropica** infections. Mixed infections are probably common. Less common respiratory pathogens include *Klebsiella pneumoniae, Corynebacterium kutscheri,* Bordetella bronchiseptica*, mouse hepatitis virus, and ectromelia virus.* Lymphocytic choriomeningitis virus* and *Streptobacillus moniliformis* (Mackie, Van Rooyen and Gilroy 1933) have been associated with a conjunctivitis in mice.

REPRODUCTIVE SIGNS

Infertility

Reduced fertility or abortion in mice may be caused by chemicals (Les 1968), estrogenic stimulation (Fredericks et al. 1981), improper light cycles (Newton 1978), im-

maturity or senescence, overcrowding, noise (Newton 1978), *Pasteurella pneumotropica** or *P. ureae* (Ackerman and Fox 1981), *Mycoplasma pulmonis,* Klebsiella oxytoca* (Davis et al. 1987; Rao et al. 1987), *Streptobacillus moniliformis* infections (Sawicki, Bruce and Andrews 1962), inbreeding, pine needle ingestion (Anderson and Lozano 1977), dichlorvos, low temperature, and nutritional restriction (Hoag et al. 1966; Marsteller and Lynch 1987a).

Litter Desertion or Death

Mouse litters are abandoned or destroyed if nests are skimpy, the litter is small, the young are injured or abnormal, the dam is malnourished, (Marsteller and Lynch 1987b) or has ectoparasitism or agalactia, the nest or young are disturbed or handled, the litter or dam is diseased, or the young become dehydrated or cold because of inadequate nest, humidity, or temperature. A wide variety of environmental disturbances, such as high-pitched sounds associated with cage washers and steam hoses, will cause maternal distress and litter abandonment. One of the most common causes of death in suckling mice is epizootic MHV infection.

Prolapsed Uterus

Uterine prolapse is uncommon but may follow dystocias or abortions of large fetuses.

NEUROMUSCULAR SIGNS

Incoordination or Convulsions

Otitis interna (Ediger, Rabstein and Olson 1971; Halliwell, McCune and Olson 1974), polioencephalomyelitis (GD VII) (Maurer 1958), mouse hepatitis virus infection,* encephalitozoonosis,* bacterial neuropathies, trauma, audiogenic seizures (Iturrian and Fink 1968; Seyfried 1979), poisonings, neoplasia and other viral infections may cause nervous system disorders in mice.

MISCELLANEOUS SIGNS

Weight Loss

Dehydration, starvation, ectoparasitism,

cestodiasis, neoplasia, malocclusion, and a wide range of infectious and noninfectious diseases can lead to weight loss in mice. Access to and consumption of food and water should always be thoroughly evaluated in cases of weight loss or death without other evident cause. A few milliliters of isotonic fluid given subcutaneously or orally, a few drops of 5% glucose solution or other caloric supplement orally, and a few wetted pellets on the cage floor can revive severely dehydrated, starving, and depressed rodents.

Death

Mice die unexpectedly from dehydration (see above), heat stress, trauma, organophosphates (Wagner and Johnson 1970),

chloroform (Deringer, Dunn and Heston 1953), antibiotics (Galloway 1968), toxemias, candidiasis (Goetz and Taylor 1967), and other infectious diseases. Salmonellosis,* ectromelia,* pseudomoniasis (Beck 1963; Brownstein and Johnson 1982), Sendai virus,* and mouse hepatitis virus* are particularly insidious killers of mice. Many times it is difficult to separate experimental factors from environmental and etiologic factors as a cause of disease.

Anemia

Anemia in mice may be associated with pediculosis,* leukemia, autoimmune disease (DeHeer and Edgington 1976), and *Eperythrozoon coccoides* infection (Riley 1964).

THE RAT

CUTANEOUS SIGNS

Alopecia

Hair loss or thinning without dermatitis is seen in some strains of rats or hair loss may occur with accompanying dermatitis. Dermatophytosis,* abrasion on cage or bedding (Andrews, 1977), and mutual grooming when on a 20% fat diet (Beare-Rogers and McGowan 1973) produce hair loss with little or no grossly evident dermatitis.

Rough Hair Coat

Older rats, especially those individually housed, or rats debilitated by malnutrition, ectoparasitism, or disease have rough hair coats. A leaking water bottle, soiled bedding, and diarrhea can also produce ruffled fur.

Dermatitis

Staphylococcus aureus, * *Polyplax spinulosa,* * and *Trichophyton mentagrophytes* * are occasionally implicated in rat dermatitis. *Staphylococcus aureus* and self-traumatization cause ulcerative lesions on the neck and anterior trunk. Self-mutilation of the tail head occurs rarely with oxyuriasis. Tail biting may be seen among group-housed rats. Other agents that rarely

cause dermatopathies in rats are *Notoedres muris* (Flynn 1960) and poxvirus (Marennikova, Shelukhina and Fimina, 1978).

Cutaneous or Subcutaneous Swelling

The most common subcutaneous mass in rats is the mammary tumor.* *Pasteurella pneumotropica* * and *Klebsiella pneumoniae* (Jackson et al. 1980) abscesses uncommonly occur, as do arthropathies (Lerner and Sokoloff 1959; Skold 1961) and salivary glands enlarged from sialodacryoadenitis virus* infection. *Pasteurella pneumotropica* may cause mastitis.*

Limb or Tail Necrosis

If unweaned rats 7 to 15 days old are housed in low (20% or less) relative humidity, they may develop one or more annular lesions of the tail (Dikshit and Sriramachari 1958; Njaa, Utne and Braekkan 1957; Totton 1958). These lesions progress through edema, inflammation, and possibly necrosis and may cause the tail distal to the lesion to slough. In many cases the animal is left with a permanent annular constriction of the tail. If the tail is sloughed, the stump heals without complications.

Low humidity and possibly artificially elevated ambient temperatures (heating) appear to result in an aberrant response of the temperature-regulating vessels in the tail of the neonatal or weanling rat. Conditions that cause lowered ambient humidity, such as wire cages, hygroscopic bedding, and excessive ventilation, predispose to ringtail. In some instances, ringtail can be prevented by providing a solid-bottom plastic cage with both adequate bedding and nesting material. The relative humidity of the room should be approximately 50%. In the northern hemisphere, ringtail is most often seen between the months of November and April, when facilities are heated and interior humidity may drop. At variance from the aforementioned scenario, ringtail may occur occasionally in barrier-sustained weanling rats at any time of the year in a rigidly controlled standard environment.

Special points of interest regarding ringtail are the possible effects of endotoxins on vascular endothelium, the effects of increased dietary lipids, and interactions among endotoxins, dietary lipids, humidity/temperature, and vascular structure of the tail.

Foot injuries, probably from loss of a toenail, laceration caused by the caging, or self-mutilation, are common among rats. Such injuries, which often result in considerable blood loss, usually heal without complications.

GASTROINTESTINAL SIGNS

Diarrhea

Diarrhea is seldom seen in rats. An epizootic diarrhea resembling rotavirus infection in mice has been reported in suckling rats.* Cestodiasis,* ileitis (Geil, Davis and Thompson 1961), and salmonellosis* are other possible enteric infections that could produce abnormal feces in rats.

Pendulous Abdomen

Pregnancy, lactation, ascites, mammary neoplasia,* and megaloileitis (Geil, Davis and Thompson 1961; Hottendorf, Hirth and Peer 1969) produce a pendulous abdomen.

Ptyalism

Slobbering in rats is associated with suffocation, heat stress, and malocclusion.

RESPIRATORY SIGNS

Nasal Discharge

Sniffling and a nasal discharge are associated with respiratory mycoplasmosis commonly and Corynebacterium kutscheri,* Pasteurella pneumotropica,* Streptococcus pneumoniae,* S. zooepidemicus,* Salmonella, and sialodacryoadenitis virus* infections uncommonly. A red ocular and nasal discharge, rich in lacrimal lipids and fluorescent porphyrins of harderian gland origin, accompanies many debilitating diseases and stress conditions (Harkness and Ridgway 1980).

Dyspnea

Pneumonia in rats, often precipitated in ammonia-laden environments or under stressful or otherwise debilitating conditions, can be caused by Mycoplasma pulmonis,* the CAR bacillus, Streptococcus pneumoniae,* and Pasteurella pneumotropica.* Less often Corynebacterium kutscheri,* Aspergillus (Singh and Chawla 1974), Streptococcus zooepidemicus,* and Haemophilus (Harr, Tinsley and Weswig 1969) are involved. Uncomplicated Sendai virus infection* can cause deaths in pregnant or aged rats. These rats develop rough hair coats and dyspnea, but not rhinitis. Chylothorax in rats has been reported (Gupta and Faith 1977). Trichosomoides crassicauda,* a nematode that infects the urinary bladder of rats, may be associated with dyspnea and lung lesions because larvae migrate through the lungs.

Ocular Discharge

Local bacterial infections (Jones 1959; Roberts and Gregory 1980; Young and Hill 1974), and sialodacryoadenitis virus* produce excessive blinking, swelling, lacrimation, and possibly an inflammatory discharge. The common "red tears" (chromodacryorrhea) are associated with several diseases and stressful situations (Harkness and Ridgway 1980). The red pigments are protoporphyrin IX and coproporphyrin III, synthesized *de novo* in the harderian gland and expelled under parasympathetic discharge. The pigments are identified in dried tears by the orange-red fluorescence under ultraviolet light. The eyelids, nares, and forepaws may be smeared with the pigment. Unilateral involvement probably indicates atrophy of the secretory cells of one harderian gland or blockage of a nasolacrimal duct.

Enucleation of the Eye

Weanling rats with acute coronavirus (RCV or SDAV) infections of the harderian glands may self-enucleate or otherwise severely damage the globe of the eye by scratching the orbital area with their rear feet.

REPRODUCTIVE SIGNS

Infertility

Rats younger than 70 days or older than 500 days of age may be sexually incompetent. Other causes of real or apparent infertility include vitamin E deficiency (Jager 1972), protein deficiency (Brown et al. 1960), Sendai virus,* rat virus (Kilham and Margolis 1966) and *Mycoplasma pulmonis* uterine* infections, organophosphate poisoning (Timmons 1975), elevated temperature (Pucak, Lee and Zaino 1977), elevated humidity (Njaa, Utne and Braekken 1957), constant light (Weihe 1976), ectoparasitism (lice),* prenatal stress (Herrenkohl 1979), and systemic disease.

Litter Desertion or Death

Factors that may lead to destruction of the neonates include excessive environmental noise, deformed or dead young, overcrowding, injured young, presence of the male parent, a small litter, dirty cage, lack of privacy, lack of nesting material (Nolen and Alexander 1966), agalactia, old age (Mohan 1974), and the use of cedar wood (Burkhart 1978) or sawdust bedding (Smith et al. 1968).

NEUROMUSCULAR SIGNS

Torticollis

Otitis interna, pituitary adenoma,* and encephalitis may cause a head tilt in rats. Otitis has been caused by *Mycoplasma pulmonis,* possibly *Streptobacillus moniliformis* (Nelson 1930), and *Pseudomonas aeruginosa* (Olson and McCune 1968).

Incoordination or Convulsions

Trauma, encephalitis (Rapp and McGrath 1975), pituitary neoplasia,* the rat virus (Kilham and Margolis 1966), and otitis interna can lead to incoordination in rats.

Paresis or Paralysis

Brain lesions, trauma of the spinal cord, malnutrition, and arthritis can produce limb weakness or paralysis. Abscesses at the head of the tail or in the perineum often extend into the spinal canal.

MISCELLANEOUS SIGNS

General Illness

Rats often exhibit nonspecific signs of illness. These signs may include anorexia, lethargy, rough hair coat, red ocular discharge, and weight loss. Among the diagnostic possibilities are mycoplasmosis, salmonellosis, streptococcal infection, *Corynebacterium* * or *Pasteurella* pneumonia,* rat virus (Jacoby, Bhatt and Jonas 1979), Sendai virus,* chronic renal disease,* Tyzzer's disease,* parasitism, water deprivation, overheating, and neoplasia.

Weight Loss

Weight loss is a nonspecific sign of illness. Overcrowding and food competition, malocclusion, pediculosis,* cestodiasis,* malnutrition (protein deficiency or water deprivation), megaesophagus (Harkness and Ferguson 1979), and nephrosis* are possible causes of weight loss in rats.

Death

Death of neonates has been described. Adults can die from pneumonias and degenerative diseases or from trauma, malnutrition, stress (Lynch and Katcher 1974), overheating or chilling, bedding (Burkhart 1978; Smith et al. 1968), megaloileitis (Hottendorf, Hirth and Peer 1969), or cecal torsion (Pollock and Hagan 1972). Cannibalism suggests death caused by starvation or water deprivation. Heat stress or suffocation that occurs during shipping may produce hyperemia of the extremities and ptyalism.

Anemia

Blood is lost through ectoparasitism and the actions of the blood protozoan *Haemobartonella muris* (Ford and Murray 1959).

Eye Lesions

Corneal opacities can be related to irritation from food or bedding dusts or to injury. Sialodacryoadenitis associated with rat coronal virus infections may lead to a transient or permanent corneal opacity.*

REFERENCES

The Rabbit

Abitol, M.M., Driscoll, S.G., and Ober, W.B.: Placental lesion in experimental toxemia in the rabbit. Am. J. Obstet. Gynecol., *125*:942–948, 1976.

Adams, C.E.: Early embryonic mortality in the rabbit. J. Reprod. Fertil., *1*:315–316, 1960.

Adams, C.E.: Prenatal mortality in the rabbit. J. Reprod. Fertil., *1*:36–44, 1960.

Adams, C.E.: Maintenance of pregnancy relative to the presence of few embryos in the rabbit. J. Endocrinol., *48*:243–249, 1970.

Adams, C.E.: Studies on prenatal mortality in the rabbit. *Oryctolagus cuniculus*: the amount and distribution of loss before and after implantation. J. Endocrinol., *19*:325–344, 1960.

Adams, C.E., and Singh, M.M.: Semen characteristics and fertility of rabbits subjected to exhaustive use. Lab. Anim., *15*:157–161, 1981.

Arendar, G.M., and Milch, R.A.: Splay leg—a recessively inherited form of femoral neck anteversion, femoral shaft torsion and subluxation of the hip in the laboratory lop rabbit. Clin. Orthop., *44*:221–229, 1966.

Baird, C.R.: Biology of *Cuterebra lepuscule* Townsent (Diptera: Cuterebridae) in cottontail rabbits in Idaho. J. Wildl. Dis., *19*:214–218, 1983.

Baxter, J.S.: Posterior paralysis in the rabbit. J. Small Anim. Pract., *16*:267–271, 1975.

Beattie, J.M., Yates, A.G., and Donaldson, R.: An epidemic disease in rabbits resembling that produced by *B. necrosis* (Schmorl) but caused by an aerobic bacillus. J. Pathol. Bacteriol., *18*:34–46, 1913.

Boro, B.R., et al.: A case of abortion in rabbits due to *Aspergillus fumigatus* infection. Vet. Rec., *103*:287–288, 1978.

Buckley, P., and Lowman, D.M.R.: Chronic noninfective conjunctivitis in rabbits. Lab. Anim., *13*:69–73, 1979.

Burrows, H.: Spontaneous uterine and mammary tumors in the rabbit. J. Pathol. Bacteriol., *51*:385–390, 1940.

Cardy, R.H.: A case of spontaneous congenital pyloric stenosis in a rabbit. Lab. Anim. Sci., *23*:588–589, 1973.

Cerna, Z., et al.: Spontaneous and experimental infections of domestic rabbits by *Sarcocystis cuniculi* Brumpt, 1913. Folia Parasit. (PRAHA), *28*:313–318, 1981.

Cheeke, P.R.: Feed preferences of adult male Dutch rabbits. Lab. Anim. Sci., *24*:601–604, 1974.

Cheeke, P.R., and Patton, N.M.: Effect of alfalfa and dietary fiber on the growth performance of weanling rabbits. Lab. Anim. Sci., *28*:167–172, 1978.

Christensen, L.R., Bond, E., and Matanic, B.: Pockless rabbit pox. Lab. Anim. Care, *17*:281–296, 1967.

Confer, A.W., Ward, B.C., and Hines, F.A.: Arsanilic acid toxicity in rabbits. Lab. Anim. Sci., *30*:234–236, 1980.

Cosgrove, M., Wiggins, J.P., and Rothenbacher, H.: *Sarcocystis* spp. in the Eastern cottontail (*Sylvilagus floridanus*). J. Wildl. Dis., *18*:37–40, 1982.

Dade, A.W., et al.: An epizootic of cerebral nematodiasis in rabbits due to *Ascaris columnaris*. Lab. Anim. Sci., *25*:65–69, 1975.

Daniels, J.J.H.M.: Enteral infections with *Pasteurella pseudotuberculosis*. Br. Med. J., *2*:997, 1961.

Doggett, V.C.: Periodicity in the fecundity of male rabbits. Am. J. Physiol., *187*:445–450, 1956.

Doggett, V.C.: Libido and the relationship to periodicity in the fecundity of male rabbits. Fed. Proc., *17*:36, 1958.

Dollahite, J.W., and Rowe, L.D.: Nitrate and nitrite intoxication in rabbits and cattle. Southwestern Veterinarian, *27*:246–248, 1974.

Dubey, J.P.: A review of *Sarcocystis* of domestic animals and of other coccidia of cats and dogs. J. Am. Vet. Med. Assoc., *169*:1061–1078, 1976.

Farrell, G., Powers, D., and Otani, T.: Inhibition of ovulation in the rabbit: Seasonal variation and the effects of indoles. Endocrinology, *83*:599–603, 1968.

Fox, J.G., et al.: Congenital entropion in a litter of rabbits. Lab. Anim. Sci., *29*:509–511, 1979.

Gray, M.L., Singh, C., and Thorp, F.: Abortion, stillbirth, early death of young in rabbits by *Listeria monocytogenes*. II. Oral exposure. Proc. Soc. Exp. Biol. Med., *89*:169–175, 1955.

Greene, H.S.N.: Rabbit pox. I. Clinical manifestations and course of the disease. J. Exp. Med., *60*:427–440, 1934.

Harcourt, R.A.: Toxoplasmosis in rabbits. Vet. Rec., *81*:191–192, 1967.

Harkins, M.J., and Saleeby, E.: Spontaneous tuberculosis of rabbits. J. Infect. Dis., *43*:554–556, 1928.

Hart, M.M., et al.: Distribution and effects of DDT in the pregnant rabbit. Xenobiotica, *2*:567–574, 1972.

Haust, M.D., and Geer, J.C.: Mechanism of calcification

*A number of general references on rabbits and rodents are cited in Chapter 2. These include the ACLAM/Academic Press monographs on the mouse (4 volumes), rat (2 volumes), rabbit, guinea pig, and hamster; the 6th edition of the UFAW Handbook (1987); and Volume 2 of the Canadian Council on Animal Care, Guide to the Care and Use of Experimental Animals (1984).

in spontaneous aortic arteriosclerotic lesions of the rabbit. Am. J. Pathol., *60*:329–346, 1970.

Hellmann, W.: Effects of fever and hyperthermia on the embryonic development of rabbits. Vet. Rec., *104*:389–390, 1979.

Holmes, R.G.: Listeriosis in rabbits. Vet. Rec., *73*:791, 1961.

Jacobson, H.A., McGinnes, B.S., and Catts, E.P.: Bot fly myiasis of the cottontail rabbit, *Sylvilagus floridanus mallurus* in Virginia with some biology of the parasite, *Cuterebra buccata.* J. Wildl. Dis., *14*:56–66, 1978.

Jones, T., et al.: Diagnostic exercise. Lab. Anim. Sci., *32*:488–489, 1982.

Kaufmann, A.F., Quist, K.D., and Broderson, J.R.: Pseudopregnancy in the New Zealand white rabbit: necropsy findings. Lab. Anim. Sci., *21*:865–869, 1971.

Kendall, K.A., Salisbury, G.W., and Vandermark, N.L.: Sterility in the rabbit associated with soybean hay feeding. J. Nutr., *42*:487–500, 1950.

Kruckenberg, S.M.: Nitrate induced abortions in rabbits: observations of field and laboratory cases. Abst. #23, 25th Annual Session, AALAS, Cincinnati, 1974.

Kunkel, H.O., and Pearson, P.B.: Magnesium in the nutrition of the rabbit. J. Nutr., *36*:657–666, 1948.

Kuristyn, I., Naumann, S., and Kaup, F.J.: Torticollis in rabbits: Etiology, pathology, diagnosis, and therapy. Berliner and Munchener, Tieraerztliche Wochenschrift., *99*:14–19, 1986.

Lainson, R.: Toxoplasmosis in England. I. The rabbit (*Oryctolagus cuniculus*) as a host of *Toxoplasma gondii.* Ann. Trop. Med. Parasitol., *49*:384–396, 1955.

Lopushinsky, T.: Myiasis of nesting cottontail rabbits. J. Wildl. Dis., *6*:98–100, 1970.

Low, R.C.: An investigation into scabies in laboratory animals. J. Pathol. Bacteriol., *15*:333–348, 1911.

Maurer, R.R., and Foote, R.H.: Maternal aging and embryonic mortality in the rabbit. J. Reprod. Fertil., *31*:15–22, 1972.

McDonald, R.A., and Pinheiro, A.F.: Water chlorination controls *Pseudomonas aeruginosa* in a rabbitry. J. Am. Vet. Med. Assoc., *151*:863–864, 1967.

McNitt, J.I.: Effect of frequency of service of male rabbits on fertility. J. Appl. Rabbit Res., *4*:18–20, 1981.

Mendlowski, B.: Neuromuscular lesions in restrained rabbits. Vet. Pathol., *12*:378–386, 1975.

Millen, J.W., and Dickson, A.D.: The effect of vitamin A upon the cerebrospinal fluid pressures of young rabbits suffering from hydrocephalus due to maternal hypovitaminosis A. Br. J. Nutr., *11*:440–446, 1957.

Mykytowycz, R., and Fullagar, P.J.: Effect of social environment on reproduction in the rabbit, *Oryctolagus cuniculus* (L.). J. Reprod. Fertil. Suppl., *19*:503–522, 1973.

Nettles, V.F., et al.: An epizootic of cerebral nematodiasis in cottontail rabbits. J. Am. Vet. Med. Assoc., *167*:600–604, 1975.

O'Donoghue, P.N., and Whatley, B.F.: *Pseudomonas aeruginosa* in rabbit fur. Lab. Anim., *5*:251–255, 1971.

Parker, H.D., Hawkings, W.W., and Brenner, E.: Epizootiologic studies of ovine virus abortion. Am. J. Vet. Res., *27*:869–877, 1966.

Patton, N.M.: Cutaneous and pulmonary aspergillosis in rabbits. Lab. Anim. Sci., *25*:347–350, 1975.

Patton, N.M., and Cheeke, P.R.: Etiology and treatment of young doe syndrome. J. Appl. Rabbit Res., *3*:23–24, 1980.

Rabbit rabies. Quoted from CDC Veterinary Public Health Notes. J. Am. Vet. Med. Assoc., *179*:84, 1981.

Ringler, D.H., and Abrams, G.D.: Laboratory diagnosis of vitamin E deficiency in rabbits fed a faulty commercial ration. Lab. Anim. Sci., *21*:383–388, 1971.

Sawin, P.B., Anders, M.V., and Johnson, R.B.: "Ataxia," a hereditary nervous disorder of the rabbit. Proc. Natl. Acad. Sci. U.S.A., *28*:123–127, 1942.

Sawin, P.B., et al.: Maternal behavior in the rabbit: hair loosening during gestation. Am. J. Physiol., *198*:1099–1102, 1960.

Sawin, P.B., and Grary, D.D.: Genetics of skeletal deformities in the domestic rabbit (*Oryctolagus cuniculus*). Clin. Orthop. Relat. Res., *33*:71–90, 1964.

Shaw, N.A., et al.: Zinc deficiency in female rabbits. Lab. Anim., *8*:1–7, 1974.

Sittmann, D.B.: Seasonal variation in reproductive traits of New Zealand white rabbits. J. Reprod. Fertil., *8*:29–37, 1964.

Stevenson, R.G., Palmer, N.C., and Findley, G.G.: Hypervitaminosis D in rabbits. Can. Vet. J., *17*:54–57, 1976.

Templeton, G.S.: Treatment for paralyzed hindquarters. Am. Rabbit J., *16*:155, 1946.

Thigpen, J.E., Clements, M.E., and Gupta, B.N.: Isolation of *Pasteurella aerogenes* from the uterus of a rabbit following abortion. Lab. Anim. Sci., *28*:444–447, 1978.

Thilsted, J.P., et al.: Fatal diarrhea in rabbits resulting from the feeding of antibiotic-contaminated feed. J. Am. Vet. Med. Assoc., *179*:360–361, 1981.

Traub, E.: Über eine mit Listerella-ähnlichen Bakterien vergesellschaftete Meningo-Encephalomyelitis der Kaninchen. Zentrabl. Bakteriol. [Orig. A] *149*:38–49, 1942.

Vetesi, F., and Kemenes, F.: Studies on listeriosis in pregnant rabbits. Acta Vet. Acad. Sci. Hung., *17*:27–37, 1967.

Ward, G.S., Crumrine, M.H., and Mattloch, J.R.: Inflammatory exostosis and abscessation associated with *Fusobacterium nucleatum* in a rabbit. Lab. Anim. Sci., *31*:280–281, 1981.

Watson, G.L., and Evans, M.G.: Listeriosis in a rabbit. Vet. Pathol., *22*:191–193, 1985.

Weber, H.W., and Van der Walt, J.J.: Cardiomyopathy in crowded rabbits. Recent Adv. Stud. Cardiac Struct. Metab., *6*:471–477, 1975.

Weisbroth, S.H., and Scher, S.: Naturally occurring hypertrophic pyloric stenosis in the domestic rabbit. Lab. Anim. Sci., *25*:355–360, 1975.

Weisbroth, S.H., Wang, R., and Scher, S.: Immune and pathologic consequences of spontaneous *Cuterebra* myiasis in domestic rabbits (*Oryctolagus cuniculus*). Lab. Anim. Sci., *23*:241–247, 1973.

Williams, C.S.F., and Gibson, R.B.: Sore dewlap: *Pseudomonas aeruginosa* on rabbit fur and skin. Vet. Med. Small Anim. Clin., *70*:954–955, 1975.

Wilson, W.K., and Dudley, F.J.: Fat colour and fur colour in different varieties of rabbit. Genetics, *47*:290–294, 1946.

Zondek, B., and Tamari, I.: Effect of audiogenic stimulation on genital function and reproduction. Am. J. Obstet. Gynecol., *80*:1041–1048, 1960.

The Guinea Pig

Aldred, P., Hill, A.C., and Young, C.: The isolation of *Streptobacillus moniliformis* from the cervical abscesses of guinea pigs. Lab. Anim., *8*:275–277, 1974.

Bender, A.D., Schottelius, D.D., and Schottelius, B.A.: Effect of short-term vitamin E deficiency on guinea pig skeletal muscle myoglobin. Am. J. Physiol., *197*:491–493, 1959.

Bishop, L.M.: Study of an outbreak of pseudotuberculosis in guinea pigs (cavies) due to *B. pseudotuberculosis rodentium*. Cornell Vet., *22*:1–9, 1932.

Bostrum, R.E., et al.: A typical fatal pulmonary butyromycosis in two guinea pigs due to *Pseudomonas aeruginosa*. J. Am. Vet. Med. Assoc., *155*:1195–1199, 1969.

Correa, W.M., and Pacheco, A.C.: Naturally occurring histoplasmosis in guinea pigs. Can. J. Comp. Med. Vet. Sci., *31*:203–206, 1967.

Dennig, H.K., and Eidmann, E.: Klebsielleninfektionen bein meerschweinchen. Berl. Tieraerztl. Wochenschr., *73*:273–274, 1960.

Farrar, W.E., Jr., and Kent, T.H.: Enteritis and coliform bacteremia in guinea pigs given penicillin. Am. J. Pathol., *47*:629–642, 1965.

Goy, R.W., Hoar, R.M., and Young, W.C.: Length of gestation in the guinea pig with data on the frequency and time of abortion and stillbirth. Anat. Rec., *128*:747–757, 1957.

Grant, E.C., and MacKintosh, J.H.: A comparison of the social postures of some common laboratory rodents. Behaviorism, *21*:246–259, 1963.

Griffiths, H.J.: Some common parasites of small laboratory animals. Lab. Anim., *5*:123–135, 1971.

Gupta, B.N., Conner, G.H., and Meyer, D.B.: Osteoarthritis in guinea pigs. Lab. Anim. Sci., *22*:362–368, 1972.

Gupta, B.N., Langham, R.F., and Conner, G.H.: Mastitis in guinea pigs. Am. J. Vet. Res., *31*:1703–1707, 1970.

Haines, G.: A statistical study of the relationship between various expressions of fertility and vigor in the guinea pig. J. Agric. Res., *42*:123–164, 1931.

Hard, G.C., and Atkinson, F.F.V.: The aetiology of "slobbers" (chronic fluorosis) in the guinea pig. J. Pathol. Bacteriol., *94*:103–112, 1967.

Juhr, N.C., and Obi, S.: Uterine infection in guinea pigs. Z. Versuchstierkd., *12*:383–387, 1970.

Lee, K.J., Johnson, W.D., and Lang, C.M.: Acute gastric dilatation associated with gastric volvulus in the guinea pig. Lab. Anim. Sci., *27*:685–686, 1977.

Lee, K.J., Johnson, W.D., and Lang, C.M.: Preputial dermatitis in male guinea pigs (*Cavia porcellus*). Lab. Anim. Sci., *28*:99, 1978.

Murray, E.S.: Guinea pig inclusion conjunctivitis virus. I. Isolation and identification as a member of the Psittacosis-Lymphogranuloma-Trachoma group. J. Infect. Dis., *114*:1–12, 1964.

Olson, G.A.: Malocclusion of the cheek teeth of a guinea pig. Lab. Anim. Dig., *7*:12–14, 1971.

Plank, S.J., and Irwin, R.: Infertility of guinea pigs on sawdust bedding. Lab. Anim. Care, *16*:9–11, 1966.

Porter, D.A., and Otto, G.F.: The guinea pig nematode, *Paraspidodera uncinata*. J. Parasitol., *20*:323, 1934.

Rehg, J.E., Yarbrough, B.A., and Pakes, S.P.: Toxicity of cecal filtrates from guinea pigs with penicillin-associated colitis. Lab. Anim. Sci., *30*:524–531, 1980.

Rossoff, I.S.: Handbook of Veterinary Drugs. New York, Springer, 1974.

Saunders, L.Z.: Myositis in guinea pigs. J. Natl. Cancer Inst., *20*:899–901, 1958.

Seidel, J.C., and Harper, A.E.: Some observations on vitamin E deficiency in the guinea pig. J. Nutr., *70*:147–155, 1960.

Sparschu, G.L., and Christie, R.J.: Metastatic calcification in a guinea pig colony: a pathological survey. Lab. Anim. Care, *18*:520–526, 1968.

Spink, R.R.: Urolithiasis in a guinea pig (*Cavia porcellanus*). Vet. Med. Small Anim. Clin., *73*:501–502, 1978.

Taylor, J.L., et al.: Chronic pododermatitis in guinea pigs, a case report. Lab. Anim. Sci., *21*:944–945, 1971.

Tzipori, S., et al.: Diarrhea due to *Cryptosporidium* infection in artificially reared lambs. J. Comp. Med., *14*:100–105, 1981.

Ward, G.S., et al.: Myopathy in guinea pigs. J. Am. Vet. Med. Assoc., *171*:837–838, 1977.

Webb, J.N.: Naturally occurring myopathy in guinea pigs. J. Pathol., *100*:155–159, 1970.

Wood, M.: Cystitis in female guinea pigs. Lab. Anim., *15*:141–143, 1981.

Wright, J.F., and Seibold, H.R.: Estrogen contamination of pelleted feed for laboratory animals: effect on guinea pig reproduction. J. Am. Vet. Med. Assoc., *132*:258–261, 1958.

The Hamster

Alspaugh, M.A., and Van Hoosier, G.C., Jr.: Naturally occurring and experimentally induced arthritides in rodents: a review of the literature. Lab. Anim. Care, *23*:724–742, 1973.

Barthold, S.W., Jacoby, R.O., and Pucak, G.J.: An outbreak of cecal mucosal hyperplasia in hamsters. Lab. Anim. Sci., *28*:723–727, 1978.

Bartlett, J.G., et al.: Antibiotic-induced lethal enterocolitis in hamsters: studies with eleven agents and evidence to support the pathogenic role of toxin-producing clostridia. Am. J. Vet. Res., *39*:1525–1530, 1978.

Chesterman, F.C., and Pomerance, A.: Cirrhosis and liver tumors in a closed colony of golden hamsters. Br. J. Cancer, *19*:802–811, 1965.

Davis, A.J., and Jenkins, S.: Cryptosporidosis and proliferative ileitis in a hamster. Vet. Pathol., *23*:632–633, 1986.

Floody, O.R., and Pfaff, D.W.: Aggressive behavior in female hamsters: The hormonal basis for fluctuations in female aggressiveness correlated with estrous state. J. Comp. Physiol. Psychol., *91*:443–464, 1977.

Frisk, C.S., Wagner, J.E., and Kusewitt, D.F.: Unusual aggressive behavior in the male golden hamster. Lab. Anim. Sci., *27*:682–684, 1977.

Frisk, C.S., Wagner, J.E., and Owens, D.R.: Streptococcal mastitis in golden hamsters. Lab. Anim. Sci., *26*:97, 1976.

Fulton, J.D.: The treatment of *Notoedres* infections in golden hamsters (*Cricetus auratus*) with dimethyl diphenylene disulphide (Mitigal) and tetraethylthiuram monosulphide. Vet. Rec., *55*:219, 1943.

Galton, M., and Slater, S.M.: Naturally occurring fatal disease of the pregnant golden hamster. Proc. Soc. Exp. Biol. Med., *120*:873–876, 1965.

Gleiser, C.A., et al.: Amyloidosis and renal paramyloid in a closed hamster colony. Lab. Anim. Sci., *21*:197–202, 1971.

Granados, H.: Nutritional studies on growth and reproduction of the golden hamster (*Mesocricetus auratus*). Acta Physiol. Scand. (Suppl.), *24*:1–138, 1951.

Gregson, J.D.: Tick paralysis in groundhogs, guinea pigs, and hamsters. Can. J. Comp. Med., *23*:266–268, 1959.

Grindeland, R.E., and Folk, G.E., Jr.: Effects of cold exposure on the oestrous cycle of the golden hamster (*Mesocricetus auratus*). J. Reprod. Fertil., *4*:1–6, 1962.

Haley, J.: Cannibalism in hamsters. Am. Small Stock Farmer, *35*:10–11, 1965.

Harkness, J.E., et al.: Weight loss and impaired reproduction in the hamster attributable to an unsuitable feeding apparatus. Lab. Anim. Sci., *27*:117–118, 1977.

Hoffman, R.A., Hester, L.J., and Townes, C.: Effect of light and temperature on the endocrine system of the golden hamster. Comp. Biochem. Physiol., *15*:525–533, 1965.

Kummeneje, K., Nesbakken, T., and Mikkelsen, T.: *Streptococcus agalactiae* infection in a hamster. Acta Vet. Scand., *16*:554–556, 1975.

Lisk, R.D., Langenber, K.K., and Buntin, J.D.: Blocked sexual receptivity in grouped female golden hamsters: independence from ovarian function and continuous group maintenance. Biol. Reprod., *22*:237–242, 1980.

Marques, D.M., and Valenstein, E.S.: Individual differences in aggressiveness of female hamsters: response to intact and castrated males and to females. Anim. Behav., *25*:131–139, 1977.

McKenna, J.M., South, F.E., and Musacchia, X.: *Pasteurella* infections in irradiated hamsters. Lab. Anim. Care, *20*:443–446, 1970.

McMartin, D.N.: Morphologic lesions in aging Syrian hamsters. J. Gerontol., *34*:502–511, 1979.

Meshorer, A.: Leg lesions in hamsters caused by wood shavings. Lab. Anim. Sci., *26*:827–829, 1976.

Nelson, W.B.: Fatal hairball in a long-haired hamster. Vet. Med. Small Anim. Clin., *70*:1193, 1975.

Nixon, C.W.: Hereditary hairlessness in the Syrian golden hamster. J. Hered., *63*:215–217, 1972.

Pearson, H.E., and Eaton, M.D.: A virus pneumonia of Syrian hamsters. Proc. Soc. Exp. Biol. Med., *45*:677–679, 1940.

Pollock, W.B.: Prolapse of invaginated colon through the anus in golden hamsters (*Mesocricetus auratus*). Lab. Anim. Sci., *25*:334–336, 1975.

Pour, P., et al.: Spontaneous tumors and common diseases in two colonies of Syrian hamsters. I. Incidence and sites. J. Natl. Cancer Inst., *56*:931–935, 1976.

Printz, R.H., and Greenwald, G.S.: A neural mechanism regulating follicular development in the hamster. Neuroendocrinology, *7*:171–182, 1971.

Printz, R.H., and Greenwald, G.S.: Effects of starvation on follicular development in the cyclic hamster. Endocrinology, *86*:290–295, 1970.

Purdy, D.M., and Hillemann, H.H.: Prenatal mortality in the golden hamster (*Cricetus auratus*). Anat. Rec., *106*:577–583, 1950.

Raymond, R.H., Chesterman, F.C., and Sebesteny, A.: Inguinal hernias in female white hamsters. J. Inst. Anim. Tech., *18*:134–138, 1967.

Rowell, T.E.: Maternal behavior in non-maternal golden hamsters (*Mesocricetus auratus*). Anim. Behav., *9*:11–15, 1961.

Rudeen, P.K., and Reiter, R.J.: Influence of a skeleton photoperiod on reproductive organ atrophy in the male golden hamster. J. Reprod. Fertil., *60*:279–283, 1980.

Schneienson, S.S., and Perlman, E.: Toxicity of penicillin for the Syrian hamster. Proc. Soc. Exp. Biol. Med., *91*:229–230, 1950.

Silberberg, R., et al.: Degenerative joint disease in Syrian hamsters. Fed. Proc., *11*:427, 1952.

Small, J.D.: Fatal enterocolitis in hamsters given lincomycin hydrochloride. Lab. Anim. Care, *18*:411–420, 1968.

Soderwall, A.L., and Britenbaker, A.L.: Reproductive capacities of different-age hamsters (*Cricetus auratus, Waterhouse*). J. Gerontol., *10*:469–470, 1955.

Swanson, L.J., and Campbell, C.S.: Maternal behavior in the primiparous and multiparous golden hamster. Z. Tierpsychol., *50*:96–104, 1979.

Vandenbergh, J.G.: The effects of gonadal hormones on the aggressive behaviour of adult golden hamsters (*Mesocricetus auratus*). Anim. Behav., *19*:589–594, 1971.

Wantland, W.W., and Lichtenstein, E.P.: Intestinal myiasis in the Syrian hamster. J. Parasitol., *140*:365, 1954.

Waterman, J.M., Myers, H.I., and Flannagan, V.D.: Effect of hyaluronidase on hemolytic *Staphylococcus* abscesses in hamsters (Abst.). J. Dent. Res., *37*:49, 1958.

Weinstein, L., et al.: Experimental streptococcal pneumonia. I. The pathologic and bacteriologic features of the untreated disease. J. Immunol., *67*:173–181, 1951.

Wise, D.A.: Aggression in the female golden hamster: Effects of reproductive state and social isolation. Horm. Behav., *5*:235–250, 1974.

Yoon, C.H., Peterson, J.S., and Corrow, D.: Spontaneous seizures: a new mutation in Syrian golden hamsters. J. Hered., *67*:115–116, 1976.

Gerbil

Ahroon, J.K., and Fidura, F.G.: The influence of the male on maternal behaviour in the Mongolian gerbil (*Meriones unguiculatus*). Anim. Behav., *24*:372–375, 1976.

Bresnahan, J.F., et al.: Nasal dermatitis in the Mongolian gerbil. Lab. Anim. Sci., *33*:258–263, 1983.

Collins, T.F.X., Hansen, W.H., and Keeler, H.V.: The effect of carbaryl (Sevin) on reproduction of the rat and the gerbil. Toxicol. Appl. Pharmacol., *19*:202–216, 1971.

Dunstone, J.J., Krupski, G.M., and Weiss, C.S.: Weight loss in gerbils (*Meriones unguiculatus*) continuously deprived of food and water. Psychol. Rep., *29*:931–936, 1971.

Elwood, R.W.: Paternal and maternal behavior in the Mongolian gerbil. Anim. Behav., *23*:766–772, 1975.

Farrar, P.L., et al.: Experimental nasal dermatitis in the Mongolian gerbil: Effect of bilateral harderian gland adenectomy on development of facial lesions. Lab. Anim. Sci., *38*:77–78, 1988.

Garibaldi, B.A., et al.: Hematuria in rabbits. Lab. Anim. Sci., *37*:769–772, 1987.

Ginsburg, H.J., and Braud, W.G.: A laboratory investigation of aggressive behavior in the Mongolian gerbil (*Meriones unguiculatus*). Psychon. Sci., *22*:54–55, 1971.

Kaplan, H., and Miezejeski, C.: Development of seizures in the Mongolian gerbil (*Meriones unguiculatus*). J. Comp. Physiol. Psychol., *81*:267–273, 1972.

Loew, F.M.: A case of overgrown mandibular incisors in a Mongolian gerbil. Lab. Anim. Care, *17*:137–139, 1967.

Loskota, W.J., Lomas, P., and Rich, S.T.: The gerbil as a model for the study of epilepsies. Seizure patterns and ontogenesis. Epilepsia, *15*:109–119, 1974.

Marston, J.H., and Chang, M.C.: The breeding, management and reproductive physiology of the Mongolian gerbil (*Meriones unguiculatus*). Lab. Anim. Care, *15*:34–48, 1965.

Norris, M.L., and Adams, C.E.: Aggressive behavior and reproduction in the Mongolian gerbil (*Meriones unguiculatus*) relative to age and sexual experience at pairing. J. Reprod. Fertil., *31*:447–450, 1972.

Norris, M.L., and Adams, C.E.: Incidence of cystic ovaries and reproductive performance in the Mongolian gerbil, *Meriones unguiculatus*. Lab. Anim. *6*:337–342, 1972.

Norris, M.L., and Adams, C.E.: Mortality from birth to weaning in the Mongolian gerbil. Lab. Anim., *6*:49–53, 1972.

Norris, M.L., and Adams, C.E.: Suppression of aggressive behaviour in the Mongolian gerbil (*Meriones unguiculatus*). Lab. Anim., *6*:295–299, 1972.

Rieder, C.A., and Reynierse, J.H.: The effects of maintenance condition on aggression and marking behavior of the Mongolian gerbil (*Meriones unguiculatus*). J. Comp. Physiol. Psychol., *75*:471–475, 1971.

Robbins, M.E.C.: Seizure resistance in albino gerbils. Lab. Anim., *10*:233–235, 1976.

Ross, C.R., et al.: Experimental transmission of *Syphacia muris* among rats, mice, hamsters and gerbils. Lab. Anim. Sci., *30*:35–37, 1980.

Spatz, C., and Granger, W.R.: Foot-thumping in the gerbil: the effect of establishing a home cage. Psychon. Sci., *19*:53–54, 1970.

Thiessen, D.D., Lindzey, G., and Friend, H.C.: Spontaneous seizures in the Mongolian gerbil, *Meriones unguiculatus*. Psychon. Sci., *11*:227–228, 1968.

Vincent, A.L., Rodrick, G.E., and Sodeman, W.A., Jr.: The pathology of the Mongolian gerbil (*Meriones unguiculatus*): A review. Lab. Anim. Sci., *29*:645–651, 1979.

Wallace, P., Owen, K., and Thiessen, D.D.: The control and function of maternal scent marking in the Mongolian gerbil. Physiol. Behav., *10*:463–466, 1973.

Wightman, S.R., Mann, P.C., and Wagner, J.E.: Dihydrostreptomycin toxicity in the Mongolian gerbil (*Meriones unguiculatus*). Lab. Anim. Sci., *30*:71–75, 1980.

Wightman, S.R., Pilitt, P.A., and Wagner, J.E.: *Dentostomella translucida* in the Mongolian gerbil (*Meriones unguiculatus*). Lab. Anim. Sci., *28*:290–296, 1978.

Yahr, P., and Kessler, S.: Suppression of reproduction in water-deprived Mongolian gerbils (*Meriones unguiculatus*). Biol. Reprod., *12*:249–254, 1975.

Yahr, P., et al.: Effects of castration on aggression between male Mongolian gerbils. Behav. Biol., *19*:189–205, 1977.

Mouse

Ackerman, J.I., and Fox, J.G.: Isolation of *Pasteurella ureae* from reproductive tracts of congenic mice. J. Clin. Microbiol., *13*:1049–1053, 1981.

Alspaugh, M.A., and Van Hoosier, G.L., Jr.: Naturally-occurring and experimentally induced arthritides in rodents: a review of the literature. Lab. Anim. Sci., *23*:724–742, 1973.

Anderson, C.K., and Lozano, E.A.: Pine needle toxicity in pregnant mice. Cornell Vet., *67*:229–235, 1977.

Beck, R.W.: The control of *Pseudomonas aeruginosa* in a mouse breeding colony by the use of chlorine in the drinking water. Lab. Anim. Care, *13*:41–45, 1963.

Bell, J.F., et al.: Dry gangrene of the ear in white mice. Lab. Anim., *4*:245–254, 1970.

Brownstein, D.G., and Johnson, E.: Experimental nasal infection of normal and leukopenic mice with *Pseudomonas aeruginosa*. Vet. Pathol., *19*:169–178, 1982.

Davis, J.K., et al.: The role of *Klebsiella oxytoca* in utero-ovarian infection of B6C3F1 mice. Lab. Anim. Sci., *37*:159–166, 1987.

DeHeer, D.H., and Edgington, T.S.: Cellular events associated with the immuno-genesis of anti-erythrocyte autoantibody responses of NZB mice. Transplant. Rev., *31*:116–155, 1976.

Deringer, M.K., Dunn, T.B., and Heston, W.E.: Results of exposure of strain C3H mice to chloroform. Proc. Soc. Exp. Biol., *83*:474–479, 1953.

Ediger, R.D., Rabstein, M.M., and Olson, L.D.: Circling in mice caused by *Pseudomonas aeruginosa*. Lab. Anim. Sci., *21*:845–848, 1971.

Fisk, S.K.: Iatrogenic pneumothorax and pneumoderma in a mouse. Lab. Anim. Sci., *26*:648, 1976.

Fredericks, G.R., et al.: Ovulation rates and embryo degeneracy in female mice fed the phytoestrogen coumestrol. Proc. Soc. Exp. Biol. Med., *167*:237–241, 1981.

Freudt, E.A.: Arthritis caused by *Streptobacillus moniliformis* and pleuropneumonia-like organisms in small rodents. Lab. Invest., *8*:1358–1375, 1959.

Galloway, J.H.: Antibiotic toxicity in white mice. Lab. Anim. Care, *18*:421–425, 1968.

Goetz, M.E., and Taylor, D.O.N.: A naturally occurring outbreak of *Candida tropicalis* infection in a laboratory mouse colony. Am. J. Pathol., *50*:361–369, 1967.

Halliwell, W.H., McCune, E.L., and Olson, L.D.: *Mycoplasma pulmonis*-induced otitis media in gnotobiotic mice. Lab. Anim. Sci., *24*:57–61, 1974.

Hoag, W.G., et al.: The effect of nutrition on fertility of inbred DBA/2J mice. Lab. Anim. Care, *16*:228–236, 1966.

Iturrian, W.B., and Fink, G.B.: Effect of noise in the animal house on seizure susceptibility and growth of mice. Lab. Anim. Care, *18*:557–560, 1968.

Krakower, C., and Gonzales, L.M.: Mouse leprosy. Arch. Pathol., *30*:308–329, 1940.

Les, E.P.: A disease related to cage population density: tail lesions of C3H/HeJ mice. Lab. Anim. Sci., *22*:56–60, 1972.

Les, E.P.: Effect of acidified chlorinated water on reproduction of C3H/HeJ and C57BL/6J mice. Lab. Anim. Care, *18*:210–213, 1968.

Litterst, C.L.: Mechanically self-induced muzzle alopecia in mice. Lab. Anim. Sci., *24*:806–809, 1974.

Mackie, T.J., Van Rooyen, C.E., and Gilroy, E.: An epizootic disease occurring in a breeding stock of mice: bacteriological and experimental observations. Br. J. Exp. Pathol., *14*:132–136, 1933.

Maronpot, R.R., and Chavannes, J-M.: Dacryoadenitis, conjunctivitis, and facial dermatitis of the mouse. Lab. Anim. Sci., *27*:277–278, 1977.

Marsteller, F.A., and Lynch, C.B.: Reproductive responses to variation in temperature and food supply by house mice: I. Mating and pregnancy. Biol. Reprod., *37*:838–43, 1987a.

Marsteller, F.A., and Lynch, C.B.: Reproductive responses to variation in temperature and food supply by house mice: II. Lactation. Biol. Reprod., *37*:844–50, 1987.

Maurer, F.D.: Mouse poliomyelitis or Theiler's mouse encephalomyelitis. J. Natl. Cancer Inst., *20*:871–874, 1958.

Mullink, J.W.: A case of actinomycosis in a male NZW mouse. Z. Versuchstierkd., *10*:225–227, 1968.

Nelson, J.B.: Association of group A streptococci with an outbreak of cervical lymphadenitis in mice. Proc. Soc. Exp. Biol. Med., *86*:542–545, 1954.

Newton, W.M.: Environmental impact on laboratory animals. Adv. Vet. Sci. Comp. Med., *22*:1–28, 1978.

Rao, G.N., et al.: Utero-ovarian infection in aged B6C3F1 mice. Lab. Anim. Sci., *37*:153–158, 1987.

Riley, V.: Synergism between a lactate dehydrogenase-elevating virus and *Eperythrozoon coccoides.* Science, *146*:921–923, 1964.

Sawicki, L., Bruce, H.M., and Andrews, C.H.: *Streptobacillus moniliformis* infection as a probable cause of arrested pregnancy and abortion in laboratory mice. Br. J. Exp. Pathol., *43*:194–197, 1962.

Schneemilch, H.S.: A naturally acquired infection of laboratory mice with *Klebsiella* capsule type 6. Lab. Anim., *10*:305–310, 1976.

Sebesteny, A.: Pathogenicity of intestinal flagellates in mice. Lab. Anim., *3*:71–77, 1969.

Seyfried, T.N.: Audiogenic seizures in mice. Fed. Proc., *38*:2399–2404, 1979.

Simpson, W., and Simmons, D.J.C.: Two *Actinobacillus* species isolated from laboratory rodents. Lab. Anim., *14*:15–16, 1980.

Stewart, D.D., et al.: An epizootic of necrotic dermatitis in laboratory mice caused by Lancefield group G streptococci. Lab. Anim. Sci., *25*:296–302, 1975.

Stowe, H.D., Wagner, J.L., and Pick J.R.: A debilitating fatal murine dermatitis. Lab. Anim. Sci., *21*:892–897, 1971.

Taylor, D.M., and Neal, D.L.: An infected eczematous condition in mice: methods of treatment. Lab. Anim., *14*:325–328, 1980.

Thornburg, L.P., Stowe, H.D., and Pick, J.R.: The pathogenesis of the alopecia due to hair chewing in mice. Lab. Anim. Sci., *23*:843–850, 1973.

Wagner, J.E., and Johnson, D.R.: Toxicity of dichlorvos for laboratory mice LD 50 and effect on serum cholinesterase. Lab. Anim. Care, *20*:45–47, 1970.

Weisbroth, S.H., Scher, S., and Boman, I.: *Pasteurella pneumotropica* abscess syndrome in a mouse colony. J. Am. Vet. Med. Assoc., *155*:1206–1210, 1969.

The Rat

Andrews, E.J.: Muzzle trauma in the rat associated with the use of feeding cups. Lab. Anim. Sci., *27*:278, 1977.

Beare-Rogers, J.L., and McGowan, J.E.: Alopecia in rats housed in groups. Lab. Anim., *7*:237–238, 1973.

Brown, A.M., et al.: Influence of nutrition on reproduction in laboratory rodents. Proc. Nutr. Soc., *19*:32–37, 1960.

Burkhart, C.A.: High rat pup mortality attributed to the use of cedar-wood shavings as bedding. Lab. Anim., *12*:221–222, 1978.

Dikshit, P.K., and Sriramachari, S.: Caudal necrosis in suckling rats. Nature, *181*:63–64, 1958.

Flynn, R.J.: *Notoedres muris* infestation of rats. Proc. Anim. Care Panel, *10*:69–70, 1960.

Ford, A.C., Jr., and Murray, T.J.: Studies on *Haemobartonella* infection in the rat. Can. J. Microbiol., *5*:345–350, 1959.

Geil, R.G., Davis, C.L., and Thompson, S.W.: Spontaneous ileitis in rats—a report of 64 cases. Am. J. Vet. Res., *22*:932–936, 1961.

Gupta, B.N., and Faith, R.E.: Chylothorax in a rat. J. Am. Vet. Med. Assoc., *171*:973–974, 1977.

Harkness, J.E., and Ferguson, F.G.: Idiopathic megaesophagus in a rat (*Rattus norvegicus*). Lab. Anim. Sci., *29*:495–498, 1979.

Harkness, J.E., and Ridgway, M.D.: Chromodacryorrhea in laboratory rats (*Rattus norvegicus*): etiologic considerations. Lab. Anim. Sci., *30*:841–844, 1980.

Harr, J.R., Tinsley, I.J., and Weswig, P.H.: *Haemophilus* isolated from a rat respiratory epizootic. J. Am. Vet. Med. Assoc., *155*:1126–1130, 1969.

Herrenkohl, L.R.: Prenatal stress reduces fertility and fecundity in female offspring. Science, *206*:1097–1099, 1979.

Hottendorf, G.H., Hirth, R.S., and Peer, R.L.: Megaloileitis in rats. J. Am. Vet. Med. Assoc., *155*:1131–1135, 1969.

Jackson, N.N., et al.: Naturally acquired infections of *Klebsiella pneumoniae* in Wistar rats. Lab. Anim., *14*:357–361, 1980.

Jacoby, R.O., Bhatt, P.N., and Jonas, A.M.: Viral diseases. *In* The Laboratory Rat. Edited by H.J. Baker, J.R. Lindsey, and S.H. Weisbroth. New York, Academic Press, 1979, pp. 271–306, Vol. I.

Jager, F.C.: Long-term dose-response effects of vitamin E in rats. Nutr. Metab., *14*:1–7, 1972.

Jones, L.P.: Purulent panophthalmitis in laboratory rats. J. Am. Vet. Med. Assoc., *135*:502–503, 1959.

Kilham, L., and Margolis, G.: Spontaneous hepatitis and cerebellar hypoplasia in suckling rats due to congenital infection with rat virus. Am. J. Pathol., *49*:457–475, 1966.

Lerner, E.M., II, and Sokoloff, L.: The pathogenesis of bone and joint infection produced in rats by *Streptobacillus moniliformis.* Arch. Pathol., *67*:20/364–28/372, 1959.

Lynch, J.J., and Katcher, A.H.: Human handling and sudden death in laboratory rats. J. Nerv. Ment. Dis., *159*:362–365, 1974.

Marennikova, S.S., Shelukhina, E.M., and Fimina, V.A.: Pox infection in white rats. Lab. Anim., *12*:33–36, 1978.

Mohan, C.: Age-dependent cannibalism in a colony of albino rats. Lab. Anim., *8*:83–84, 1974.

Nelson, J.B.: The bacteria of the infected middle ear in adult and young albino rats. J. Infect. Dis., *46*:64–75, 1930.

Njaa, L.R., Utne, F., and Braekkan, O.R.: Effect of relative humidity on rat breeding and ringtail. Nature, *180*:290–291, 1957.

Nolen, G.A., and Alexander, J.C.: Effects of diet and type of nesting material on the reproduction and lactation of the rat. Lab. Anim. Sci., *16*:327–336, 1966.

Olson, L.D., and McCune, E.L.: Histopathology of chronic otitis media in the rat. Lab. Anim. Care, *18*:478–485, 1968.

Pollock, W.B., and Hagan, T.R.: Two cases of torsion of the cecum and ileum in rats. Lab. Anim. Sci., 22:549–551, 1972.

Pucak, G.J., Lee, C.S., and Zaino, A.S.: Effects of prolonged high temperature on testicular development and fertility in the male rat. Lab. Anim. Sci., 27:76–77, 1977.

Rapp, J.P., and McGrath, J.T.: Mycotic encephalitis in weanling rats. Lab. Anim. Sci., 25:477–480, 1975.

Roberts, S.A., and Gregory, B.J.: Facultative *Pasteurella* ophthalmitis in hooded lister rats. Lab. Anim., 14:323–324, 1980.

Singh, B., and Chawla, R.S.: A note on an outbreak of pulmonary aspergillosis in albino rat colony. Indian J. Anim. Sci., 44:804–807, 1974.

Skold, B.H.: Chronic arteritis in the laboratory rat. J. Am. Vet. Med. Assoc., 138:204–207, 1961.

Smith, P.C., et al.: Intestinal obstruction and death in suckling rats due to sawdust bedding. Lab. Anim. Care, 18:224–228, 1968.

Timmons, E.H.: Dichlorvos effects on estrous cycle onset in the rat. Lab. Anim. Sci., 25:45–47, 1975.

Totton, M.: Ringtail in newborn Norway rats. A study of the effect of environmental temperature and humidity on incidence. J. Hyg., 56:190–196, 1958.

Weihe, W.H.: *In* Control of the Animal House Environment. Edited by T. McSheeby, London, Lab. Anim. Ltd., Lab. Anim. Handbook No. 7, 1976.

Young, C., and Hill, A.: Conjunctivitis in a colony of rats. Lab. Anim., 8:301–304, 1974.

Chapter 5
Specific Diseases and Conditions

The diseases and conditions of rabbits and rodents described in this chapter were selected for inclusion because of their prevalence in pet and laboratory animals, public health significance, or potential importance to biomedical research. This listing is certainly not complete. Tuberculosis, listeriosis, yersiniosis, mycotic infections, many viral diseases, congenital abnormalities, and several nutritional conditions are referenced in chapter 4 but are not described further. References in this chapter are included at the end of each discussion of a disease or condition.

ACARIASIS

Hosts. Mites commonly infect rabbits and rodents. Species of mites are usually host specific or limited to a narrow host range.

Etiology. Mites are arachnids and as adults possess eight legs. Mites commonly found on rabbits and rodents are members of the suborder Prostigmata, superfamily Cheyletidae. They are confined to the host by anatomic and feeding requirements except when adults are brushed onto the bedding or when mites leave because of overpopulation or host death. Eggs are attached to hair shafts.

With the exception of *Liponyssus bacoti,* the mites described live on the skin surface and obtain nourishment from epithelial debris and tissue fluids. Life cycles are approximately 10 to 21 days long, and eggs hatch in about 8 days, although cycle lengths and intervals vary with the hosts' condition and environment.

Representative species of rodent and rab-

bit mites include these more commonly encountered ectoparasites:

In the rabbit, *Psoroptes cuniculi* is the ear mite, and *Cheyletiella parasitivorax* and *Listrophorus gibbus* are fur mites. The life cycle of *Psoroptes cuniculi* is approximately 21 days.

In the guinea pig, *Chirodiscoides caviae* is a fur mite, and *Trixacarus caviae* is a sarcoptid mite.

Demodex mites affect gerbils and hamsters. Two species have been described in hamsters: *D. criceti,* which is found in epidermal pits, and *D. aurati,* which is found in the pilosebaceous system and hair follicles.

In the mouse, *Myobia musculi, Myocoptes musculinus,* and *Radfordia affinis* are fur mites, and *Psorergates simplex* is a rarely seen follicular mite.

In the rat, *Radfordia ensifera* is a fur mite, and *Notoedres muris* is the rare mange mite.

Another mite that affects rodents is *Liponyssus bacoti,* the tropical rat mite and a representative of the suborder Mesostig-

111

mata. This blood-sucking parasite feeds on the host and lives off the host in nests and on bedding.

Transmission. Mites are spread by direct contact with infected hosts, pelage debris, or bedding. Adult female mites may survive a week or more off the host.

Predisposing Factors. Mite infestations are often more severe in debilitated or chronically diseased older hosts, in young hosts, or in animals housed in unsanitary environments. Pet or laboratory rodents attract wild rodents of similar species; this contact provides the opportunity for the transfer of many diseases, including acariasis. Also, the black strains of mice are particularly susceptible to the pruritic effects of mite infestations, which may result in self-mutilation. This is possibly due to variations in grooming behavior or immune response. Low temperature and high humidity favor survival of the mites off the host.

Clinical Signs. *Psoroptes cuniculi* in rabbits causes accumulation of serum and brown crust in the external ear canal, pinna, and, rarely, the adjacent head and throat or other parts of the body. Beneath the crusts the skin is hairless, moist, and raw. The inflammation and pruritus, particularly pronounced in uncommon secondary bacterial infections, may cause scratching and head shaking. Extension of an uncomplicated mite infection to the middle and inner ears is rare, if it occurs at all, but a bacterial otitis externa or media could extend to the inner ear and cause torticollis.

Even in heavy infestations, *Myobia musculi* may cause signs no more dramatic than scratching, but in susceptible mice, scratching and a roughened hair coat are often seen even with low parasite populations. Alopecia, epidermal scaling, and scant to extensive ulceration may occur; however, these signs are probably caused by a scratching or rubbing response to the allergic, pruritic stimulus induced by the mites. It is presumed that *Myocoptes musculinus* and *Radfordia affinis* (mice) and *Radfordia ensifera* (rats) may cause similar signs.

Demodex in hamsters and gerbils has been associated with alopecia, rough hair coat, and scaly, scabby dermatitis of the rump and back. Demodecosis is particularly common in hamsters older than 1.5 years and is frequently secondary to chronic renal disease, intercurrent systemic disease, and malnutrition.

Trixacarus cavaie of guinea pigs frequently produces a severe alopecia, dermatitis, and pruritus on the body and legs, which may result in self-mutilation. The skin is thickened, dry, and scaly.

Cheyletiella parasitivorax in rabbits may cause no signs, or it may cause loose hair, which can be pulled out in clumps, and reddened, oily, hairless, scaly patches over the back and head. Rabbits with *Cheyletiella* infections appear to have increased "dandruff" in the fur.

The small *Listrophorus gibbus* mites of rabbits are relatively nonpathogenic even though they may be present in large numbers.

Psorergates simplex inhabits hair follicles and sebaceous glands of mice. As mites and debris accumulate in the skin, follicular invaginations are visible as small (2-mm) white nodules on the skin of the head and neck. These nodules are best seen on the back of reflected skin at necropsy.

Liponyssus bacoti differs from the other mites described because most of the life cycle is spent off the host. Nymphal and adult forms intermittently suck blood from the host, and the loss of blood causes anemia, decreased fertility, weakness, and death. As with fur mites, mammalian hosts may develop an allergic dermatitis to these mites, and the pruritic response may be intense.

Histologic examination of a mite-caused lesion reveals hyperkeratosis, acanthosis, and cutaneous ulcers, scabs, and crusts.

Diagnosis. A skin scraping, deep if *Demodex* is suspected, with an oil- or glycerin-wetted scalpel blade will collect mites or lice. Scraped material is placed on a slide with a few drops of warmed 10% KOH, covered, allowed to sit 10 minutes, and then examined microscopically.

The pelt, especially the head, ears, neck, and margins of lesions, can be examined with a hand lens or dissecting microscope for adult or immature ectoparasites. Engorged *Liponyssus* and other mites moving in the fur can be seen with the unaided eye. If the animal is dead, examination with the hand lens will provide a more certain diagnosis if the pelt is cooled in the refrigerator for 30 or more minutes, removed for 10 or more minutes, and then examined with a lens. The parasites migrate off the cold pelt and toward the warmer tips of the hair. If the pelt is placed on a dark-colored paper and surrounded by a frame of double-gummed cellophane tape or petroleum jelly, the mites will become trapped after they leave the animal.

Psoroptes cuniculi mites may be identified with an otoscope. Examination of a mineral oil scraping of debris from the ear canal reveals oval-shaped mites with well-developed legs and bell-shaped suckers on the ends of jointed pedicles. In the case of *Cheyletiella parasitivorax* infections, examination of the fur may reveal white mites about 0.3-mm long with piercing chelicerae and large curved papal hooks. *Listrophorus gibbus* is found predominantly on the abdomen and back of rabbits.

Ectoparasites of mice or rats often are seen at the time of microscopic examination of perianal cellophane tape impressions for pinworm ova. Additionally, strips of heavy cellophane tape can be pressed firmly against the live animal and examined microscopically for ectoparasites.

The pelt can be dissolved overnight in 5% trypsin solution and then heated until dissolved in 10% potassium hydroxide solution. The cloudy liquid is then passed through an 80-mesh brass filter. A xylene wash may be needed to remove excessive fat. The mites are retained on the filter and easily identified and counted by microscopic examination.

The pelage can also be examined for ectoparasites by vigorously brushing (toothbrush) or combing (a very fine-tooth comb) over a dark background. The collected material is carefully examined with a hand lens for live mites. Additionally, the collected debris can be transferred to a glass microscope slide, coverslipped, and examined microscopically for eggs, small mites, or immature forms.

Treatment.   Mites, once established in a colony, are more realistically controlled than eliminated. *Psoroptes cuniculi,* the rabbit ear mite, is treated with one of several available oil-based insecticide preparations. One or 2 ml of the liquid are massaged into the ear canal. Acaricides are usually not ovicidal, and treatment is repeated 7 to 10 days later to eliminate immature mites. In severe cases the use of a sebumlytic compound, antiseptic soap, and topical or systemic antibiotic may be indicated. A single intramuscular or subcutaneous injection of ivermectin (Ivomec—Merck, Sharp & Dohme, Rahway, NJ) at 400 μg/kg eliminated all *P. cuniculi* mites (Wright and Riner 1985). Ivermectin has been used to treat ascariasis in small rodents, but efficacious and safe dosages are not known. A dose of 200 μg/kg given subcutaneously or orally is recommended for many small mammals (Sikarskie 1986).

Cheyletiella, Chirodiscoides, Myobia, Radfordia, Myocoptes, Psorergates, and *Liponyssus* can be reduced by dusting the adults, weanlings, and bedding with a carbaryl or permethrin powder at weekly intervals. Control of *Myobia* in mice was achieved by mixing 4.0 g of 0.25% permethrin dust (Ectiban—ICI Americas, Goldsboro, NC) into the bedding weekly (Bean-Knudsen, Wagner and Hall 1986). Individual animals may be dipped into 0.025% lindane solution or other ectoparasite dip. Two percent malathion, used as a dip, is toxic to rabbits (Jones 1984). Preweanling animals should not be dipped. Silicate dusts alone or compounded with insecticides abrade and desiccate ectoparasites and provide a mechanical alternative to potentially toxic chemicals. This material is mixed with contact bedding in the affected animals' cages. *Trixacarus* is treated by wetting with 1:40 lime sulfur in water repeated weekly for 6 weeks (McDonald and Lavoipierre 1980), or with 1% lindane solution repeated weekly for 3 weeks (Manning, Wagner and Harkness 1984).

Demodex in hamsters and gerbils burrow into the skin and are difficult to detect. Mitaban (Amitraz—Upjohn, Kalamazoo, MI) is widely used for treating generalized demodicosis of dogs and, when package directions are followed, should be highly effective for treating hamsters. Clip and bathe the hamster with a mild soap and water and towel dry before treating the entire animal topically by wetting with a mitaban-water mixture and allowing to air dry. Three to six topical treatments fourteen days apart are recommended.

Dichlorvos, an organophosphate and cholinesterase inhibitor, has often been demonstrated as efficacious in reducing mite infestations. Resin strips impregnated with dichlorvos and placed on or near a cage for 48-hour periods at several weekly intervals will markedly reduce ectoparasite populations. Entire rooms of animals can be treated by restricting air flow and using multiple resin strips (Wagner 1969). Exact dosages or effective concentrations of the vapors have not been established. Effective doses are likely to temporarily curtail breeding and may result in increased cannibalism of newborn litters. Pelleted dichlorvos scattered in the bedding likewise can be used to treat ectoparasites (Fraser et al. 1974). Treatment must be undertaken only after one is aware of the consequences such treatment may have on subsequent use of the animals, i.e., the nature of the research for which the animals are destined determines whether or not they can be treated and how long one must wait between treating and using the animals. People must be protected from excessive exposure to the vapors.

Elimination of ectoparasites from premises requires treatment and removal of animals, thorough mechanical scrubbing, and fumigation of the room with formaldehyde gas generated from paraformaldehyde. Fomites (feed bags, carts, trash cans, clothing) moved from room to room must also be cleaned to remove ectoparasites. Rooms should be repopulated with animals proven free of ectoparasites.

Prevention.   Mite infestations are prevented by placing clean stock into clean premises. Infected animals should be separated and treated, and the premises should be disinfected. Mites off the host die within 3 weeks. Wild rodents must be excluded from the colony.

Public Health Significance. Psoroptes, Chirodiscoides, Psorergates, Myobia, Radfordia, Myocoptes, Listrophorus, and *Demodex* mites of rabbits and rodents are not known to affect man.

Cheyletiella parasitivorax can, in rare cases, cause a dermatitis in humans, especially children. *Cheyletiella* can transfer the rabbit myxoma virus.

Liponyssus bacoti bites man and can serve as a vector for murine typhus, Q fever, and plague.

REFERENCES

Arlian, L.G., et al.: Infestivity of *Psoroptes cuniculi* in rabbits. Am. J. Vet. Res., *42*:1782–1784, 1981.

Baies, A., Suteu, I., and Klemm, W.: *Notoedres* scabies of the golden hamster. Z. Versuchsteirkd., *1*:251–257, 1968.

Barr, A.R.: A case of "mange" of the domestic rabbit due to *Cheyletiella parasitivorax* (Megnin). J. Parasitol., *41*:323, 1955.

Bean-Knudsen, D.E., Wagner, J.E., and Hall, R.D.: Evaluation of the control of *Myobia musculi* infestations on laboratory mice with permethrin. Lab. Anim. Sci., *36*:268–270, 1986.

Bjotvedt, G., and Geib, L.W.: Otitis media associated with *Staphylococcus epidermidis* and *Psoroptes cuniculi* in a rabbit. Vet. Med. Small Anim. Clin., *76*:1015–1016, 1981.

v. Bronswijk, J.E.M.H., and de Kreek, E.J.: *Cheyletiella* (Acari: Cheyletiellidae) of dog, cat and domesticated rabbit, a review. J. Med. Entomol., *13*:315–327, 1976.

Campbell, D.J.: Parasitic diseases of laboratory animals. Can. Med. Assoc. J., *98*:908–910, 1968.

Cloyd, G.G., and Moorhead, D.P.: Facial alopecia in the rabbit associated with *Cheyletiella parasitivorax*. Lab. Anim. Sci., *26*:801–803, 1976.

Constantin, M.L.: Effects of insecticides on acariasis in mice. Lab. Anim., *6*:279–286, 1972.

Csiza, C.K., and McMartin, D.N.: Apparent acaridal dermatitis in a C57BL/6 Nya mouse colony. Lab. Anim. Sci., *26*:781–787, 1976.

Desch, C.E., Jr.: Redescription of *Demodex nanus* (Acari:Demodicidae) from *Rattus norvegicus* and *R. rattus* (Rodentia). J. Med. Entomol., *24*:19–23, 1987.

Estes, P.C., Richter, C.B., and Franklin, J.A.: Demodectic mange in the golden hamster. Lab. Anim. Sci., *21*:825–828, 1971.

Flatt, R.E., and Wiemers, J.: A survey of fur mites in domestic rabbits. Lab. Anim. Sci., *26*:758–761, 1976.

Flynn, R.J.: The diagnosis of some forms of ectoparasitism of mice. Lab. Anim. Care, *13*:111–125, 1963.

Fraser, J., et al.: The use of pelleted dichlorvos in the

control of murine acariasis. Lab. Anims., *8*:271–274, 1974.

Friedman, S., and Weisbroth, S.H.: The parasitic ecology of the rodent mite, *Myobia musculi.* II. Genetic factors. Lab. Anim. Sci., *25*:440–445, 1975.

Friedman, S., and Weisbroth, S.H.: The parasitic ecology of the rodent mite, *Myobia musculi.* IV. Life cycle. Lab. Anim. Sci., *27*:34–37, 1977.

Gibson, T.E.: Parasites of laboratory animals transmissible to man. Lab. Anims., *1*:17–24, 1967.

Green, C.J., and Needham, J.R.: Control of mange mites in a large mouse colony. Lab. Anims., *8*:245–251, 1974.

Griffiths, H.J.: Some common parasites of small laboratory animals. Lab. Anims., *5*:123–135, 1971.

Henderson, J.D.: Treatment of cutaneous acariasis in the guinea pig. J. Am. Vet. Med. Assoc., *163*:591–592, 1973.

Hsu, C-K.: Parasitic diseases. *In* The Laboratory Rat. Vol. I. Biology and Diseases. Edited by H.J. Baker, J.R. Lindsey, and S.H. Weisbroth. Orlando, Academic Press, 1979, pp. 307–331.

Jones, J.M.: Organophosphate poisoning in two Rex rabbits. N.Z. Vet. J., *32*:9–10, 1984.

Keefe, T.J., Scanlon, J.E., and Wetherald, L.D.: *Ornithonyssus bacoti* (Hirst) infestation in mouse and hamster colonies. Lab. Anim. Care, *14*:366–369, 1964.

Kraus, A.L.: Arthropod parasites. *In* The Biology of the Laboratory Rabbit. Edited by S.H. Weisbroth, R.E. Flatt, and A.L. Kraus. Orlando, Academic Press, 1974, pp. 287–315.

Kummel, B.A., Estes, S.A., and Arlian, L.G.: *Trixacarus caviae* infestation of guinea pigs. J. Am. Vet. Med. Assoc., *177*:903–908, 1980.

Levine, J.L., and Lage, A.L.: House mouse mites infesting laboratory rodents. Lab. Anim. Sci., *34*:393–394, 1984.

Lin, S.L., Pinson, D.M., and Lindsey, J.R.: Diagnostic exercise. Lab. Anim. Sci., *34*:353–555, 1984.

Lund, E.L.: Ear mange in domestic rabbits. Am. Rabbit J., *21*:67–69, 1951.

Manning, P., Wagner, J., and Harkness, J.: Biology and diseases of guinea pigs. *In* Laboratory Animal Medicine. Edited by J. Fox, B. Cohen, and F. Loew. Orlando, Academic Press, 1984.

McDonald, S.E., and Lavoipierre, M.M.J.: *Trixacarus caviae* infestation in two guinea pigs. Lab. Anim. Sci., *30*:67–70, 1980.

Nutting, W.B.: *Demodex aurati* sp. nov. and *D. criceti,* ectoparasites of the golden hamster (*Mesocricetus auratus*). Parasitol., *51*:515–522, 1961.

Nutting, W.B., and Rauch, H.: Distribution of *Demodex aurati* in the host (*Mesocricetus auratus*) skin complex. J. Parasitol., *49*:323–329, 1963.

Owen, D., and Young, C.: The occurrence of *Demodex aurati* and *Demodex criceti* in the Syrian hamster (*Me-*

socricetus auratus) in the United Kingdom. Vet. Rec., *92*:282–284, 1973.

Pence, D.B.: Diseases of laboratory animals. *In* Mammalian Diseases and Arachnids, Vol. II, Medico-Veterinary, Laboratory and Wildlife Diseases, and Control. Edited by W.B. Nutting. Boca Raton, FL, CRC Press, 1984.

Ronald, N.C., and Wagner, J.E.: The arthropod parasites of the genus *Cavia. In* The Biology of the Guinea Pig. Edited by J.E. Wagner and P.J. Manning. Orlando, Academic Press, 1976, pp. 201–209.

Schwarzbrott, S.S., Wagner, J.E., and Frisk, C.S.: Demodicosis in the Mongolian gerbil (*Meriones unguiculatus*): A case report. Lab. Anim. Sci., *24*:666–668, 1974.

Sikarski, J.G.: The use of invermectin in birds, reptiles, and small mammals. *in:* Current Veterinary Therapy, Kirk, R.W., Ed. Vol IX, p. 743–745, 1986.

Tarshis, I.B.: The use of silica aerogel compounds for the control of ectoparasites. Proc. Anim. Care Panel, *12*:217–258, 1962.

Thoday, K.L., and Beresford-Jones, W.P.: The diagnosis and treatment of mange in the guinea-pig caused by *Trixacarus* (*Caviacoptes*) *caviae.* J. Small Anim. Pract., *18*:591–595, 1977.

Wagner, J.E.: Control of mouse ectoparasites with resin vaporizer strips containing vapona. Lab. Anim. Care, *19*:804–807, 1969.

Wagner, J.E., Al-Rabiai, S., and Rings, R.W.: *Chirodiscoides caviae* infestation in guinea pigs. Lab. Anim. Sci., *22*:750–752, 1972.

Wagner, J.E.: Parasitic diseases. *In* The Laboratory Hamster. Edited by G.L. Van Hoosier, Jr., and C.W. McPherson. Orlando, Academic Press, 1987, pp. 135–156.

Walberg, J.A., et al.: Demodicidosis in laboratory rats (*Rattus norvegicus*). Lab. Anim. Sci., *31*:60–62, 1981.

Weisbroth, S.H., et al.: The parasitic ecology of the rodent mite *Myobia musculi.* I. Grooming factors. Lab. Anim. Sci., *24*:510–516, 1974.

Weisbroth, S.H., Friedman, S., and Scher, S.: The parasitic ecology of the rodent mite, *Myobia musculi.* III. Lesions in certain host strains. Lab. Anim. Sci., *26*:725–735, 1976.

Wright, F.C., and Riner, J.C.: Comparative efficacy of injection routes and doses of ivermectin against *Psoroptes* in rabbits. Am. J. Vet. Res., *46*:752–754, 1985.

Yunker, C.: Mites. *In* Parasites of Laboratory Animals. Edited by R.J. Flynn. Ames, The Iowa State University Press, 1973, pp. 425–492.

Zajac, A., Williams, J.F., and Williams, C.S.F.: Mange caused by *Trixacarus caviae* in guinea pigs. J. Am. Vet. Med. Assoc., *177*:900–903, 1980.

Zenoble, R.D., and Greven, J.H.: Sarcoptid mite infestation in a colony of guinea pigs. J. Am. Vet. Med. Assoc., *177*:898–900, 1980.

ALLERGIES TO LABORATORY ANIMALS

From 11 to 15% of veterinarians, animal handlers, and pet owners are allergic to laboratory animals. The allergic reactions among people in frequent contact with animals include anaphylaxis and involvement of the upper airways, eyes, lower airways, and skin. Usually months or years pass before people in contact with animals become sensitized. Just as humans develop allergies to allergens of a variety of animals, it is possible that animals develop sensitivities to other animal species.

Predisposing Factors.   Predisposing factors to animal allergens include atopy, intensity and directness of contact, and lack of use of safety factors or devices. The species of animal is also a factor. Allergies are usually species specific, i.e., particular to one species of animal or another. Among common laboratory animal species, cats are the most common source of antigen to which people develop allergic reactions. Others in rank order (most common first) are rats, guinea pigs, rabbits, mice, hamsters, and gerbils. These rankings, no doubt, are skewed to overemphasize proportionally the more common species of laboratory animals— mice and rats. Allergies to aerosolized rat urine (serum) proteins (detected by ammonia smell) are particularly dangerous because they are associated with severe pulmonary congestion, the symptoms of which develop rapidly after sensitized persons enter facilities with poor ventilation or infrequent cage cleaning.

Clinical Signs.   Clinical signs of an allergy to laboratory animals are runny eyes and nose, a persistent cough, especially at night, asthma or shortness of breath, or various skin manifestations. Reactions may occur 15 to 20 minutes after exposure or several hours later, e.g., at night after the person leaves work where he or she has had animal contact. An exception is acute pulmonary congestion, which develops instantly following inhalation of rat urine proteins carried with ammonia gas, which results from urea-splitting bacterial growth in dirty cage bedding.

Treatment.   Pulmonary congestion does not respond well to antihistamines commonly used for eye, nose, and skin allergic reactions. Prophylactically, cromolyn sodium, a bronchodilator, reduces asthma attacks by inhibiting the degranulation of sensitized mast cells that occurs after exposure to specific antigens. A physician should be consulted in troublesome cases of animal allergy.

Prevention.   Precautions to reduce exposure to animal allergens include the use of special hygiene procedures, specially constructed exhaust ventilated cages, ceiling-to-floor air flow (does not place allergen-containing air at breathing zone of person), protective clothing (gloves, gown, mask), special care in handling excreta, control of traffic flow, and the creation of other barriers, such as arranging as much work as possible in clean areas where there is no animal exposure. The matter of sensitivities to animal allergens should be considered in pre-employment medical examinations of potential animal technicians and in establishing policies on employment.

REFERENCES

Brown, F.R., and Wolfe, H.I.: Observations on animal dander hyposensitization. Ann. Allergy, *26*:305–308, 1968.

Carroll, K.B., et al.: Extrinsic allergic alveolitis due to rat serum proteins. Clin. Allergy, *5*:443–456, 1975.

Dewdney, J.M.: Allergy induced by exposure to animals. J. R. Soc. Med., *74*:928–932, 1981.

Edwards, R.G., Beeson, M.F., and Dewdney, J.M.: Laboratory animal allergy: The measurement of airborne urinary allergens and the effects of different environmental conditions. Lab. Anims., *17*:235–239, 1983.

Gilday, F.J.: Bronchial asthma due to rat hair. Del. Med. J., *28*:110–111, 1956.

Lincoln, T.A., Bolton, N.E., and Garret, A.S., Jr.: Allergy to animal dander and sera. J. Occup. Med., *16*:465–469, 1974.

Lutsky, I., and Neuman, I.: Laboratory animal dander allergy: I. An occupational disease. Ann. Allergy, *35*:201–205, 1975.

Lutsky, I., and Toshner, D.: A review of allergic respiratory disease in laboratory animal workers. Lab. Anim. Sci., *28*:751–756, 1978.

McGivern, D., Longbottom, J., and Davies, D.: Allergy to gerbils. Clin. Allergy, *15*:163–165, 1985.

Neuman, I., and Lutsky, I.: Laboratory animal dander allergy: II. Clinical studies and the potential protective effect of disodium cromoglycate. Ann. Allergy, *36*:23–29, 1976.

Ohman, J.L., Jr., Lowell, F.C., and Block, K.J.: Allergens of mammalian origin. II. Characterization of allergens extracted from rat, mouse, guinea pig, and rabbit pelts. J. Allergy Clin. Immunol., *55*:16–24, 1975.

Ohman, J.L., Jr.: Allergy in man caused by exposure to mammals. J. Am. Vet. Med. Assoc., *172*:1403–1406, 1978.

Rajka, G.: Ten cases of occupational hypersensitivity to laboratory animals. Acta Allergol., *16*:168–176, 1961.

Sorrell, A.H., and Gottesman, J.: Mouse allergy. A case report. Ann. Allergy, *15*:662–663, 1957.

Wilson, J.A.: Hamster-hair hypersensitivity in adults of low atopic status. Br. Med. J., *4*:341, 1971.

ANOREXIA

Etiology. Anorexia, or a reduction or cessation of food intake, may result from neophobia, water deprivation, climatic extremes, an unpalatable or improperly compounded diet, malocclusion, pain, loss of olfaction, metabolic disorders, toxemia, gastric hairball, neoplasia, territorial and behavioral traits, and mechanical factors that prevent access to feed. Guinea pigs and rats are particularly neophobic, and a change from one food or flavor to another should be done gradually by using transitional mixtures. Excessively hard pellets and elevated or nonfunctional sipper tubes frequently result in weanling deaths.

Clinical Signs. Clinical manifestations of anorexia are weight loss or failure to gain, cannibalism, increased susceptibility to disease, dehydration, loss of litter, and death. Guinea pigs may rapidly develop a fatal, irreversible ketosis in spite of a resumption of feeding.

Treatment. Some problems involving anorexia can be overcome through using sweetened or preferred feeds (corn, oats, calf manna, sunflower seeds, vegetables, hay, apples); by changing feeds or feeders and waterers; by giving B vitamins or anabolic steroids; by treating abnormalities; by reducing crowding or separating incompatible individuals; by mixing feeds; and by reducing obesity. Oral administration of a high caloric supplement, yogurt, ground feed, and 50% glucose solution may provide an adequate interim feed during a postoperative period.

A commonly noted relationship exists between obesity in sedentary rabbits, anorexia, and eventual death. Rabbits rarely live 3 full weeks after feces no longer appear in the pan. Limited feeding (3 to 5 oz per day) or feeding a high-fiber diet to nonpregnant adult rabbits usually prevents this problem.

REFERENCES

Harkness, J.E., et al.: Weight loss and impaired reproduction in the hamster attributable to an unsuitable feeding apparatus. Lab. Anim. Sci., *27*:117–118, 1977.

Morris, M.L., Jr.: Index of dietetic management. *In* Current Veterinary Therapy. VI. Small Animal Practice. Edited by R.W. Kirk. Philadelphia, W.B. Saunders Co., 1977, pp. 59–73.

Plaut, S.M., et al.: Maternal deprivation in the rat: Prevention of mortality by nonlactating adults. Psychosom. Med., *36*:311–320, 1974.

Siegmund, O.H., et al., editors: The Merck Veterinary Manual. 5th Ed. Rahway, NJ, Merck & Co., Inc., 1979, pp. 86–89, 1387–1389.

BORDETELLA BRONCHISEPTICA INFECTION

Hosts. Clinical infection with *Bordetella bronchiseptica* is relatively common in guinea pigs, dogs, and swine. Rats, rabbits, cats, birds, and primates may also develop clinical infections, but these animals usually become carrier hosts. Epizootic respiratory disease caused by *B. bronchiseptica* can have disastrous consequences in a guinea pig colony.

Etiology. *Bordetella bronchiseptica*, a small, motile, gram-negative bacillus or coccobacillus, is the causative organism. After incubation for 48 hours at 37° C on blood agar, *Bordetella* colonies are 1 to 2 mm in diameter, yellowish-brown, and variably hemolytic. Carbohydrates are not fermented.

Transmission. Transmission of *B. bronchiseptica* is by direct contact with clinically affected animals, carrier hosts, contaminated fomites, and respiratory aerosol. Subclinical infections or carrier animals are common, and *B. bronchiseptica* can be cultured from the respiratory tracts of clinically normal animals. Uterine infections with *Bordetella* have been reported in guinea pigs.

Predisposing Factors. Many outbreaks of *B. bronchiseptica* are precipitated by stressors such as nutritional imbalances, temperature changes, crowding, feed changes, experimental procedures, and marginal diets, especially those deficient in vitamin C in the case of guinea pigs. Guinea pigs, especially the young, pregnant, anorectic, ketotic, and old, are particularly susceptible. Contact with carrier species, such as rabbits, dogs, cats, and subhuman primates, and mixing of guinea pigs from multiple sources should be avoided.

Clinical Signs.   Clinical signs of *B. bronchiseptica* infection in guinea pigs are usually associated with pneumonia and vary from no signs to anorexia, inappetence, nasal and ocular discharge, dyspnea, and death. The incubation period is 5 to 7 days. The acute or epizootic form of the infection has a sudden onset and lasts 2 to 3 days. High mortality, abortions, and stillbirths are noted in affected guinea pigs during epizootics, which occur when the agent is introduced into susceptible colonies or when the immunity of individual animals in enzootically infected colonies drops below an effective level.

Rabbits frequently harbor *B. bronchiseptica* in their upper respiratory passages, but the usual consequence, through ciliary and epithelial damage, is predisposition to other infections, particularly pasteurellosis.

Necropsy Signs.   The most common finding is discrete consolidation of a portion of a lobe or of an entire lobe of a lung. The affected portion is firm and dark red or reddish-tan. Otitis media, rhinitis, tracheitis, and a red-brown pleural exudate may accompany the bronchopneumonia.

The histologic features of epizootic bronchopneumonia that characterize *B. bronchiseptica* infection in guinea pigs are suppuration, exudation, and hemorrhage. There is a marked purulent bronchitis. The lumens of air passageways contain substantial amounts of polymorphonuclear leukocytes, mucus, and desquamated bronchiolar epithelium. The exudate may extend to terminal bronchioles. Alveoli are characteristically packed with degenerating, protein-rich masses composed largely of cellular debris, primarily neutrophils, and fibrin. Metritis may also occur.

Diagnosis.   A definitive diagnosis is based on clinical signs and culture on blood agar of exudate from the lower trachea, bronchial lumen, or middle ear. Slow growth results in very small colonies at 24 hours. The pearl-like, hemolytic colonies reach maximal size at 72 hours. When nasal swabs are used to screen animal colonies for

carriers, MacConkey agar is the preferred culture medium. Frequently, pure cultures of *B. bronchiseptica* can be recovered from animals found to have suppurative otitis media at necropsy. Fresh isolants can be used in serodiagnosis with an agglutination test using known immune serum. Suppurative otitis media can be detected in living guinea pigs by radiographic examination of the tympanic bullae. Otitis media can be detected as tympanic membrane cloudiness visible through an otoscope.

Treatment.   Treatment of *Bordetella* bronchopneumonia in guinea pigs is usually not practical except in individual pets, and even then the prognosis is poor. A variety of injectable and oral Tribrissen® products are available (Coopers Inc. Box 167, Kansas City, Mo 64141). A number of people have informally reported success in treating a variety of rabbit and rodent infections with these products. Chloramphenicol palmitate oral suspension may also be used. The dose recommended for dogs is 25 mg/lb. every 6 hours. Sulfamethazine in the drinking water at 4.0 ml of a 12.5% stock solution per 500 ml for 1 to 2 weeks may suppress, but rarely cures, an active infection. Affected guinea pigs should be given ascorbic acid orally or by injection.

Prevention.   Good husbandry, clean stock, and separation of carrier animals from healthy guinea pigs are essential. A *Bordetella*-free colony must be managed on the closed-colony basis with entry restricted to guinea pigs from colonies known to be free of the organism. Generally, guinea pigs with enzootic *B. bronchiseptica* infections are unsatisfactory for research because they tend to succumb to infection when stressed.

A single intramuscular injection of formalin-killed bacterin with incomplete Freund's adjuvant into guinea pigs results in protective titers lasting 4 to 6 months. Prolonged use of the bacterin will eliminate the carrier state from an affected colony. A live, attenuated bacterin prepared from the heat-sensitive (34° C), porcine ts-S34 mutant *B. bronchiseptica* grows in the guinea pig's nasal passage but not in the lungs. Protective antibodies are produced following intranasal administration. Use of commercially available porcine *B. bronchiseptica* bacterins protect guinea pigs against fatal pneumonias (Matherne, Steffen and Wagner, 1987).

Public Health Significance. The importance of *B. bronchiseptica* infection in man is minimal, although the organism is occasionally recovered from the human nasopharynx; it can cause a whooping cough syndrome, and has caused bronchopneumonia in an elderly person.

REFERENCES

Baskerville, M., Baskerville, A., and Wood, M.: A study of chronic pneumonia in a guinea pig colony with enzootic *Bordetella bronchiseptica* infection. Lab. Anims., *16*:290–296, 1982.

Bemis, D.A., Greisen, H.A., and Appel, M.J.G.: Bacteriological variation among *Bordetella bronchiseptica* isolates from dogs and other species. J. Clin. Microbiol., *5*:471–480, 1977.

Burek, J.D., et al.: The pathology and pathogenesis of *Bordetella bronchiseptica* and *Pasteurella pneumotropica* infection in conventional and germ-free rats. Lab. Anim. Sci., *22*:844–849, 1972.

Ganaway, J.R., Allen, A.M., and McPherson, C.W.: Prevention of acute *Bordetella bronchiseptica* pneumonia in a guinea pig colony. Lab. Anim. Care, *15*:156–162, 1965.

Ghosh, H.K., and Trauter, J.: *Bordetella bronchicanis* (*bronchiseptica*) infection in man: review and a case report. J. Clin. Pathol., *32*:546–548, 1979.

Manning, P., Wagner, J., and Harkness, J.: Biology and disease of guinea pigs. *In* Laboratory Animal Medicine. Edited by J. Fox, B. Cohen, and F. Loew. Orlando, Academic Press, 1984.

Matherne, C.M., Steffen, E.K., and Wagner, J.E.: Efficacy of commercially available vaccines in protecting guinea pigs against *Bordetella bronchiseptica* pneumonia. Lab. Anim. Sci., *37*:191–194, 1987.

Matsuyama, T., and Takino, T.: Scanning electromicroscopic studies of *Bordetella bronchiseptica* on the rabbit tracheal mucosa. J. Med. Microbiol., *13*:159–161, 1980.

Nakagawa, M., et al.: Some observations on diagnosis of *Bordetella bronchiseptica* infection in guinea pigs. Exp. Anim., *18*:105–116, 1969.

Nakagawa, M., et al.: Prophylaxis of *Bordetella bronchiseptica* infection in guinea pigs by vaccination. Jpn. J. Vet. Sci., *36*:33–42, 1974.

Nikkels, R.J., and Mullink, J.W.M.A.: *Bordetella bronchiseptica* pneumonia in guinea pigs. Description of the disease and elimination by vaccination. Z. Versuchstierkd., *13*:105–111, 1971.

Roudebush, P., and Fales, W.H.: Antibacterial susceptibility of *Bordetella bronchiseptica* isolates from small companion animals with respiratory disease. J. Am. Anim. Hosp. Assoc., *17*:793–797, 1981.

Shimizu, T.: Prophylaxis of *Bordetella bronchiseptica* infection in guinea pigs by intranasal vaccination with live strain ts-S34. Infect. Immun., *22*:318–321, 1978.

Simpson, W., and Simmons, D.J.C.: Problems associated with the identification of *Bordetella bronchiseptica*. Lab. Anims., *10*:47–48, 1976.

Trahan, C.J., et al.: Airborne-induced experimental *Bordetella bronchiseptica* pneumonia in Strain 18 guinea pigs. Lab. Anims., *21*:226–232, 1987.

Wagner, J.E., et al.: Otitis media of guinea pigs. Lab. Anim. Sci., *26*:902–907, 1976.

Watson, W.T., et al.: Experimental respiratory infection with *Pasteurella multocida* and *Bordetella bronchiseptica* in rabbits. Lab. Anim. Sci., *25*:459–464, 1975.

Winsser, J.: A study of *Bordetella bronchiseptica*. Proc. Anim. Care Panel, *10*:87–104, 1960.

Woode, G.N., and McLeod, N.: Control of acute *Bordetella bronchiseptica* pneumonia in a guinea-pig colony. Lab. Anims., *1*:91–94, 1967.

CESTODIASIS

Hosts. Cestodes, or tapeworms and their intermediate forms, infect a wide range of species, including rabbits and rodents.

Etiology. *Hymenolepis nana*, the dwarf tapeworm, is found in rodents, especially hamsters, and primates. *Hymenolepis diminuta* is the rat tapeworm but may also be found in other rodents. Adults of *Taenia pisiformis* are found in carnivores, and the larvae occur in rabbits. *Taenia taeniaformis*, as the larval *Cysticercus fasciolaris*, occurs in mice and rats, and the adult form inhabits the intestines of felines and other carnivores. The larval stages of *Multiceps serialis* are found in rabbits. The adult occurs in canines.

Transmission. *Hymenolepis nana* has three cycle variations. In the direct cycle, ova are passed in the feces from one definitive host and are ingested by another. The prepatent period is 15 to 30 days. Tissue migration by the parasites occurs in this cycle, as it does in autoinfection, in which ova mature in the intestinal lumen of the definitive host. In the indirect life cycle, which confers minimal host reaction because tissue migration does not occur, the ova are passed in the feces from a definitive host through beetles, cockroaches, or fleas to another definitive host.

Hymenolepis diminuta eggs appear in the feces of the definitive host and are ingested by insects. When the insects are eaten, the eggs are passed to another definitive host. Because of the need for an intermediate host (insect), *H. diminuta* infections are rarely seen in contemporary laboratory rodents. *Taenia* and *Multiceps* ova are passed in carnivore feces and then are ingested by the rabbit or rodent. Oncospheres pass from the intestines to internal organs. The cycle is completed when a carnivore eats the rabbit.

Predisposing Factors. General debility and absence of previous exposure to cestodes predispose to the more serious effects of infection. Contamination of feed and bedding with carnivore and rodent feces contributes to the spread of cestodes to rabbits and domestic rodents. Susceptibility and reaction to cestode infection vary with sex, age, strain of animal, and virulence of the parasite. For example, the DBA/2 strain of mice is more susceptible to *Hymenolepis* infection than is the C3H strain.

Clinical Signs. Hymenolepids may cause no clinical signs or, especially when present in large numbers, may cause constipation, catarrhal diarrhea, weight loss, and death. *Taenia* infestations are, with rare exception, subclinical in rabbits and rodents, but in heavy infestations, abdominal distention, lethargy, and weight loss may occur. *Multiceps serialis* infestation in the rabbit results in cyst formation under the skin or in other tissues, including the muscles and brain. Clinical signs depend on tissues affected and displaced.

Necropsy Signs. Hymenolepid infestations are usually asymptomatic or may cause a catarrhal enteritis, emaciation, and, indirectly, abscessation of mesenteric lymph nodes. The adult worms are found in the small intestine and occasionally in the pancreatic and biliary ducts of rodents. Intestinal obstruction and impaction can occur. Worm masses migrate into the anterior two thirds of the small intestine following ingestion of a meal by the host; when the stomach is empty, the worms retreat to the posterior two thirds.

Taenia pisiformis cysts, containing a single scolex and measuring up to 2 cm in diameter, are found attached to the mesenteries

of the rabbit. The hepatic lesions, which result from hepatic migration of the larvae, are usually white areas with sharp boundaries. These lesions may appear as pale foci or streaks in the liver, especially beneath the capsule. They grossly resemble the hepatic lesions of *Eimeria stiedae* infection, but *Eimeria* lesions are usually concentric with fuzzy edges.

Taenia taeniaformis cysticerci usually number one to ten, white to clear, thick-walled cysts in a rodent's liver. The cysts, which grossly resemble abscesses, are several millimeters in diameter, contain a coiled strobila, and are protected against rejection by a coating of blocking antibodies.

Multiceps serialis cysts, containing multiple scolices and measuring up to 5 cm in diameter, occur most often in the subcutaneous tissues and connective tissue of skeletal muscle in rabbits, but they may occur in other tissues.

Diagnosis. *Hymenolepis nana* is identified by the oval ovum (44 to 62 $\times$ 30 to 55 mm), which contains an embryo with 3 pairs of small hooks. The adult tapeworms (20 to 30 mm $\times$ 1 mm) occur in the small intestine or pancreatic and biliary ducts and are found at necropsy. Onchospheres (16 to 25 $\times$ 24 to 30 μm) occur in the intestinal villi. *Hymenolepis diminuta* adults measure 20 to 60 μm $\times$ 4 mm, and the ova 52 to 81 $\times$ 62 to 88 μm. Because *Hymenolepis* ova reach the environment in proglotids shed in the feces, fecal flotation techniques to find ova in the feces are not reliable diagnostic tests.

Taenia pisiformis larvae are recognized on necropsy in the liver as scattered white foci or in the abdominal cavity as cysts with a single scolex. The larval cysts are usually between 0.5 and 2.0 cm in diameter. Histologically, the *Taenia* migration lesion in the hepatic parenchyma consists of focal granulomas containing scattered polymorphonuclear leukocytes. The *Eimeria* lesion involves the bile duct epithelium (destruction, hyperplasia, oocyst development) and adjacent portal tissue (fibrosis with chronic inflammatory cell infiltration).

Multiceps serialis larvae may be detected by

observation, palpation, and aspiration of cyst fluid, or by demonstration of cysts on necropsy. *Taenia taeniaformis* larvae in rodents are detected in the liver on necropsy examination.

Treatment. *Hymenolepis* infestation may be treated with niclosamide (Yomesan) (Hughes, Barthel and Lang 1973) (Ronald and Wagner 1975) but the drug is no longer available in the U.S. Thiabendazole (0.3% in the feed for 7 to 14 days) or uredofos (25 mg/kg in the drinking water and 125 ppm in the feed for 6 days) has also been used successfully (Taffs 1976). Praziquantel (Droncit) in the feed at 140 ppm for 1 week eliminated *Hymenolepis* in mice (Arther, Cox and Schmidl 1981). *Taenia* infestations in rabbits are not treated. *Multiceps serialis* cysts can be removed surgically unless they are numerous or inaccessible. Treatment should be accompanied by changes to prevent reinfection.

Prevention. Cestode infestations are prevented by excluding carnivores, wild rodents, arthropod vectors, and insects from rabbit and rodent colonies. Food and bedding should be clean, and dogs and cats in the immediate area of rabbits and rodents should be treated for tapeworms. High humidity favors the survival of cestode eggs in the environment.

Public Health Significance. *Hymenolepis nana* is pathogenic for man and can cause enteric disease. *Taenia pisiformis* does not affect man, but *Taenia taeniaformis* may, if the rodent is eaten. *Multiceps serialis* larvae can affect man as they affect other intermediate hosts.

REFERENCES

Ambu, S., and Kwa, B.H.: Susceptibility of rats to *Taenia taeniaformis* infection. J. Helminthol., *54*:43–44, 1980.

Arther, R.G., Cox, D.D., and Schmidl, J.A.: Praziquantel for control of *Hymenolepis nana* in mice. Lab. Anim. Sci., *31*:301–302, 1981.

Ayuya, J.M., and Williams, J.F.: The immunological response of the rat to infection with *Taenia taeniaformis.* VII. Immunizations by oral and parenteral administration of antigens. Immunology, *36*:825–834, 1979.

Balk, M.W., and Jones, S.R.: Hepatic cysticercosis in a mouse colony. J. Am. Vet. Med. Assoc., *157*:678–679, 1970.

Coman, B.J.: The survival of *Taenia pisiformis* eggs under laboratory conditions and in the field environment. Aust. Vet. J., *51*:560–565, 1975.

Duwel, D., and Brech, K.: Control of oxyuriasis in rabbits by fenbendazole. Lab. Anims., *15*:101–105, 1981.

Flatt, R.E., and Campbell, W.W.: Cysticercosis in rabbits: Incidence and lesions of the naturally occurring disease in young domestic rabbits. Lab. Anim. Sci., *24*:914–918, 1974.

Flatt, R.E., and Moses, R.W.: Lesions of experimental cysticercosis in domestic rabbits. Lab. Anim. Sci., *25*:162–167, 1975.

Ghazal, A.M., and Avery, R.A.: Observations on coprophagy and the transmission of *Hymenolepis nana* infections in mice. Parasitology, *73*:39–45, 1976.

Heath, D.D., and Chevis, R.A.F.: Duration of immunity to *Taenia pisiformis* larvae in rabbits. J. Parasitol., *64*:252, 1978.

Hopkins, C.A.: Diurnal movement of *Hymenolepis diminuta* in the rat. Parasitology, *60*:255–271, 1970.

Hughes, H.C., Jr., Barthel, C.H., and Lang, C.M.: Niclosamide as a treatment for *Hymenolepis nana* and *Hymenolepis diminuta* in rats. Lab. Anim. Sci., *23*:72–73, 1973.

Insler, G.D., and Roberts, L.S.: *Hymenolepis diminuta*: lack of pathogenicity in the healthy rat host. Exp. Parasitol., *39*:351–357, 1976.

Ito, A.: A simple method for collecting infective cysticercoids of *Hymenolepis nana* from the mouse intestine. J. Parasitol., *63*:167–168, 1977.

Kwa, B.H., and Liew, F.Y.: Studies on the mechanism of long term survival of *Taenia taeniaformis* in rats. J. Helminthol., *52*:1–6, 1978.

Leiper, R.T.: Specimens illustrating the larval development of *Hymenolepis* in the wall of the small intestine of the gerbillae. Trans. Ry. Soc. Trop. Med. Hyg., *26*:319–320, 1933.

Lussier, G., and Loew, F.M.: Natural *Hymenolepis nana* infection in Mongolian gerbils (*Meriones unguiculatus*). Can. Vet. J., *11*:105–107, 1970.

Mitchell, G.F., Rajasekariah, G.R., and Rickard, M.D.: A mechanism to account for mouse strain variation in resistance to the larval cestode *Taenia taeniaformis.* Immunology, *39*:481–489, 1980.

Peeters, J.E., et al.: Clinical and pathological changes after *Eimeria intestinalis* infection in rabbits. Zentralbl. Vet. Med. B., *31*:9–24, 1984.

Read, C.P.: *Hymenolepis diminuta* in the Syrian hamster. J. Parasitol., *37*:324, 1951.

Ronald, N.C., and Wagner, J.E.: Treatment of *Hymenolepis nana* in hamsters with Yomesan® (niclosamide). Lab. Anim. Sci., *25*:219–220, 1975.

Simmons, D.J.C., and Walkey, M.: *Capillaria* and *Hymenolepis* in a wild rat: Hazards to barrier-maintained laboratory animals. Lab. Anim., *5*:49–55, 1971.

Taffs, L.F.: Further studies on the efficacy of thiabendazole given in the diet of mice infected with *H nana, S obvelata* and *A tetraptera.* Vet. Rec., *99*:143–144, 1976.

Tucek, P.C., Woodard, J.C., and Moreland, A.F.: Fibrosarcoma associated with *Cysticercus fasciolaris.* Lab. Anim. Sci., *23*:401–407, 1973.

Voge, M.: Cestodes. *In* Parasites of Laboratory Animals. Edited by R.J. Flynn. Ames, The Iowa State University Press, 1973, pp. 155–202.

Worley, D.E.: Quantitative studies on the migration and development of *Taenia pisiformis* larvae in laboratory rabbits. Lab. Anim. Sci., *24*:517–522, 1974.

COCCIDIOSIS (HEPATIC)

Hosts. Coccidia are highly specific as to host and anatomic location within that host. Hepatic coccidiosis is a protozoal disease of rabbits and wild lagomorphs and was, in fact, the first protozoal disease and protozoan identified (Van Leeuwenhoek in 1678).

Etiology. *Eimeria stiedae* is a highly pathogenic protozoan commonly found in domestic rabbits. Sporulated oocysts excyst in the duodenum and pass via blood and lymph to the liver and other organs. Schizogony and gametogony occur in the biliary epithelium, and unsporulated oocysts pass via bile ducts to the intestine. The prepatent period is 15 to 18 days.

Transmission. Transmission of *E. stiedae* is by ingestion of sporulated oocysts passed in the feces. Sporulation requires 2 or more days outside a host. Coprophagy does not allow time for sporulation, and reinfection by this route is unlikely. Also, a long-lasting immunity develops following initial exposure. Coccidial oocysts are extremely resistant and may be found in soil and feed and on personnel, vegetables, caging, and utensils. Oocysts may remain infectious in the environment for several months.

Predisposing Factors. Factors predisposing to clinical coccidiosis include lack of immunity to a specific coccidium, the number of oocysts ingested, and the general disease resistance of the host. Although 3- to 4-week-old rabbits have large numbers of oocysts in their feces, the disease is usually more severe clinically in weanling (5 to 8 weeks) rabbits, especially those in unsanitary or contaminated environments.

Clinical Signs. Infection with *Eimeria stiedae* is usually subclinical, but consequences in an animal without previous exposure are dependent on the number of ingested oocysts. Anorexia, failure to gain weight, weight loss, enlarged abdomen, icterus, diarrhea, debilitation, and death can occur. In young susceptible animals exposed to many oocysts, mortality may reach 50% or more. Liver coccidiosis can be expected to alter many liver and serum components and other physiologic functions so that infected rabbits are frequently unsuitable for research. Therefore, managers of animal facilities and investigators should not buy infected rabbits when uninfected rabbits from other sources are available.

Necropsy. The most common sign of hepatic coccidiosis is few to numerous yellow-white nodules or cords on and within the liver. These foci are irregularly shaped, raised, fuzzy-edged, and, when cut, may ooze a yellow-green fluid. The cords are several millimeters in diameter and course into the body of the liver following the path of affected bile ducts. The gallbladder and major extrahepatic ducts may be thickened, distended, and contain the yellow-green fluid. The number of foci and the degree of hepatomegaly are related to the number of infective oocysts ingested. Hepatic fibrosis may be evident on cutting the liver.

The microscopic lesion is primarily a papillary hyperplasia of the biliary epithelium, which usually contains coccidia intracellularly in various stages of development. Occasionally infection and compression of hepatocytes adjacent to portal areas result in a mixed inflammatory cell infiltration and fibrosis. Sometimes enlarged bile ducts will rupture releasing contents into surrounding tissue, which initiates a severe granulomatous response. Histopathologic examination of livers from rabbits recovered from active hepatic coccidiosis frequently reveals periportal fibrosis and varying amounts of monomorphonuclear cell infiltration. Pathologists examining livers from research rabbits must be aware that the lesion is not associated with the experimental manipulation of the rabbit but rather is a residual lesion of a naturally occurring disease. Remnants of lesions may remain for the life of the rabbit.

Diagnosis. Diagnosis is by recognition of characteristic lesions at necropsy. Also, each species of *Eimeria* has oocysts that appear in the feces and have characteristic features. Those of *E. stiedae* are elongated ellipsoidal bodies, about 37 × 20 μm. Flotation and concentration-flotation methods are used to detect oocysts in the feces. They are smooth, light-yellow, and walled with a wide thin micropyle and residual body; the sporocyst has a terminal knob (Stiedae body). Histologically, developmental coccidial stages can be demonstrated in the bile duct epithelium. Bile duct hyperplasia may be so extreme as to resemble neoplasia. It should be remembered that in heavy primary infections, especially in young rabbits, it may be possible for death to occur before oocysts appear in the feces. Direct smears prepared from liver lesions at necropsy are an excellent diagnostic technique.

Treatment. Hepatic coccidiosis is best prevented or controlled through rigid sanitation practices. Effective treatment is possible only during a short, inapparent stage early in the asexual part of the protozoan life cycle. For all practical purposes, hepatic coccidiosis, ordinarily a subclinical disease, is untreatable. As a last resort, treatments used for intestinal coccidiosis may be of value in controlling hepatic coccidiosis.

Prevention. Prevention of coccidiosis rests primarily in the use of good husbandry and sanitation: clean stock, proper equipment, and vermin control. The use of coccidiostats often results in frustration because of confusion about dosages and regimens and the reappearance of the disease when the drug is removed.

Because immunity develops following exposure to *Eimeria,* rabbits often become carriers. This self-limiting factor, combined with the difficulty of eliminating oocysts from caging, bedding, and soil, makes coccidiosis very difficult to eliminate from a rabbit colony. Control of coccidiosis requires screening of resident and incoming animals for oocysts in the feces, culling of infected animals, and strict cleaning and disinfecting procedures. It is important to note that oocysts are commonly found in the feces of clinically healthy rabbits and a "clean" rabbit rarely occurs.

Detergents, disinfectants, and a good scrub brush properly used remove the organic material in which oocysts remain viable, but no commonly used disinfectant, used as directed, kills coccidial oocysts. Hot 2% lye solution, flaming, heat sterilization (120° C for 20 minutes), and such small and hazardous molecules as ammonia (10% solution), methylbromide, and carbon disulfide are means of killing oocysts. Soil removal, where appropriate, may also be used. The use of hanging wire cages, sippertube waterers, and hopper feeders is extremely important in preventing coccidiosis, as is the elimination of vermin, education of handlers, and early weaning of the young.

Drugs, especially sulfonamides, are widely used to prevent coccidiosis; however, there are several cautions associated with this practice. The only drug product approved for use in rabbits by the U.S. Food and Drug Administration is a specific preparation of sulfaquinoxaline (Merck Co., Sulfaquinoxaline Premix 40%–181.6 gm per pound), which, though approved, is no longer marketed. Other sulfaquinoxaline products, though unapproved for rabbits, are available. Rabbits so treated should not be slaughtered for food until the required 10 day withdrawal period has passed. The responsibilities for residues remaining at slaughter lie with the owner or veterinarian. Sulfonamides also have a variety of side effects that may adversely affect medical research. These range from inhibition of phagocytosis to nephrotoxicity to blood clotting disorders.

Rabbit feeds supplemented with 0.025 to 0.03% sulfaquinoxaline (Joyner, Catchpole, and Berrett 1983; Kraus et al. 1984.) and fed continuously during the critical weaning period (3 to 8 weeks) reduce coccidiosis (and acute pasteurellosis) in young rabbits. Sulfaquinoxaline may also be added to the drinking water at 0.025% to 0.1% (Hagen 1974). This drug is given continuously for 4 to 8 weeks, for alternating "on-off-on" 2-week periods, or on a wide range of other schedules; the intention is to have the drug present at adequate levels when oocysts excyst.

Other prophylactic regimens include daily ingestion of 0.625 g succinylsulfathiazole; sulfamethazine 0.05% to 1.0% in the feed; and sulfadimethoxine at 75 mg/kg for

7 days (Pakes 1974). Sulfamethazine in the drinking water at 0.2% on alternate days was recommended by Norton and colleagues in 1977. These drugs may control the protozoan until the body's own defenses overcome the infection.

Monensin, robenidine (66 ppm in the feed) (Peeters and Halen 1980; Peeters, Halen and Meulemans 1979), and other coccidiostats have also been used in rabbits with good effect, but withdrawal times have not been established. Avatec® (Hoffmann-LaRoche, 340 Kingsland Street, Nutley, NJ 07110) a premix containing 20% lasalocid, a coccidiostat for chickens, or Bovatec® (15% lasalocid) for cattle, may be worth considering and FDA approval for use in rabbits is pending. Perhaps the best that can be said about the use of coccidiostats in rabbitries is that they provide a short-term shield against coccidiosis until the owner or manager controls coccidia through good husbandry and stock selection. The most important consideration in controlling coccidiosis is to prevent rabbits from contacting infective feces or contaminated food and water containers by using self-cleaning cages (wire or barred floors) and external feeders and waterers.

Public Health Significance. *Eimeria* of rabbits does not affect man.

REFERENCES

Barriga, O.O., and Arnoni, J.V.: *Eimeria stiedae:* weight, oocyst output, and hepatic function of rabbits with graded infections. Exp. Parasitol., *48*:407–414, 1979.

Chapman, M.P.: The use of sulfaquinoxaline in the control of liver coccidiosis in domestic rabbits. Vet. Med., *43*:375–379, 1948.

Hagen, K.: Colony husbandry, Chapter 2. *In* The Biology of the Laboratory Rabbit. Edited by S. Weisbroth. Orlando, Academic Press, 1974.

Horton-Smith, C., Taylor, E.L., and Turtle, E.E.: Ammonia fumigation for coccidial disinfection. Vet. Rec., *52*:829–832, 1940.

Joyner, L.P., Catchpole, J., and Berrett, S.: *Eimeria stiedae* in rabbits: The demonstration of responses to chemotherapy. Res. Vet. Sci., *34*:64–67, 1983.

Kraus, A.L., et al.: Biology and diseases of rabbits, Chapter 8. *In* Laboratory Animal Medicine. Edited by J.G. Fox, B.J. Cohen, and F.M. Loew. Orlando, Academic Press, 1984.

Levine, N.D.: Sporozoans and neosporans. *In* Parasites of Laboratory Animals. Edited by R.J. Flynn. Ames, Iowa State University Press, 1973, pp. 48–113.

Long, P.L., Burns Brown, W., and Goodship, G.: The effect of methyl bromide on coccidial oocysts determined under controlled conditions. Vet. Rec., *90*:562–566, 1972.

Norton, C.C., Catchpole, J., and Rose, M.E.: *Eimeria stiedae* in rabbits: the presence of an oocyst residuum. Parasitology, *75*:1–7, 1977.

Owen, D.: Life cycle of *Eimeria stiedae.* Nature (London), *227*:304, 1970.

Pakes, S.P.: Protozoal diseases, Chapter 8. *In* The Biology of the Laboratory Rabbit. Edited by S. Weisbroth. Orlando, Academic Press, 1974.

Peeters, J.E., Halen, P., and Meulemans, G.: Efficacy of robenidine in the prevention of rabbit coccidiosis. Br. Vet. J., *135*:349–354, 1979.

Peeters, J.E., and Halen, P.: Robenidine treatment of rabbits naturally infected with coccidia. Lab. Anim., *14*:53–54, 1980.

Peeters, J.E., and Geeroms, R.: Efficacy of Toltrazuril against intestinal and hepatic coccidiosis in rabbits. Vet. Parasitol., *22*:21–35, 1986.

Smetana, H.: Coccidiosis of the liver in rabbits. I. Experimental study on the excystation of oocysts of *Eimeria stiedae.* Arch. Pathol., *15*:176–192, 1933.

Varga, I.: Large scale management systems and parasite populations: Coccidia in rabbits. Vet. Parasitol., *11*:69–84, 1982.

Yvore, P., et al.: The effects in the rabbit of hepatic coccidiosis on the digestibility of fat and energy. Ann. Rech. Vet., *7*:343–348, 1976.

COCCIDIOSIS (INTESTINAL)

Hosts. Intestinal coccidiosis occurs in rabbits, and the disease may occasionally occur in rodents.

Etiology. Several protozoal genera can be included as intestinal coccidia, but the genus of concern in rabbit colonies is *Eimeria*. The most frequently encountered *Eimeria* species in the rabbit intestine are: *E. flavescens, E. irresidua, E. neoleporis, E. intestinalis,* and *E. magna,* highly pathogenic inhabitants of the small intestine, *E. perforans* of low pathogenicity in the small intestine, and *E. media* of moderate pathogenicity in the large and small intestines. *Eimeria falciformis* is the most common agent infecting the large intestine of the mouse, and *E. caviae* infects guinea pigs. Prepatent periods range from 7 to 10 days, and mixed infections are common. Host immunity develops against each species; immunity against one species will not prevent disease caused by another species later in life.

The transmission, predisposing factors, treatment, prevention, and public health significance of intestinal coccidiosis are similar to those of hepatic coccidiosis that was discussed in the previous section of this chapter. The diarrhea of intestinal coccidiosis can be treated symptomatically, but usually to no avail. In addition to being a direct cause of disease in rabbits, coccidia may predispose to bacterial enteritis, particularly clostridial infections and colibacillosis.

Clinical Signs. Coccidia frequently reside in the intestine more as innocuous flora than as pathogenic organisms. This characteristic is common in immune, adult animals; however, the introduction of new *Eimeria* species in young or old can produce a wide range of clinical signs.

Signs may include failure to gain weight, weight loss, diarrhea with soft to watery feces, and mild to severe dehydration. Affected rabbits tend to show intense thirst. Feces may contain mucus or blood and have a variety of colors and odors. Usually the feces are watery, have a foul odor, are brown-green, and smear the belly, rear quarters, and legs of the weakened animal.

Mortality may be high or low, depending on the *Eimeria* species involved, immune status of the host, and the amount of inoculum. In severe outbreaks, following ingestion of large numbers of pathogenic organisms, all or most young will weaken and die over a few days. Even if drugs are used at the first sign of an outbreak, the disease will probably progress unabated for several days within the exposed colony.

Diagnosis. Diagnosis involves finding characteristic forms of the protozoa on microscopic examination of intestinal scrapings together with lesions in the intestine or finding oocysts in the feces. It may be difficult to decide if oocysts found in feces are associated with primary clinical disease or if they are the product of an incidental, low-grade infection. *E. flavescens* has oocysts smaller and more ovoidal than the yellow,

broadly ellipsoidal or ovoidal *E. irresidua* oocysts.

Necropsy. Animals may die of dehydration and secondary bacteremia. Gross lesions seen with intestinal coccidiosis may be inapparent or include edema, reddening, and white streaking of the intestinal wall. A common sign is an obviously red intestinal tract filled with fluid. The mucosa may slough, and blood may accumulate in the intestinal lumen. The severity of the disease varies with the depth of schizont penetration into the epithelial cells, the lamina propria, and submucosa, and with the contribution of enteric bacteria to the inflammatory process. Death is due to a combination of diarrhea and tissue invasion by bacteria.

Treatment. Treatments recommended for liver coccidiosis can generally be applied to intestinal coccidiosis.

REFERENCES

Ellis, P.A., and Wright, A.E.: Coccidiosis in guinea-pigs. J. Clin. Pathol., *14*:394–396, 1961.
Gallazzi, D.: Cyclical variations in the excretion of intestinal coccidial oocysts in the rabbit. Folia Vet. Lat., *7*:371–380, 1977.
Gregory, M.W., and Catchpole, J.: Coccidiosis in rabbits: The pathology of *Eimeria flavescens* infection. Int. J. Parasitol., *16*:131–145, 1986.
Hankinson, G.J., Murphy, J.C., and Fox, J.G.: Diagnostic exercise. Lab. Anim. Sci., *32*:35–36, 1982.
Kleeberg, H.H., and Steeken, W., Jr.: Severe coccidiosis in guinea-pigs. J. S. Afr. Vet. Med. Assoc., *34*:49–52, 1963.
Levine, N.D.: Sporozoans and neosporans. *In* Parasites of Laboratory Animals. Edited by R.J. Flynn. Ames, Iowa State University Press, 1973, pp. 48–113.
McPherson, C.W., et al.: Eradication of coccidiosis from a large breeding colony of rabbits. Proc. Anim. Care Panel, *12*:133–140, 1962.
Niilo, L.: Acquired resistance to reinfection of rabbits with *Eimeria magna*. Can. Vet. J., *8*:201–208, 1967.
Norton, C.C., Catchpole, J., and Joyner, L.P.: Redescription of *Eimeria irresidua* Kessel & Jankiewicz, 1931 and *E. flavescens* Marotel & Guilhon, 1941 from the domestic rabbit. Parasitol., *79*:231–248, 1979.
Owen, D.: Effects of pelleting and sterilization of diet on 2 strains of rabbit coccidia. Lab. Anims., *12*:49–50, 1978.
Peeters, J.E., and Geeroms, R.: Efficacy of Toltrazuril against intestinal and hepatic coccidiosis in rabbits. Vet. Parasitol., *22*:21–35, 1986.
Rutherford, R.L.: The life cycle of four intestinal coccidia of the domestic rabbit. J. Parasitol., *29*:10–32, 1943.
Ryley, J.F.: Recent developments in coccidian biology: Where do we go from here? Parasitol., *80*:189–209, 1980.

COLIBACILLOSIS

Hosts. Colibacillosis is a common vertebrate disease, especially in neonatal and weanling animals. Both rabbits and rodents have diarrheal conditions during which the usually scant intestinal coliforms proliferate in huge numbers, but whether this proliferation is the cause of the diarrhea, a contributing factor, or a secondary phenomenon is not known.

Etiology. The name *colibacillosis* confines this disease to an *Escherichia coli* causation, but in rabbits the action of this organism is poorly defined and often unrelated to the clinical signs observed. These observations give rise to suggestions that coliforms act with other microbial organisms and changes in the intestinal milieu in an etiologic complex or proliferate secondarily in an otherwise debilitated intestine. The *E. coli* enterotoxin is associated with increased intestinal secretion and hypermotility.

Pathogenic serotypes of *E. coli* cultured from rabbits with acute diarrhea include 085, 0153, 0128, and an unclassified serogroup 0X1; however, other types, some thought to be nonpathogenic, have also been cultured during diarrheal outbreaks. The gastrointestinal tracts of rabbits, guinea pigs, and gerbils normally have few or no *E. coli,* whereas other rodents carry the organism in low numbers.

Other probable coliform diseases in small rodents include diarrhea in gerbils and enteritis in hamsters, which may occur alone or in combination with a proliferative or hyperplastic lesion of the ileum. Following antibiotic injections, guinea pigs and hamsters may die within 2 to 6 days; coliform or clostridial overgrowth has been implicated as the cause.

Because the etiology of acute enteritis of rabbits and rodents is so poorly understood, it is likely that new causes will be discovered and clarified. The association between clostridial toxins and acute fatal enteritis and antibiotic-induced colitis of rabbits and other species is becoming more firmly established. It is likely that some of the enteropathies of laboratory animals called colibacillosis in the past were in fact caused by clostridial toxins.

Transmission. Transmission of *E. coli* is by the fecal-oral route. The clinical disease, however, probably develops from activation of a pre-existing component of the intestinal flora.

Predisposing Factors. Predisposing factors for colibacillosis include change of diet, entrance of pathogenic organisms into the relatively uncolonized gut of the weanling rabbit, high starch-low fiber diets, stress, other disease, or alteration of the microflora with antibiotics active against gram-positive or nonresistant bacteria. Factors predisposing to diarrhea or other enteropathies in rabbits and rodents are discussed in more detail under "Enteropathy, Nonspecific," in this chapter.

Clinical Signs. Colibacillosis, as the other components of the enteropathy complex, frequently involves weanling animals. Signs include acute onset, fever, and either slight diarrhea and recovery or watery diarrhea or a massive colonic expulsion of a fecal puddle, salivation, bloat, and death within 6 to 72 hours. Mortality, however, is usually high.

Necropsy Signs. Gross signs of colibacillosis include cecal hemorrhage and edema and a watery or mucous brown, fetid, luminal content. Other changes include enlarged mesenteric nodes and congested kidneys. Microscopically, there are hemorrhages and edema of the gut wall, loss of epithelium, and inflammatory infiltration of the cells of the submucosa and lamina.

Diagnosis. Diagnosis is through clinical signs and culture.

Treatment. Treatment of colibacillosis and other diarrheal conditions is discussed under "Enteropathy, Nonspecific," in this chapter.

Prevention. Prevention of colibacillosis is discussed under "Enteropathy, Nonspecific" and otherwise follows the recommendations for enterotoxemia.

Public Health Significance. Some strains of *E. coli* are enteropathogenic for both man and rabbits, but the likelihood of encountering and transmitting such an organism is slight.

REFERENCES

Cantey, J.R., and Blake, R.K.: Diarrhea due to *Escherichia coli* in the rabbit: A novel mechanism. J. Infect. Dis., *135*:454–462, 1977.

Cantey, J.R., O'Hanley, P.D., and Blake, R.K.: A rabbit model of diarrhea due to invasive *Escherichia coli*. J. Infect. Dis., *136*:640–648, 1977.

Coussement, W.: Pathology of experimental colibacillosis in rabbits. Zentralbl. Vet. Med. B., *31*:64–72, 1984.

Farrar, W.E., Jr., and Kent, T.H.: Enteritis and coliform bacteremia in guinea pigs given penicillin. Am. J. Pathol., *47*:629–642, 1965.

Farrar, W.E., Jr., Kent, T.H., and Elliott, V.B.: Lethal gram-negative bacterial superinfection in guinea-pigs given bacitracin. J. Bacteriol., *92*:496–501, 1966.

Frisk, C.S., Wagner, J.E., and Owens, D.R.: Enteropathogenicity of *Escherichia coli* isolated from hamsters (*Mesocricetus auratus*) with hamster enteritis. Infect. Immun., *20*:319–320, 1978.

Glantz, P.J.: Serotypes of *Escherichia coli* associated with colibacillosis in neonatal animals. Ann. N.Y. Acad. Sci., *176*:67–79, 1971.

Glantz, P.J.: Unclassified *Escherichia coli* serogroup 0X1 isolated from fatal diarrhea of rabbits. Can. J. Comp. Med., *34*:47–49, 1970.

Peeters, J.E., et al.: Pathogenic properties of *Escherichia coli* strains isolated from diarrheic commercial rabbits. J. Clin. Microbiol., *20*:34–39, 1984.

Peeters, J.E., Charlier, G.J., and Halen, P.H.: Pathogenicity of attaching effacing enteropathogenic *Escherichia coli* isolated from diarrheic suckling and weanling rabbits for newborn rabbits. Infect. Immun., *46*:690–696, 1984.

Prescott, J.F.: *Escherichia coli* and diarrhoea in the rabbit. Vet. Pathol., *15*:237–248, 1978.

Prohászka, L.: Antibacterial effect of volatile fatty acids in enteric *E. coli*-infections of rabbits. Zentralbl. Vet. Med. B, *27*:631–639, 1980.

Prohászka, L.: Study of pathogenesis of enteric *Escherichia coli* infection in model experiments on rabbits. Zentralbl. Bakt. Hyg., I. Abt. Orig. A, *221*:314–321, 1972.

Richter, C.B., and Hendren, R.L.: The pathology and epidemiology of acute enteritis in captive cottontail rabbits (*Sylvilagus floridanus*). Pathol. Vet., *6*:159–175, 1969.

Savage, N.C., and Sheldon, W.G.: An epizootic of diarrhea in a rabbit colony. Pathology and bacteriology. Can. J. Comp. Med., *37*:313–319, 1973.

Schiff, L.J., et al.: Enteropathogenic *Escherichia coli* infections: increasing awareness of a problem in laboratory animals. Lab. Anim. Sci., *22*:705–708, 1972.

Schiff, L.J., et al.: The use of BSVR/SrCr mice for detection of murine strains of toxigenic *Escherichia coli*. Lab. Anim. Sci., *24*:752–756, 1974.

Small, J.D.: Fatal enterocolitis in hamsters given lincomycin hydrochloride. Lab. Anim. Care, *18*:411–420, 1968.

Smith, H.W.: Observations on the flora of the alimentary tract of animals and factors affecting its composition. J. Pathol. Bacteriol., *89*:95–122, 1965.

Vetesi, F., and Kutas, F.: Mucoid enteritis in the rabbit associated with *E. coli*. Changes in carbohydrate metabolism. Acta Vet. Acad. Sci. Hung., *24*:303–311, 1974.

Vetesi, F., and Kutas, F.: Mucoid enteritis in the rabbit associated with *E. coli*. Changes in water, electrolyte and acid-base balance. Acta Vet. Acad. Sci. Hung., *23*:381–388, 1973.

Yuill, T.M., and Hanson, R.P.: Coliform enteritis of cottontail rabbits. J. Bacteriol., *89*:1–8, 1965.

CORYNEBACTERIUM KUTSCHERI INFECTION

Hosts. Mice and rats are affected by *Corynebacterium kutscheri*. Rare isolations have been reported in the guinea pig.

Etiology. *Corynebacterium kutscheri (murium)* is a gram-positive, nonmotile, diphtheroid bacillus. After 48-hour aerobic incubation on 5% blood agar, colonies are circular, 1 to 4 mm in diameter, translucent, gray to yellow, smooth, and non-hemolytic.

Transmission. *Corynebacterium kutscheri* is carried in the oral cavity and regional lymph nodes (submaxillary) of rats. Transmission is speculated to be by the oral-nasal and fecal-oral routes and possibly by respiratory aerosol. Prenatal transmission may occur. The organism is an opportunistic pathogen.

Predisposing Factors. *Corynebacterium kutscheri* infections in mice and rats are often in-apparent or latent and become overt disease following stressful manipulations. Nutritional deficiencies, concomitant infections, cortisone injections, pregnancy, and radiation exposure are predisposing stresses. There is a great variation in susceptibility to the bacterium among inbred strains of mice. A strain mice, for example, are highly susceptible. Generally, strains of mice sensitive to *C. kutscheri* infection are relatively resistant to *Salmonella* infections and vice versa. Rats are more resistant to the spontaneous disease than are mice.

Clinical Signs. The infection is usually latent and subclinical. The acute clinical disease, with high morbidity and mortality, is characterized by rough hair coat, emaciation, rapid respiration, hunched posture, abnormal gait, nasal and ocular discharges, septic swollen joints, and lethargy. In cutaneous infections, the skin is abscessed, ulcerated, and underlaid with fistulous tracts. Loss of necrotic extremities may mimic some of the lesions associated with mousepox. Death usually occurs within a week. A chronic infection, with low morbidity and mortality, may be inapparent or produce nonspecific signs.

Necropsy Signs. The extension of the organism from the oral cavity or regional lymph nodes through the circulatory system results in focal embolic abscessation in a variety of organs, most notably the lungs, kidneys, heart, and liver. The brain, middle ear, lymphatic system, joints, and skin may also be affected. The lesions are scattered, gray to light yellow, and raised with caseopurulent foci up to 15 mm in diameter. There are usually no splenic lesions as in salmonellosis. Rats may develop a severe pneumonia with a striking pattern of white, multifocal abscesses against a dark red background. The abscesses are infiltrated with a mixed cellular exudate and are surrounded by aggregations of macrophages and neutrophils. Clumps of the gram-positive bacteria are evident on smear preparations. Lung lesions eventually become granulomatous, thus the reason for calling the disease pseudotuberculosis.

Diagnosis. Giemsa or gram staining of impression smears of affected tissues or examination of tissue sections may reveal the typical "Chinese character" arrangement of pleomorphic gram-positive rods. Definitive diagnosis requires culture on blood agar. Cortisone acetate provocation to trigger latent infections to active disease with subsequent culture can be used in diagnostic screening of rat and mouse colonies. The agent is difficult to recover from animals in latently infected colonies. Possibly it is carried as a low-grade infection. Microtiter and tube agglutination tests have been used. Refined diagnostic techniques, such as the enzyme-linked immunosorbent assay (ELISA) and DNA hybridization assays, used to reveal presence of organisms in tissue, have

been described. Use of these tests in conjunction with eliciting an anamnestic response may be productive.

Treatment. The rapid course of the clinical disease renders treatment of individual animals difficult. The bacterium is sensitive to a wide variety of antibiotics, including ampicillin, chloramphenicol, and tetracycline.

Prevention. Prevention of a *C. kutscheri* outbreak involves selection of clean stock, good husbandry practices, and the removal of affected animals. Potentially infected animals should not be mixed with "clean" animals. Unfortunately culture is unreliable as a screening tool because latently infected or carrier animals are difficult to detect, i.e., they are generally culture negative. Immunologically compromised mice and rats should be isolated and carefully observed for signs of the disease.

Public Health Significance. *Corynebacterium kutscheri* is a rodent pathogen, but a case of chorioamnionitis has been observed in man.

REFERENCES

Ackerman, J.I., Fox, J.G., and Murphy, J.C.: An enzyme linked immunosorbent assay for detection of antibodies to *Corynebacterium kutscheri* in experimentally infected rats. Lab. Anim. Sci., *34*:38–43, 1984.

Brownstein, D.G., et al.: Experimental *Corynebacterium kutscheri* infection in rats: Bacteriology and serology. Lab. Anim. Sci., *35*:135–138, 1985.

Fauve, R.M., Pierce-Chase, C.H., and Dubos, R.: Corynebacterial pseudotuberculosis in mice. II. Activation of natural and experimental latent infections. J. Exp. Med., *120*:283–304, 1964.

Fitter, W.F., DeSa, D.J., and Richardson, H.: Chorioamnionitis and funisitis due to *Corynebacterium kutscheri*. Arch. Dis. Child., *55*:710–712, 1979.

Fox, J.G., et al.: Comparison of methods to diagnose an epizootic of *Corynebacterium kutscheri* pneumonia in rats. Lab. Anim. Sci., *37*:72–75, 1987.

Giddens, W.E., Jr., et al.: Pneumonia in rats due to infection with *Corynebacterium kutscheri*. Pathol. Vet., *5*:227–237, 1968.

Hirst, R.G., and Wallace, M.E.: Inherited resistance to *Corynebacterium kutscheri* in mice. Infect. Immun., *14*:475–482, 1976.

Lawrence, J.J.: Infection of laboratory mice with *Corynebacterium murium*. Aust. J. Sci., *20*:147, 1957.

LeMaistre, C., and Tompsett, R.: The emergence of pseudotuberculosis in rats given cortisone. J. Exp. Med., *95*:393–408, 1952.

Saltzgaber-Muller, J., and Stone, B.A.: Detection of *Corynebacterium kutscheri* in animal tissues by DNA-DNA hybridization. J. Clin. Microbiol., *24*:759–763, 1986.

Suzuki, E., Mochida, K., and Nakagawa, M.: Naturally occurring subclinical *Corynebacterium kutscheri* infection in laboratory rats: Strain and age related antibody response. Lab. Anim. Sci., *38*:42–45, 1988.

Weisbroth, S.H., and Scher, S.: *Corynebacterium kutscheri* infection in the mouse. I. Report of an outbreak, bacteriology, and pathology of spontaneous infections. Lab. Anim. Sci., *18*:451–458, 1968.

Weisbroth, S.H., and Scher, S.: *Corynebacterium kutscheri* infection in the mouse. II. Diagnostic serology. Lab. Anim. Sci., *18*:459–468, 1968.

Yokoiyama, S., Mizuno, K., and Fujiwara, K.: Antigenic heterogeneity of *Corynebacterium kutscheri* from mice and rats. Exp. Anim., *26*:263–266, 1977.

CRYPTOSPORIDIOSIS

Hosts.   Cryptosporidiosis may be an important cause of enterocolitis and diarrhea in guinea pigs, nonhuman primates, young ruminants (calves and lambs), foals, snakes, birds, and man. Asymptomatic infections have been reported in mice, rats, hamsters, and rabbits.

Etiology.   *Cryptosporidium* sp. was first identified by Tyzzer in 1907 in the pyloric glands of laboratory mice. Protozoans of the genus *Cryptosporidium* are small (2 to 6 μm) intracellular coccidian parasites that typically infect the absorptive surface of epithelial cells covering intestinal villi. Less often, epithelial cells of the respiratory mucosa may be infected. During infection, cryptosporidia fuse with host membranes, invaginate between microvilli, and ultimately lie within the apex of absorptive epithelial cells or, on occasion, within the cytoplasm of M cells in Peyer's patches. Cryptosporidia produce sporulated infective oocysts. Like other enteric coccidia, only one host is required to complete the life cycle (monoxenous).

Originally, cryptosporidia were considered highly host-specific, as are most coccidia, and each new species was named or speciated based on the host species from which it was recovered. More recent experimental and naturally occurring cross-species transmissions among mammals, however, indicate that *Cryptosporidium* is not always a host-specific parasite. Some investigators have proposed that there should be a single species within the genus *Cryptosporidium.* Levine (1984) has suggested designating four species of *Cryptosporidium,* one for each host class: mammal, *C. muris;* reptile, *C. crotali;* fish, *C. nasorum;* and birds, *C. meleagridis.*

Transmission.   Transmission of crypto-sporidia occurs with ingestion of mature and infective sporulated oocysts. Oocysts in the feces of animals with cryptosporidiosis are immediately infective for other animals. Indirect transmission may occur through contact with oocyst-contaminated food, water, or fomites. Excystation is believed to occur by digestion of the oocyst wall in the gastrointestinal tract of the host.

Predisposing Factors.   Clinical cryptosporidiosis is more likely to occur in young animals and in immunoincompetent animals. Recently weaned animals, particularly guinea pigs, are most likely to show clinical signs. Stress caused by shipping and overcrowding may contribute to the severity of disease. Heavily contaminated environments may increase intestinal microbial load and augment clinical signs.

Clinical Signs.   Cryptosporidial infections are frequently subclinical and asymptomatic. Lethargy, rough hair coat, failure to gain or maintain weight, and weight loss have been reported in natural and experimental infections in mice, rats, and rabbits. In guinea pigs, failure to gain weight, watery diarrhea, and occasionally death have been associated with cryptosporidiosis.

Necropsy Signs.   Usually there are no macroscopic lesions associated with subclinical cryptosporidiosis in rodents and rabbits. In guinea pigs with clinical signs, however, emaciation, hyperemia of the small intestine, serosal edema of the cecal wall, and watery ingesta throughout the intestines may be seen.

Depending on severity and duration of infection, the primary microscopic lesions are villus atrophy and blunting, villus fusion or bridging, and metaplasia of the mucosal epithelium from normal or tall columnar to cuboidal, absorptive epithelial cells. Small oval-to-round cryptosporidial forms may be observed microscopically within or on the brush border of mucosal epithelial cells. Developmental stages of *Crypotsporidium* spp. may be seen from the pylorus or duodenum to the cecum, with the highest concentration in the ileum.

Diagnosis.   Diagnosis is made by recognition of clinical signs, characteristic lesions, and parasites in microscopic sections or by identification of oocysts in the feces. Cryptosporidial oocysts frequently are overlooked in fecal sample analysis because of their small size (2 to 6 μm) and their morphologic similarity to yeasts. Diagnosis of cryptosporidiosis may be optimized by concentration techniques, such as oocyst flotation from fecal specimens (Sheather's sugar solution), and subsequent examination of meniscal fluid by light or phase-contrast microscopy. Examination of fresh or formalin-fixed fecal smears stained with acid-fast or Giemsa stains may also detect oocysts. Phase-contrast examination of impressions or scrapings of the ileal mucosa made at necropsy will also reveal the small, mature birefringent oocysts.

Treatment.   No effective treatment has been identified.

Prevention.   Prevention of transmission of cryptosporidia is based on the use of husbandry and sanitation procedures that preclude exposure to carrier animals and contaminated objects. Many animal species, especially lambs and calves, may be carriers and sources of infection. Therefore, contact with these species must be minimized. Oocysts in the environment may be destroyed by treatment with 5% ammonia solution. Exposure to temperatures below 0° C or above 65° C kills oocysts.

Public Health Significance.   Because of their infectious nature and tendency to cross species lines, *Cryptosporidium* spp. are a potential public health hazard. Caution is warranted in the handling of potentially infected animals, specimens, and material from the suspected carrier's environment. Cryptosporidial infections have been documented in a veterinary student and in animal han-

dlers caring for calves with cryptosporidiosis. In addition, cryptosporidia isolated from man have been experimentally transmitted to several animal species. Lack of host specificity suggests that animals may be a zoonotic reservoir of cryptosporidia for man.

The course of cryptosporidiosis in man is largely determined by the host's immune status. Infection may be asymptomatic or cause profuse watery diarrhea without gross or microscopic blood, abdominal pain, and anorexia. Diarrhea is self-limiting in immunocompetent people, whereas immunosuppressed individuals develop chronic, severe diarrhea, which may contribute to the death of the patient. Cryptosporidia cause severe irreversible diarrhea in human patients with immunologic disorders, especially acquired immune deficiency syndrome (AIDS).

REFERENCES

Anderson, B.: Cryptosporidiosis: A review. J. Am. Vet. Med. Assoc., *180*:1455–1457, 1982.

Angus, K.W., Hutchison, G., and Munro, H.M.C.: Infectivity of a strain of *Cryptosporidium* found in the guinea-pig (*Cavia porcellus*) for guinea-pigs, mice and lambs. J. Comp. Pathol., *95*:151–165, 1985.

Angus, K.W., et al.: Prophylactic effects of anticoccidial drugs in experimental murine cryptosporidiosis. Vet. Rec., *114*:166–168, 1984.

Campbell, I., et al.: Effect of disinfectants on survival of cryptosporidium oocysts. Vet. Rec., *111*:414–415, 1982.

Davis, A.J., and Jenkins, S.J.: Cryptosporidiosis and proliferative ileitis in a hamster. Vet. Pathol., *23*:632–633, 1986.

Gibson, S.V., and Wagner, J.E.: Cryptosporidiosis in guinea pigs: A retrospective study. J. Am. Vet. Med. Assoc., *189*:1033–1034, 1986.

Inman, L.R., and Takeuchi, A.: Spontaneous cryptosporidiosis in an adult female rabbit. Vet. Pathol., *16*:89–95, 1979.

Levine, N.D.: Taxonomy and review of the coccidian genus *Cryptosporidium* (Protozoa, apicomplexa). J. Protozool., *31*:94–98, 1984.

Marcial, M.A., and Madara, J.L.: *Cryptosporidium*: Cellular localization, structural analysis of absorptive cell-parasite membrane-membrane interactions in guinea pigs, and suggestion of protozoan transport by M cells. Gastroenterol., *90*:583–594, 1986.

Navin, T.R., and Juranek, D.D.: Cryptosporidiosis: clinical, epidemiologic, and parasitologic review. Rev. Infect. Dis., *6*:313–327, 1984.

Pohjola, S., Jokipii, L., and Jokipii, A.M.M.: Dimethylsulphoxide-Ziehl-Neelsen staining technique for detection of cryptosporidial oocysts. Vet. Rec., *115*:442–443, 1985.

Reese, N.C., et al.: Cryptosporidiosis of man and calf: A case report and results of experimental infections in mice and rats. Am. J. Trop. Med. Hyg., *31*:226–229, 1982.

Rehg, J.E., Lawton, G.W., and Pakes, S.P.: *Cryptosporidium cuniculus* in the rabbit (*Oryctolagus cuniculus*). Lab. Anim. Sci., *29*:656–660, 1979.

Soave, R., and Ma, P.: Cryptosporidiosis: Traveler's diarrhea in two families. Arch. Intern. Med., *145*:70–72, 1985.

Tzipori, S.: Cryptosporidiosis in animals and humans. Microbiol. Rev., *47*:84–96, 1983.

Tzipori, S., Campbell, I., and Angus, K.W.: The therapeutic effect of 16 antimicrobial agents on *Cryptosporidium* infection in mice. Aust. J. Exp. Biol. Med. Sci., *60*:187–190, 1982.

Vetterling, J.M., et al.: *Cryptosporidium wrairi* sp. n. from the guinea pig *Cavia porcellus* with an emendation of the genus. J. Protozool., *18*:243–247, 1971.

Vetterling, J.M., Takeuchi, A., and Madden, P.A.: Ultrastructure of *Cryptosporidium wrairi* from the guinea pig. J. Protozool., *18*:248–260, 1971.

DERMATOPHYTOSIS

Hosts. Dermatophytes affect a wide range of animals. Most reports of ringworm or dermatophytosis in small mammals refer to rabbits, guinea pigs, and mice.

Etiology. *Trichophyton mentagrophytes* is the organism usually encountered in rabbit and rodent dermatophytoses, although *Microsporum* occurs occasionally. There are several cultural variants of *Trichophyton mentagrophytes,* a circumstance that leads to confusion in the literature and in the diagnostic laboratory.

Transmission. Although clinical infections are uncommon, asymptomatic carriers of the dermatophyte are not and pose a continuing threat to both animals and caretakers. Dermatophytes are transmitted easily by direct contact with spores on hair coats, bedding, and soil.

Predisposing Factors. Husbandry, nutritional, and environmental or internal stress factors that increase exposure or reduce resistance predispose animals to dermatophytoses. Genetic background, overcrowding, heat and humidity, ectoparasitism, youth, old age, and pregnancy have all been implicated as predisposing factors.

Clinical Signs. Most animals with *T. mentagrophytes* apparently have few or no clinical signs of infection. Lesions in guinea pigs usually arise on the face and spread over the back and limbs. Lesions are "sore spots": ovoid, hairless, and scaling with crusts or scabs over raised, pruritic areas. Lesions on the rat are generally over the back, whereas tail lesions are often seen in mice. Scant to complete focal hair loss, erythema, scaling skin, and scabs are signs of ringworm infection in mice. In young rabbits there may be (characteristically) 2- to 3-cm hairless areas with reddening and slight to heavy crust formations over the snout, face, forelimbs, or ears.

Diagnosis. Diagnosis may be established by microscopic examination of skin scrapings taken from the periphery of the lesion and mounted in 10% KOH under a petrolatum-ringed coverslip. Scrapings should be inoculated onto a suitable agar culture medium and cultivated aerobically at room temperature for at least 10 days. Dermatophyte test medium (DTM) with color indicators is commonly used. Specimens for routine cultural screenings can be collected by using a sterile toothbrush or surgical scrub brush to dislodge hair and cellular debris directly onto the agar surface, or the brush can be inoculated onto the agar. Histologic sections stained with periodic acid Schiff or silver will aid in confirming a diagnosis. Ultraviolet fluorescence is diagnostic for some forms of *Microsporum canis* only.

Treatment. Successful treatment of ringworm involves elimination of the organism, not just the lesions. Treatment should be undertaken only after consideration of the public health significance of the fungus.

Topical and systemic antifungal agents are available. Topical antifungal creams or lotions, such as Conofite (Pitman-Moore, Washington Crossing, NJ), are applied once daily for 2 to 4 weeks. Griseofulvin is administered at 25 mg/kg body weight daily in the water for 14 days (Hagen 1969) or in the feed (20 mg/kg of feed) for 25 days (Cheek, Patton, and Templeton 1987).

Prevention. Maintenance of high-level husbandry standards, particularly with the young, aged, pregnant, or otherwise stressed, is a protective measure. Cultural screening for dermatophytes, proper adjustment of temperature and humidity, removal of ectoparasites, culling carriers, and sterilization of contaminated equipment are other preventive measures.

Public Health Significance. *Trichophyton mentagrophytes* may infect the caretakers before it is noticed on the animals. The fungus is highly infectious for man, particularly children and the infirm.

REFERENCES

Alteras, I., and Cojocaru, I.: Human infection by *Trichophyton mentagrophytes* from rabbits. Mykosen, *12*:543–544, 1969.

Balsari, A., et al.: Dermatophytes in clinically healthy laboratory animals. Lab. Anims., *15*:75–77, 1981.

Banks, K.L., and Clarkson, T.B.: Naturally occurring dermatomycosis in the rabbit. J. Am. Vet. Med. Assoc., *151*:926–929, 1967.

Carroll, H.F.: Evaluation of dermatophyte-test medium for diagnosis of dermatophytosis. J. Am. Vet. Med. Assoc., *165*:192–195, 1974.

Cheeke, P.R., Patton, M.M., and Templeton, G.S.: Rabbit Production. 6th Edition. Danville, IL, Interstate Printers and Publishers, Inc., 1987.

Fischman, O., DeCamargo, Z.P., and Grinblat, M.: *Trichophyton mentagrophytes* infection in laboratory white mice. Mycopathologia, *59*:113–115, 1976.

Gentles, J.C.: Experimental ringworm in guinea pigs: Oral treatment with griseofulvin. Nature (London), *182*:476–477, 1958.

Hagen, K.W.: Ringworm in domestic rabbits: oral treatment with griseofulvin. Lab. Anim. Care, *19*:635–638, 1969.

Kuttin, E.S., Beemer, A.M., and Amani, H.: *Trichophyton mentagrophytes* infection in rabbits successfully treated with a polyvinyl iodine solution. Lab. Anim. Sci., *26*:960, 1976.

McAleer, R.: An epizootic in laboratory guinea pigs due to *Trichophyton mentagrophytes*. Aust. Vet. J., *56*:234–236, 1980.

McAlister, H.A.: Dermatophytes and dermatophytoses. *In* Diagnostic Procedures in Veterinary Microbiology. 2nd Edition. Edited by G.R. Carter. Springfield, IL. Charles C Thomas, 1973, pp. 205–219.

Pombier, E.C., and Kim, J.C.S.: An epizootic outbreak of ringworm in a guinea-pig colony caused by *Trichophyton mentagrophytes*. Lab. Anims., *9*:215–221, 1975.

Post, K., and Saunders, T.R.: Topical treatment of experimental ringworm in guinea-pigs with griseofulvin in dimethylsulfoxide. Can. Vet. J., *20*:45–48, 1979.

Povar, M.L.: Ringworm (*Trichophyton mentagrophytes*) infection in a colony of albino Norway rats. Lab. Anim. Care, *15*:264–265, 1965.

Smith, J.M.B.: Diseases of laboratory animals—Mycotic. *In* C.R.C. Handbook of Laboratory Animal Science. Edited by E.C. Melby, Jr. and N.H. Altman. Cleveland, CRC Press, Inc., 1974, Vol. II, pp. 331–344.

Sprouse, R.F.: Mycoses. *In* Biology of the Guinea Pig. Edited by J.E. Wagner and P.J. Manning. Orlando, Academic Press, 1976, pp. 153–161.

Vogel, R.A., and Timpe, A.: Spontaneous *Microsporum audouinii* infection in a guinea pig. J. Invest. Dermatol., *28*:311–312, 1957.

Weisbroth, S.H., and Scher, S.: *Microsporum gypseum* dermatophytosis in a rabbit. J. Am. Vet. Med. Assoc., *159*:629–634, 1971.

Young, C.: *Trichophyton mentagrophytes* infection of the Djungarian hamster (*Phodopus sungorus*). Vet. Rec., *94*:287–289, 1974.

ENCEPHALITOZOONOSIS (NOSEMATOSIS)

Hosts. Rabbits are the principal host among small mammals, although mice, rats, guinea pigs, and hamsters may be infected. The disease is common in rabbits and is a frequent research complication.

Etiology. The infection is caused by *Encephalitozoon cuniculi*, a microsporidium that is an obligate, intracellular, protozoan parasite.

Transmission. Urine-oral passage is the most important route of transmission in a rabbit colony. The major exchange occurs between the doe and her young. The organism enters the intestine, passes in phagocytes to the blood, and is distributed to other organs. Spores appear in the kidneys approximately 35 days postinfection.

Predisposing Factors. The severity of the infection, as with other host-parasite interaction, depends on host resistance and infective dose. Young rabbits, whose maternal antibodies subside at about 4 weeks, are at an increased risk to develop the clinical disease if they are housed in unsanitary conditions and eat or drink from crocks, which are readily contaminated with spore-bearing urine from infected dams.

Clinical Signs. Most cases of encephalitozoonosis are chronic and subclinical and therefore diagnosed only on postmortem examination. In some cases, retarded growth, tremors, torticollis, paresis, convulsions, and death occur weeks after ingestion of infective spores. Typically one or several members of a litter have progressive weight loss and become runted. Death may ensue. No signs of renal impairment have been reported. IgG decreases and IgM increases during the course of an infection.

Necropsy Signs. *Encephalitozoon* has a predilection for the kidney and later the brain during an infection, although the lungs, heart, and liver may be transiently affected during the first few weeks after ingestion. The infection is terminated between 40 and 70 days in the kidney, by which time lesions are forming in the brain.

In the earlier or more acute stages of the disease, the kidneys may be moderately enlarged and may have small white areas in the cortex. Chronic renal infection is evidenced by numerous, randomly scattered, small (1- to 3-mm) pits on the cortical surface. The pits, caused by tissue destruction and scarring, remain for the life of the animal. Brain lesions are not grossly evident.

Microscopic lesions include interstitial nephritis with granulomas, and in the brain foci of necrosis surrounded by lymphocytes, plasma cells, microglia, and epithelioid cells appear, as well as extensive perivascular cuffing, meningitis, and focal granulomas.

Diagnosis. *Encephalitozoon* infections can be detected by several diagnostic tests, including an indirect fluorescent antibody test, a complement fixation test, an india ink test, and an intradermal skin test using an *Encephalitozoon* antigen. The production of ascitic fluid (containing *E. cuniculi*) in susceptible mice following an intraperitoneal injection of infected tissue or the histologic observation of characteristic focal granulomas and perivascular cuffing in the renal interstitium and brain are also useful tests. The organisms stain well with Goodpasture's stain and appear as oval or crescent-shaped trophozoites 1 nm in length or in pseudocysts 8 to 12 nm in diameter.

An ELISA or other sensitive serologic test could be used to screen animal colonies for encephalitozoonosis. In most cases histopathologic findings and laboratory findings remain the preferred way of diagnostic monitoring for encephalitozoon infections. Lesions of encephalitozoonosis may remain throughout the life of a rabbit. Histopathologists must be acutely aware of residual lesions of encephalitozoonosis in the brain and kidney of rabbits.

Treatment. No effective antibiotic treatment exists.

Prevention. Selection of rabbits known free of *E. cuniculi* is the best method to prevent colony infection. Urine and fecal transmission are reduced if sipper tubes or automatic waterers and hopper feeders are used and cages are kept clean.

Elimination of an infection from an existing colony is difficult. Extensive serologic testing and culling of seropositive animals and good sanitary procedures using disinfectants and hot water reduce or eliminate infective spores. Spores may remain alive for 4 weeks in dry conditions.

Public Health Significance. A few cases of suspected *Encephalitozoon* encephalitis have been reported in man, but there is considerable evidence that the organism is not pathogenic for man.

REFERENCES

Bywater, J.E.C.: Is encephalitozoonosis a zoonosis? Lab. Anim., *13*:149–151, 1979.

Bywater, J.E.C., and Kellett, B.S.: *Encephalitozoon cuniculi* antibodies in a specific pathogen-free rabbit unit. Infect. Immun., *21*:360–364, 1978.

Bywater, J.E.C., and Kellett, B.S.: The eradication of *Encephalitozoon cuniculi* from a specific pathogen free rabbit colony. Lab. Anim. Sci., *28*:402–404, 1978.

Cox, J.C.: Altered immune responsiveness associated with *Encephalitozoon cuniculi* infection in rabbits. Infect. Immun., *15*:392–395, 1977.

Cox, J.C., and Gallichio, H.A.: Serological and histological studies on adult rabbits with recent, naturally acquired encephalitozoonosis. Res. Vet. Sci., *24*:260–261, 1978.

Cox, J.C., Hamilton, R.C., and Attwood, H.D.: An investigation of the route and progression of *Encephalitozoon cuniculi* infection in adult rabbits. J. Protozool., *26*:260–265, 1979.

Cox, J.C., Horsburgh, R., and Pye, D.: Simple diagnostic test for antibodies to *Encephalitozoon cuniculi* based on enzyme immunoassay. Lab. Anims., *15*:41–43, 1981.

Cox, J.C., and Pye, D.: Serodiagnosis of nosematosis by immunofluorescence using cell-culture-grown organisms. Lab. Anims., 9:297–304, 1975.

Flatt, R.E., and Jackson, S.J.: Renal nosematosis in young rabbits. Pathol. Vet., 7:492–497, 1970.

Gannon, J.: The course of infection of Encephalitozoon cuniculi in immunodeficient and immunocompetent mice. Lab. Anims., 14:189–192, 1980.

Gannon, J.: The immunoperoxidase test diagnosis of Encephalitozoon cuniculi in rabbits. Lab. Anims., 12:125–127, 1978.

Goodman, D.G., and Garner, F.M.: A comparison of methods for detecting Nosema cuniculi in rabbit urine. Lab. Anim. Sci., 22:568–572, 1972.

Howell, J. McC., and Edington, N.: The production of rabbits free from lesions associated with Encephalitozoon cuniculi. Lab. Anims., 2:143–146, 1968.

Hunt, R.D., King, N.W., and Foster, H.L.: Encephalitozoonosis: Evidence for vertical transmission. J. Infect. Dis., 126:212–214, 1972.

Kellett, B.S., and Bywater, J.E.C.: The indirect india-ink immunoreaction for detection of antibodies to Encephalitozoon cuniculi in rat and mouse serum. Lab. Anims., 14:83–86, 1980.

Kunstyr, I., and Naumann, S.: Head tilt in rabbits caused by pasteurellosis and encephalitozoonosis. Lab. Anims., 19:208–213, 1985.

Lyngset, A.: A survey of serum antibodies to Encephalitozoon cuniculi in breeding rabbits and their young. Lab. Anim. Sci., 30:558–561, 1980.

Mollet, T.: A survey on toxoplasmosis and encephalitozoonosis in laboratory animals. Z. Versuchstierkd., 10:27–38, 1968.

Owen, D.G., and Gannon, J.: Investigation into the transplacental transmission of Encephalitozoon cuniculi in rabbits. Lab. Anims., 14:35–38, 1980.

Pakes, S.P.: Protozoal diseases. In The Biology of the Laboratory Rabbit. Edited by S.H. Weisbroth, R.E. Flatt, and A.L. Kraus. Orlando, Academic Press, 1974, pp. 263–386.

Pakes, S.P., Shadduck, J.A., and Olsen, R.G.: A diagnostic skin test for encephalitozoonosis (nosematosis) in rabbits. Lab. Anim. Sci., 22:870–877, 1972.

Pakes, S.P., et al.: Comparison of tests for the diagnosis of spontaneous encephalitozoonosis in rabbits. Lab. Anim. Sci., 34:356–359, 1984.

Pye, D., and Cox, J.C.: Isolation of Encephalitozoon cuniculi from urine samples. Lab. Anims., 11:233–234, 1977.

Pye, D., and Cox, J.C.: Simple focus assay for Encephalitozoon cuniculi. Lab. Anims., 13:193–195, 1979.

Ruge, H.: [Encephalitozoon in the guinea pig.] Zentralbl. Bacteriol. Hyg., I. Abt. Orig., 156:543–544, 1951.

Shadduck, J.A.: Nosema cuniculi. In vitro isolation. Science, 166:516–517, 1969.

Shadduck, J.A., and Geroulo, M.J.: A simple method for the detection of antibodies to Encephalitozoon cuniculi in rabbits. Lab. Anim. Sci., 29:330–334, 1979.

Shadduck, J.A., and Pakes, S.P.: Encephalitozoonosis (nosematosis) and toxoplasmosis. Am. J. Pathol., 64:657–674, 1971.

Shadduck, J.A., et al.: Animal infectivity of Encephalitozoon cuniculi. J. Parasitol., 65:123–129, 1979.

Waller, T.: Growth of Nosema cuniculi in established cell lines. Lab. Anims., 9:61–68, 1975.

Waller, T.: Sensitivity of Encephalitozoon cuniculi to various temperatures, disinfectants and drugs. Lab. Anims., 13:227–230, 1979.

Waller, T.: The india-ink immunoreaction: a method for the rapid diagnosis of encephalitozoonosis. Lab. Anims., 11:93–97, 1977.

Waller, T., Morein, B., and Fabiansson, E.: Humoral immune response to infection with Encephalitozoon cuniculi in rabbits. Lab. Anims., 12:145–148, 1978.

Waller, T., and Bergquist, N.R.: Rapid simultaneous diagnoses of toxoplasmosis and encephalitozoonosis by carbon immunoassay. Lab. Anim. Sci., 32:515–517, 1982.

Wilson, J.M.: The biology of Encephalitozoon cuniculi. Med. Biol., 57:84–101, 1979.

Wosu, N.J., et al.: Diagnosis of encephalitozoonosis in experimentally infected rabbits by intradermal and immunofluorescence tests. Lab. Anim. Sci., 27:210–216, 1977.

Wright, J.H., and Craighead, E.M.: Infectious motor paralysis in young rabbits. J. Exp. Med., 36:135–140, 1922.

Yost, D.H.: Encephalitozoon infection in laboratory animals. J. Natl. Cancer Inst., 20:957–963, 1958.

ENTEROPATHY (NONSPECIFIC)

Hosts. Enteropathies of undetermined or controversial origin occur frequently in rabbits and rodents, as well as in other animals.

Etiology. Most of the common enteropathies of small domestic mammals result from alterations in the populations of intestinal microbes. These alterations often follow changes in the intestinal environment. These changes may include osmotic pressure, pH, salinity, nutritional substrate, and roughage content. Other factors that may alter microbial function and numbers include changes in the host's immune defenses, intestinal motility, and antimicrobial activity. Qualitative and quantitative changes in microbial populations may lead to hypermotility or atony, increased permeability, hypersecretion, or malabsorption, any one of which can produce a clinical abnormality.

The roles of specific microbes in the genesis of diarrhea are incompletely known. Even such well-described enteric conditions as salmonellosis, coccidiosis, Tyzzer's disease, and clostridial infections may have etiologies involving synergistic actions of two or more agents. Enterotoxemia, colibacillosis, and mucoid enteropathy are poorly defined conditions in rabbits and rodents, as are rotaviral, adenoviral, and chlamydial infections. Anaerobic infections are even less well understood. The roles of *Clostridium* and *Campylobacter* spp. are only beginning to be appreciated. Umemura et al. (1982) reported on the occurrence of histiocytic enteritis associated with *Campylobacter* spp. infections in rabbits.

Transmission. The spread of enteropathogenic agents is assumed to be by the fecal-oral route, although reports of spontaneous enteropathies in rabbits, especially those occurring in adult rabbits, do not suggest spread through a colony. These conditions may arise from circumstances within susceptible individuals.

Predisposing Factors. Enteropathies often occur in stressed and weanling animals, although stress or other causes of immunosuppression are certainly not always obvious prerequisites. Other contributory factors are age, diet, and concurrent infection. The circumstances commonly contributing to weanling diarrhea involve the establishment of a potentially pathogenic microflora in the relatively uninhabited gut of the nursing animal. Without the protective effects of milk components and benign microbes, the aberrant flora proliferates. Other predisposing factors include the winter season, pregnancy and lactation, concurrent disease, unsanitary conditions, carriers of disease agents in the colony, dietary changes, overfeeding of high starch feed, antibiotics, lack of water or feed, and dietary fiber below 6%.

Clinical Signs. Animals with enteropathies are usually 3 to 10 weeks old. Signs include anorexia, lethargy, and often diarrhea. Enteropathies in rabbits and rodents often have an acute, abrupt onset with recovery or death occurring within a few days or weeks, although prolonged intestinal disturbances occur.

In peracute cases a rabbit may die suddenly with no obvious signs, or it may pass a massive puddle of fecal slush and die. In acute cases constipation or explosive, smelly, fluid diarrhea occurs. The diarrhea, which may be brown, green, or yellow and contain blood or mucus, stains the rear quarters of the animal. In cases of longer duration, the feces may be soft or tarry, smelly, and sticky and may continue to smear the rabbit and cage for weeks.

Signs associated with the acute condition include bloat, bubbling sounds, tooth grinding, squinting, weakness, anorexia, polydipsia or the opposite, rough hair, altered temperature, and convulsions and death. Chronic cases are more a nuisance than a threat to the animal's life, although progressive weight loss does occur.

Necropsy Signs. Necropsy signs of an enteropathy vary from inapparent to a dried or fluid content in various sections of the gut. Intestinal hemorrhage, edema, hyperemia, and necrosis occur, although the most common sign is a color different from the normal gray-green of the cecum and intestine. Gas and mucous accumulation are commonly noted in the intestinal lumen, the mesenteric lymph nodes are swollen and reddened, the liver is pale, and the gallbladder is distended with bile.

Treatment. Among rabbit and rodent owners and their veterinarians there is naturally a strong desire to treat the common and often fatal enteropathies. There is an understandable imperative to "do something" rather than just stand back and allow valuable or beloved animals to die. All parties should understand, however, that diarrheal or enteric diseases in rabbits and rodents are prevented through good and appropriate husbandry and that treatments for these conditions are often ineffective.

It is common practice among rabbit raisers to use food additives prophylactically and therapeutically for enteropathies. Such substances include blackberry leaves and comfrey, vinegar in the water, yogurt, plasterboard, salt, and tea. Many breeders attest to the effectiveness of these products, but few, if any, controlled clinical trials have been conducted on the medical uses of these substances. The common practice of adding dietary fiber to rabbit diets may cause hypoglycemia, malnutrition, or caloric deprivation in the young or in dwarf breeds.

Other practices are less dramatic. Some people withhold water or feed for several hours or a few days, whereas others simply change the diet. Over-the-counter antibiotics and antimicrobials are popular and are usually given for extended periods, even if the antibiotics themselves eventually cause the diarrhea. These drugs are not approved for use in rabbits, their withdrawal times in rabbits are unknown, they are of questionable effectiveness, and they are poor substitutes for good preventive husbandry.

Symptomatic treatment of diarrhea in small pets or valuable research or breeding animals can be more easily justified, although the treated animal should be kept separate from the susceptible herd. Pectin and kaolin are useful with fluid therapy (electrolytes and glucose). Steroids for toxemia, antispasmodics for hypermotility, and even antibiotics may help (Banerjee, et al 1987).

Prevention. Prevention of enteropathies resides in good husbandry. Use of self-cleaning, wire-floored rabbit cages, hopper feeders, sipper-tube waterers, healthy stock, quality diet, fresh water, good ventilation, exclusion of vermin, proper use of disinfectants, culling of sick animals, and constant surveillance will greatly reduce enteric problems.

A plethora of preventive measures can be used, but of these additional measures, added roughage (small handful of oats or hay daily), sulfaquinoxaline at 0.025 to 0.03% in the feed and water from the time the animal is 3 to 8 weeks of age, and lactobacilli in the feed or water are the only procedures with some reputation for occasional success.

Other prophylactic regimens are mentioned only to emphasize that in almost all controlled studies involving antimicrobials morbidity is rarely lowered. Mortality, however, is sometimes decreased. Colimy-

cin (0.1 g/kg diet), tetracycline (0.15 g/kg diet), and nitrofurantoin (0.45 g/kg diet) administered alternately over several weeks have been used to prevent an enteropathy (Meshorer 1976), as has oxytetracycline at 600 ppm in the feed (Kruijt 1976). Oxytetracycline plus neomycin and chlortetracycline plus vitamin B_{12} also have adherents (Sinkovics 1978, Sinkovics 1978). Umemura et al. (1982) reported that neomycin administered at 50 mg/rabbit *per os* for 5 days was effective in reducing diarrhea associated with *Campylobacter* spp. infections.

Public Health Significance. Because the etiologies of these enteric syndromes are unknown, the significance for human handlers is also unknown.

REFERENCES

Banerjee, A.K., et al.: Acute diarrheal disease in rabbit: Bacteriological diagnosis and efficacy of oral rehydration in combination with loperamide hydrochloride. Lab. Anims., *21*:314–317, 1987.

Bryner, J.H., et al.: Infectivity of three *Vibrio fetus* biotypes for gallbladder and intestines of cattle, sheep, rabbits, guinea pigs, and mice. Am. J. Vet. Res., *32*:465–470, 1971.

Caldwell, M.B., and Walker, R.I.: Adult rabbit model for *Campylobacter* enteritis. Am. J. Pathol., *122*:573–576, 1986.

Cheeke, P.R., and Patton, N.M.: Effect of alfalfa and dietary fiber on the growth performance of weaning rabbits. Lab. Anim. Sci., *28*:167–172, 1978.

Dubluzeau, R., et al.: Digestive tract microflora in healthy and diarrheic young hares born in captivity. Effect of intake of different antibiotics. Ann. Biol. Anim. Biochem. Biophys., *15*:529–539, 1975.

Fuller, R., and Moore, J.H.: The effect on rabbit intestinal microflora of diets which influence serum cholesterol levels. Lab. Anims., *5*:25–30, 1971.

Kruijt, B.C.: Enteritis in a conventional rabbit colony. Lab. Anims., *10*:189–194, 1976.

Meshorer, A.: Histological findings in rabbits which died with symptoms of mucoid enteritis. Lab. Anims., *10*:199–202, 1976.

Nikkels, R.J., Mullink, J.W.M.A., and Van Vliet, J.C.J.: An outbreak of rabbit enteritis: Pathological and microbiological findings and possible therapeutic regime. Lab. Anims., *10*:195–198, 1976.

Prescott, J.F.: Intestinal disorders and diarrhea in the rabbit. Vet. Bull., *48*:475–480, 1978.

Sinkovics, G.: Rabbit dysentery: 1 clinical, epizootological and bacteriological studies. Vet. Rec., *103*:326–328, 1978.

Sinkovics, G.: Rabbit dysentery: 2 therapeutic experiments. Vet. Rec., *103*:328–331, 1978.

Sinkovics, G.: Rabbit dysentery: 3 diagnostic differentiation. Vet. Rec., *103*:331–332, 1978.

Smith, H.W.: The antimicrobial activity of the stomach

contents of suckling rabbits. J. Pathol. Bacteriol., 91:1–9, 1966.

Smith, H.W., and Crabb, W.E.: The faecal bacterial flora of animals and man: Its development in the young. J. Pathol. Bacteriol., 82:53–66, 1961.

SultanDosa, A.B., Bryner, J.H., and Foley, J.W.: Pathogenicity of Campylobacter jejuni and Campylobacter coli strains in the pregnant guinea pig model. Am. J. Vet. Res., 44:2175–2178, 1983.

Taylor, D.E., and Bryner, J.H.: Plasmid content and pathogenicity of Campylobacter jejuni and Campylobacter

coli strains in the pregnant guinea pig model. Am. J. Vet. Res., 45:2201–2202, 1984.

Umemura, T., et al.: Histiocytic enteritis of rabbits. Vet. Pathol., 19:326–329, 1982.

Wells, C.L., and Balish, E.: Gastrointestinal ecology and histology of rats monoassociated with anaerobic bacteria. Appl. Environ. Microbiol., 39:265–267, 1980.

Whitney, J.C.: Treatment of enteric disease in the rabbit. Vet. Rec., 95:533, 1974.

Zumpt, I.: Treatment of enteric disease in the rabbit. Vet. Rec., 96:345–346, 1975.

ENTEROTOXEMIA (CLOSTRIDIAL)

Hosts.   Sporadic outbreaks of enterotoxemia attributed to *Clostridium* spp. occur in rabbits. Hamsters and guinea pigs given certain antibiotics may develop a fatal enterotoxemia after *Clostridium* proliferation. Administration of lincomycin or clindamycin may be associated with development of enterotoxemia in rabbits.

Etiology.   Enterotoxemia of suspected clostridial origin has only recently and tentatively been distinguished from the poorly defined diarrheal conditions of rabbits grouped into the "rabbit enteropathy complex," a designation that remains useful. Although *Clostridium spiroforme* and its iota-like toxin have been implicated as the cause of the clinical disease enterotoxemia, other bacteria, protozoa, and even viruses may alone or in combinations be involved. The pathogenic action of *Clostridium* on the animal is effected by toxin-induced changes in intestinal and vascular permeability. These effects may persist for several days following the onset of diarrhea and presumably after the starch substrate has been degraded.

Outbreaks of fatal diarrhea within 3 to 6 days following antibiotic administration to guinea pigs and hamsters have been attributed to *Clostridium difficile* and *C. sordellii*, organisms known from humans, hares, and rodents. *Clostridium spiroforme* causes fatal enterotoxemia in weanling or adult rabbits with gut flora disturbed by antibiotic treatment. *C. spiroforme* may be present in the environment of many colonies. References on antibiotic-induced enterotoxemias are included at the end of this section and in the section on colibacillosis.

Transmission.   Clostridia are variably present in a benign state and in low populations in the large intestine of rabbits and rodents. Transmission would be by the fecal-oral route, although clinical cases likely arise from proliferation of organisms resident within the patient's own cecum.

Predisposing Factors.   Enterotoxemia in rabbits usually occurs in weanling fryers 4 to 8 weeks old, when passively acquired maternal immunity has decreased, body growth is rapid, and intestinal microbes are populating the intestine. The disease, however, may also affect adults on high caloric diets or does returned to full feed immediately after kindling.

The postulated, primary predisposition to the disease is dietary: digestible starches contained in high-grain feeds arrive in the large intestine in excessive quantities and provide an environment for clostridial proliferation. Other predisposing factors include tight packing of the food bolus in the stomach with sequestering and protection of starch and bacteria, excessive acid in the duodenum inhibiting amylase activity, altered intestinal pH, and hypermotility.

The success of high-fiber diets in the reduction of rabbit diarrhea may be due to the inverse relationship between cereal grains and roughage in rabbit feed and to the looser boluses formed from coarse roughage. Other factors that might contribute to the disease signs are changes in feed quality, anorexia, the use of antibiotics, environmental stress, and rotavirus infections. Breeding does are especially susceptible to clostridial enterotoxemia during the first few weeks postpartum if they change from restricted feeding during pregnancy to *ad libitum* feeding at kindling.

Clinical Signs. The principal sign of enterotoxemia is an acute onset of profuse, green or brown, watery diarrhea in young rabbits. Dehydration and death occur within 6 to 72 hours from the first diarrheal sign. Anorexia, polydipsia, depression, fever, bloat, intestinal sounds, fecal mucus, fecal soiling of fur, and a fetid odor are other signs.

The young doe syndrome occurs in primiparous and multiparous does and may be related to clostridial enterotoxemia. Affected rabbits die suddenly when their litters are 1 to 4 weeks old. A few days preceding death, the does go off feed and some salivate profusely. Fatal septicemia derived from a staphylococcal mastitis causes similar signs.

The cause of neonatal or milk enterotoxemia is uncertain, but it may be caused by a toxin passed in low amounts through the milk or by clostridia passed into the neonatal gut. Young die suddenly during the first 3 days postpartum. This problem, which can produce great economic losses in a rabbitry, is more common during the winter months.

Necropsy Signs. Signs of enterotoxemia include brain, subserosal, and submucosal petechial and ecchymotic hemorrhages and edema of the cecum and adjacent ileum and colon. The intestinal wall is thickened, and the gut lumen contains a green, brown, or yellow mucoid or fluid material and gas. Necropsy signs of the young doe syndrome either are inconclusive or are as those just described.

Diagnosis. Anaerobic culture of affected portions of the gastrointestinal tract may result in recovery of clostridial species, although this does not necessarily establish a cause-and-effect relationship. Cell-free extracts of cecal contents can be introduced intraperitoneally into mice, and the mice can be examined later for toxic effects, including death. Histopathologic examination may reveal cecal lesions.

Treatment. Treatment of rabbit diarrhea is discussed in this chapter under "Enteropathy, Nonspecific."

Prevention. General preventive measures appropriate for enterotoxemia in rabbits are discussed under "Enteropathy, Nonspecific" in this chapter. More specifically, prevention of enterotoxemia probably involves feeding growing rabbits a diet with some grain replaced by roughage; this change would require supplementing pelleted rabbit feed with hay, feeding a higher fiber feed, reducing the carbohydrate component of the feed a few percentage points, and reducing daily caloric intake. Copper sulfate mixed thoroughly into the feed at 250 ppm (2 pounds per ton) and fed continuously is a promising but relatively untested preventive measure.

The young doe syndrome can be prevented by increasing the feed after kindling by 10 to 30 g per day, rather than all at once, until *ad libitum* levels of approximately 450 g per day are reached.

Enterotoxemia in hamsters following antibiotic administration is avoided by giving only those antibiotics (neomycin, chloramphenicol, and perhaps the tetracyclines at low doses) that suppress both gram-negative and gram-positive bacteria.

Public Health Significance. Whereas *C. perfringens* affects a variety of species, *C. spiroforme* is primarily a rabbit pathogen.

REFERENCES

Bartlett, J.G., et al.: Antibiotic-induced lethal enterocolitis in hamsters: studies with eleven agents and evidence to support the pathogenic role of toxin-producing clostridia. Am. J. Vet. Res., *39*:1525–1530, 1978.

Baskerville, M., Wood, M., and Seamer, J.H.: *Clostridium perfringens* type E enterotoxaemia in rabbits. Vet. Rec., *107*:18–19, 1980.

Borriello, S.P., and Carman, R.J.: Association of iota-like toxin and *Clostridium spiroforme* with both spontaneous and antibiotic-associated diarrhea and colitis in rabbits. J. Clin. Microbiol., *17*:414–418, 1983.

Brooks, D.L.: Endemic diarrhea of domestic rabbits in California. Dissertation Abst. Inter., *39B*:5853, 1979.

Carman, R.J., and Borriello, S.P.: Laboratory diagnosis of *Clostridium spiroforme*-mediated diarrhea (iota enterotoxaemia) of rabbits. Vet. Rec., *113*:184–185, 1983.

Carman, R.J., and Borriello, S.P.: Infectious nature of

Clostridium spiroforme-mediated rabbit enterotoxaemia. Vet. Microbiol., 9:497–502, 1984.

Cheeke, P.R., and Patton, N.M.: Carbohydrate-overload of the hindgut—a probable cause of enteritis. J. Appl. Rabbit Res., 3:20–23, 1980.

Clapp, H.W., and Graham, W.R.: An experience with Clostridium perfringens in cesarean derived barrier sustained mice. Lab. Anim. Care, 20:1081–1086, 1970.

Dubos, F., et al.: Immediate postnatal inoculation of a microbial barrier to prevent neonatal diarrhea induced by Clostridium difficile in young conventional and gnotobiotic hares. Am. J. Vet. Res., 45:1242–1244, 1984.

Duncan, C.L., and Strong, D.H.: Experimental production of diarrhea in rabbits with Clostridium perfringens. Can. J. Microbiol., 15:765–770, 1969.

Eaton, P., and Fernie, D.S.: Enterotoxaemia involving Clostridium perfringens iota toxin in a hysterectomy-derived rabbit colony. Lab. Anims., 14:347–351, 1980.

Fekety, R., et al.: Antibiotic-associated colitis: effects of antibiotics on Clostridium difficile and the disease in hamsters. Rev. Infect. Dis., 1:386–396, 1979.

Harris, D.J., Cheeke, P.R., and Patton, N.M.: The effect of dietary buffers on growth of weanling rabbits and incidence of enteritis. J. Appl. Rabbit Res., 5:12–13, 1982.

Harris, I.E., and Portas, B.H.: Enterotoxaemia in rabbits caused by Clostridium spiroforme. Aust. Vet. J., 62:342–343, 1985.

Knoop, F.C.: Clindamycin-associated enterocolitis in guinea pigs: evidence for a bacterial toxin. Infect. Immun., 23:31–33, 1979.

Lelkes, L.: A review of rabbit enteric diseases: A new perspective. J. Appl. Rabbit Res., 10:55–61, 1987.

Lowe, B.R., Fox, J.G., and Bartlett, J.G.: Clostridium difficile-associated cecitis in guinea pigs exposed to penicillin. Am. J. Vet. Res., 41:1277–1279, 1980.

Madden, D.L., Horton, R.E., and McCullough, N.B.: Spontaneous infection in ex-germfree guinea pigs due to Clostridium perfringens. Lab. Anim. Care, 20:454–455, 1970.

Patton, N.M.: Young doe syndrome. J. Appl. Rabbit Res., 2:11–12, 1979.

Patton, N.M., and Cheeke, P.R.: Etiology and treatment of young doe syndrome. J. Appl. Rabbit Res., 3:23–24, 1980.

Patton, N.M., et al.: Enterotoxemia in rabbits. Lab. Anim. Sci., 28:536–540, 1978.

Peeters, J.E., et al.: Significance of Clostridium spiroforme in the enteritis-complex of commercial rabbits. Vet. Microbiol., 12:25–31, 1986.

Rehg, J.E., and Lu, Y-S.: Clostridium difficile colitis in a rabbit following antibiotic therapy for pasteurellosis. J. Am. Vet. Med. Assoc., 179:1296–1297, 1981.

Rehg, J.E., and Pakes, S.P.: Clostridium difficile antitoxin neutralization of cecal toxin(s) from guinea pigs with penicillin-associated colitis. Lab. Anim. Sci., 31:156–160, 1981.

Rehg, J.E., Yarbrough, B.A., and Pakes, S.P.: Toxicity of cecal filtrates from guinea pigs with penicillin-associated colitis. Lab. Anim. Sci., 30:524–531, 1980.

Rehg, J.E., and Pakes, S.P.: Implication of Clostridium difficile and Clostridium perfringens iota toxins in experimental lincomycin-associated colitis of rabbits. Lab. Anim. Sci., 32:253–257, 1982.

Schneierson, S.S., and Perlman, E.: Toxicity of penicillin for the Syrian hamster. Proc. Soc. Biol. Med., 91:229–230, 1956.

Thilsted, J.P., et al.: Fatal diarrhea in rabbits resulting from the feeding of antibiotic-contaminated feed. J. Am. Vet. Med. Assoc., 179:360–361, 1981.

Yonushonis, W.P., et al.: Diagnosis of spontaneous Clostridium spiroforme iota enterotoxemia in a barrier rabbit breeding colony. Lab. Anim. Sci., 37:69–71, 1987.

EPILEPSY IN GERBILS

Etiology. Pet owners, animal technicians, and others not familiar with gerbils may become alarmed on witnessing, for the first time, the spontaneous seizures so common in gerbils. These epileptiform seizures are precipitated by novel or excitement-producing experiences such as handling. There appear to be seizure-resistant and seizure-sensitive strains of gerbils. Incidence of seizure susceptibility in natural populations is approximately 20% but approaches 100% in some groups.

Clinical Signs. Seizures vary from a mild hypnotic state characterized by cessation of activity and twitching of vibrissae and pinnae to severe myoclonic convulsions followed by tonic extensor rigidity and death. Seizures last from about 30 seconds to 2 minutes. Gerbils begin having seizures at about 45 days of age, although the onset is extremely variable. A refractory period of several days usually follows a severe seizure.

Prevention and Treatment. Frequent handling from an early age may suppress development of seizures. Anticonvulsant drugs may cause fatalities and are therefore contraindicated.

REFERENCES

Cox, B., and Lomax, P.: Brain amines and spontaneous epileptic seizures in the Mongolian gerbil. Pharmacol. Biochem. Behav., 4:203–267, 1976.

Goldblatt, D.: Seizure disorder in gerbils. Neurology, 18:303–304, 1968.

Goldblatt, D., et al.: The effect of anticonvulsants on seizures in gerbils. Neurology, 21:433–434, 1971.

Harriman, A.E.: "Spontaneous" seizing in openfield

tests by Mongolian gerbils fed magnesium at different rates. Percept. Mot. Skills, *47*:1031–1035, 1978.

Kaplan, H.: What triggers seizures in the gerbil, *Meriones unguiculatus?* Life Sci., *17*:693–698, 1975.

Kaplan, H., and Miezejeski, C.: Development of seizures in the Mongolian gerbil (*Meriones unguiculatus*). J. Comp. Physiol. Psychol., *81*:267–273, 1972.

Kaplan, H., and Silverman, W.P.: Early experience affects seizure latency and post seizure recovery time in the Mongolian gerbil. Neuropsychologica, *16*:649–652, 1978.

Loskota, W.J., and Lomax, P.: Mongolian gerbil as an animal model for studies of the epilepsies: anticonvulsant screening. Proc. West. Pharmacol. Soc., *17*:40–45, 1974.

Loskota, W.J., and Lomax, P.: The Mongolian gerbil (*Meriones unguiculatus*) as a model for the study of the epilepsies: EEG records of seizures. Electroencephalogr. Clin. Neurophysiol., *38*:597–604, 1975.

Loskota, W.J., Lomax, P., and Rich, S.T.: The gerbil as a model for the study of epilepsy: seizure habituation and seizure patterns. Proc. West. Pharmacol. Soc., *15*:189–194, 1972.

Loskota, W.J., Lomax, P., and Rich, S.T.: The gerbil as a model for the study of the epilepsies. Seizure patterns and ontogenesis. Epilepsia, *15*:109–119, 1974.

Paul, L.A., Schain, R.J., and Bailey, B.G.: Structural correlates of seizure behavior in the Mongolian gerbil. Science, *213*:924–926, 1981.

Robbins, M.E.C.: Seizure resistance in albino gerbils. Lab. Anim., *10*:233–235, 1976.

Schonfeld, A.R., and Glick, S.D.: Effect of handling-induced seizures and passive avoidance learning in the Mongolian gerbil (*Meriones unguiculatus*). Behav. Biol., *24*:101–106, 1978.

Thiessen, D.D., Lindzey, G., and Friend, H.C.: Spontaneous seizures in the Mongolian gerbil (*Meriones unguiculatus*). Psychonom. Sci., *11*:227–228, 1968.

Watanabe, K.S., et al.: Effects of phenobarbitol on seizure activity in the gerbil. Pediatr. Res., *12*:918–922, 1978.

HEAT STROKE

Hosts. All animals are susceptible to heat stroke, but rabbits and guinea pigs are particularly susceptible.

Etiology. In high ambient temperatures, in combination with the predisposing factors listed, thermoregulatory mechanisms fail, and the body temperature rises beyond a level compatible with life.

Predisposing Factors. Predisposing factors to heat stroke include an ambient temperature above 28° C (85° F), high humidity (above 70%), a thick hair coat, obesity, direct sunlight, poor ventilation, insufficient or warm water, crowding, and psychologic stress.

Clinical Signs. Signs of heat stress include hyperemia of peripheral vessels, rapid respiration, hyperthermia (high rectal temperature, i.e., above 41° C or 105° F), cyanosis, prostration, and death. There may be blood-tinged fluid on the nose and mouth. Food intake drops and water intake is greatly increased. Rodents have large salivary glands and respond to overheating by profuse salivation. Excess saliva escaping from the corners of the mouth to wet the body has a cooling effect through evaporation.

Heat-stressed rabbits eat and grow less, consume more water, and at temperatures of 32° C (90° F) and above for several days or even weeks males become sterile. Younger bucks (5 to 7 months) are less susceptible to heat-induced sterility.

Necropsy Signs. Lesions of heat stroke include hyperemia of the tissues, particularly of the lungs and intestinal wall.

Diagnosis. The history of acute onset in hot weather and the absence of evidence of ketosis or infectious or toxic disease support a diagnosis of heat stroke. Rodents with acute heat stroke or suffocation will have evidence of excessive salivation from the corners of the mouth. Mouths and noses of heat-prostrated rabbits may be tinged with bloody fluid. Fur may be wet and matted, and extremities (ears, tail, scrotum, and feet) may be hyperemic and cyanotic.

Treatment. Heat-stressed animals can be sprayed with water, carefully dipped into a cool bath, or wrapped in a dampened cloth until the rectal temperature returns to normal. Supportive care and steroids may be administered.

Prevention. Prevention of heat stress includes provision of shade; adequate air circulation, feed, and water; and water sprays or a container of ice to cool cages. Obese and heavily furred animals are particularly susceptible and should be conditioned or eliminated.

REFERENCES

Adolph, E.F.: Tolerance to heat and dehydration in several species of mammals. Am. J. Physiol., 151:564–575, 1947.

Besch, E.L., and Woods, J.E.: Heat dissipation biorhythms of laboratory animals. Lab. Anim. Sci., 27:54–59, 1977.

Daily, W.M., and Harrison, T.R.: A study of the mechanism and treatment of experimental heat pyrexia. Am. J. Med. Sci., 215:42–54, 1948.

March, F.: Experiments in heatstrokes in Iran. Trans. R. Soc. Trop. Med. Hyg., 32:371–394, 1938.

Oloufa, M.M., Bogart, R., and McKenzie, F.F.: Effect of environmental temperature and the thyroid gland on fertility in the male rabbit. Fertil. Steril., 2:223–229, 1951.

Pucak, G.J., Lee, C.S., and Zaino, A.S.: Effects of prolonged high temperature on testicular development and fertility in the male rat. Lab. Anim. Sci., 27:76–77, 1977.

Rathore, A.K.: High temperature exposure of male rabbits: Fertility of does mated to bucks subjected to 1 and 2 days of heat treatment. Br. Vet. J., 126:168–172, 1970.

Schall, W.D.: Heat stroke (heat stress, hyperpyrexia). In Current Veterinary Therapy. VII. Small Animal Practice. Edited by R.W. Kirk. Philadelphia, W.B. Saunders, 1980, pp. 195–197.

Young, W.C.: The influence of high temperature on the guinea-pig testis. J. Exp. Zool., 49:459–499, 1927.

HEMORRHAGIC FEVER WITH RENAL SYNDROME

Etiology.   Hemorrhagic fever with renal syndrome (HFRS) or Korean hemorrhagic fever (KHF) is a febrile disease with renal involvement that has occurred primarily among researchers and animal technicians working with rats. It is caused by infection with one of the Hantaviruses. Between 1975 and 1981, 102 cases of HFRS were detected in 15 Japanese institutions. KHF is enzootic in wild rodents (*Apodemus* spp.) in Korea. From serologic evidence, laboratory rats with enzootic HFRS were believed to be the source of the disease in Japan. There is serologic evidence that the Hantaan virus or a related Hantavirus infects most major wild rat populations in the United States.

Clinical Signs.   Clinical features of HFRS in people include acute high fever, severe malaise, myalgia, headaches, diarrhea, nausea and vomiting, proteinuria, oliguria, and possibly hemorrhagic manifestations. Clinical signs and lesions have not been reported to occur in laboratory rats, although they develop relatively high antibody titers to the agent.

Prevention.   Routes by which wild rodents might contaminate laboratory stocks must be eliminated. Laboratory work with these agents must be done under appropriate conditions for controlling hazardous agents. Monitoring for the KHF or HFRS agent should be considered when laboratory rats, *Apodemus* spp., and possibly mice are moved from Japan and Belgium (or other areas of the world where the agent may exist) to other countries and uninfected areas. Testing should be done in advance of shipment or during quarantine of the animals after arrival. Testing of sera for HFRS antibody is done by using an indirect immunofluorescent antibody (IFA) test. Serum neutralization and hemagglutination inhibition (HAI) tests have also been used. Although testing is not done on a routine basis in the United States, inquiries regarding tests may be made through the laboratories of the U.S. Army Medical Institute for Infectious Diseases, Frederick, MD, or of the Virology Section, Special Pathogen Laboratory, Communicable Disease Center, Atlanta, GA.

Public Health Significance.   While not likely to cause clinical disease in rodents, the Hantaviruses may cause serious disease and death in man. Laboratory personnel in Japan, Belgium, Korea, and the United Kingdom have contracted disease from infected laboratory rats.

REFERENCES

Kawamata, J., et al.: Control of laboratory acquired hemorrhagic fever with renal syndrome (HFRS) in Japan. Lab. Anim. Sci., 37:431–436, 1987.

Kurata, T., Tsai, T.F., Bauer, S.P., and McCormick, J.B.: Immunofluorescence studies of disseminated Hantaan virus infection in suckling mice. Infect. Immun., 41:391–398, 1983.

LeDuc, J.W.: Epidemiology of Hantaan and related viruses. Lab. Anim. Sci., 37:413–418, 1987.

Lee, H.W., Baek, L.J., and Johnson, K.M.: Isolation of Hantaan virus, the etiologic agent of Korean hemorrhagic fever from wild urban rats. J. Infect. Dis., *146*:638–644, 1982.

Lee, H.W., and Johnson, K.M.: Laboratory-acquired infections with Hantaan virus, the etiologic agent of Korean hemorrhagic fever. J. Infect. Dis., *146*:645–651, 1982.

Lee, H.W., Lee, P.W., and Johnson, K.M.: Isolation of the etiologic agent of Korean hemorrhagic fever. J. Infect. Dis., *137*:298–308, 1978.

Lee, H.W.: Korean hemorrhagic fever. Prog. Med. Virol., *28*:96–113, 1982.

Morita, C., et al.: Age-dependent transmission of hemorrhagic fever with renal syndrome (HFRS) virus in rats. Arch. Virol., *85*:145–149, 1985.

Quimby, F.W.: Zoonotic implications of Hantaan-like viruses: An introduction. Lab. Anim. Sci., *37*:411–412, 1987.

Takahashi, Y., et al.: Comparison of immunofluorescence and hemagglutination inhibition tests and enzyme-linked immunosorbent assay for detection of serum antibody in rats infected with hemorrhagic fever with renal syndrome virus. J. Clin. Microbiol., *24*:712–715, 1986.

Tsai, T.F., et al.: Preliminary evidence that Hantaan or a closely related virus is enzootic in domestic rodents. N. Engl. J. Med., *307*:623–625, 1982.

Tsai, T.F.: Serologic and virological evidence of a Hantaan virus-related enzootic in the United States. J. Infect. Dis., *152*:126–136, 1985.

Tsai, T.F.: Hemorrhagic fever with renal syndrome: Clinical aspects. Lab. Anim. Sci., *37*:419–427, 1987.

Tsai, T.F.: Hemorrhagic fever with renal syndrome: Mode of transmission to humans. Lab. Anim. Sci., *37*:428–430, 1987.

HYPOVITAMINOSIS C (SCURVY)

Hosts.   Guinea pigs and primates, including man, are among the animal groups with an absolute requirement for dietary L-ascorbic acid, a vitamin active in certain biologic oxidation and reduction systems.

Etiology.   A gene-controlled absence or deficiency in the guinea pig and primate of the hepatic enzyme L-gulonolactone oxidase, one of the several enzymes necessary for the production of L-ascorbic acid from D-glucose, necessitates an exogenous or dietary supply of the vitamin. Ascorbic acid is involved in the formation of hydroxyproline and hydroxylysine and in the metabolism of cholesterol, amino acids, and carbohydrates. A deficiency leads to fragmentation of collagen and the intercellular ground substance.

Predisposing Factors.   Age, sex, type of diet, pregnancy, lactation, concomitant disease, and environmental conditions all affect the duration of onset and the signs of ascorbic acid deficiency; guinea pigs show some signs of deficiency within 2 weeks if vitamin C is withheld.

Ascorbic acid content of feed is reduced by dampness, light, heat, and storage. Approximately 50% of the vitamin C activity in stored feed is lost in 6 weeks. In water in an open crock, up to 50% of the vitamin C activity is lost within 24 hours.

Clinical Signs.   The pet or research guinea pig with scorbutus is lethargic, weak, and anorexic, may vocalize or bite (from pain) when restrained, and often has enlarged limb joints and costochondral junctions, a rough hair coat, diarrhea, weight loss, and ocular and nasal discharge. Young, growing animals are more susceptible to bone deformities. Death from starvation or secondary infections follows in 3 to 4 weeks.

Necropsy Signs.   Gross lesions of scorbutus are related to abnormalities of bones and blood vessels. Signs, most prominent in growing animals, include hemorrhage into the subperiosteum, subcutaneous tissues, skeletal muscle, and intestine, and separation at the epidiaphyseal junction.

Microscopically, the consequences of aberrant synthesis of collagen and connective tissue are evident. In metaphyseal areas, trabeculae of the osteoid are reduced, leaving islands of calcified cartilage. Epiphyses are narrowed, osteoblastic activity is reduced, and osteoclasts are increased. Endothelial junctional defects in skeletal muscle lead to hemorrhage and fragmentation of myofilaments.

Diagnosis.   The clinical history and necropsy signs, particularly joint stiffness and hemorrhage, provide a tentative diagnosis. Marginal deficiencies, especially in adults, and early cases are often difficult to diagnose from the nonspecific signs of weakness and anorexia. Careful questioning about and examination of the diet are often necessary. Secondary bacterial or metabolic diseases may obscure subtle signs of the vitamin deficiency.

Treatment. Active ascorbic acid supplied daily, or at least every third day, via feed, water, or parenteral injection at 10 mg/kg body weight for maintenance or 30 mg/kg during pregnancy will reverse the consequences of a deficiency (Manning, Wagner, and Harkness. 1984). Guinea pigs retain a dose of ascorbic acid for approximately 4 days. Recovery usually requires less than a week. If the abrupt change in the taste of the feed or water causes the guinea pig to refuse the diet, then oral ascorbic acid drops will be necessary. Multivitamin drops containing vitamin C should not be used because of the potential for toxic overdose of other vitamins.

Prevention. Provision of adequate, fresh, stabilized vitamin C in feed or water at recommended daily levels (200 mg/L in the drinking water) will prevent hypovitaminosis C (Manning, Wagner and Harkness. 1984). The instability of the vitamin-water solution, as well as the presence of chlorine in water, warrants daily vitamin-water changes. Commercial guinea pig feeds contain about 800 mg/kg diet of vitamin C when milled. Feed must have adequate levels of vitamin C, be stored below 22° C, and used within 90 days of milling. Newer irradiated diets have longer storage lives. One hundred grams of fresh kale contains approximately 125 mg of vitamin C. Approximately 50 g (half cup) of fresh cabbage supplies 30 mg of vitamin C. Metal, hard water, and heat cause accelerated deterioration of ascorbic acid in solution. L-Ascorbic acid phosphate is more stable against oxidation or hydrolysis in neutral or alkaline solutions.

REFERENCES

Clarke, G.L., et al.: Subclinical scurvy in the guinea pig. Vet. Pathol., *17*:40–44, 1980.

Collins, M., and Elvehjem, C.A.: Ascorbic acid requirement of the guinea pig using growth and tissue ascorbic acid concentrations as criteria. J. Nutr., *64*:503–511, 1958.

Davies, J.E.W., and Hughes, R.E.: A note on the effect of food restriction on tissue ascorbic acid in guinea-pigs. Br. J. Nutr., *38*:299–300, 1977.

Denmark, S.J.: Ascorbic acid staining of scorbutic guinea pig incisors. J. Dent. Res., *45*:762–767, 1966.

Eva, J.K., Fifield, R., and Rickett, M.: Decomposition of supplementary vitamin C in diets compounded for laboratory animals. Lab. Anims., *10*:157–159, 1976.

Fullmer, H.M., Martin, G.R., and Burns, J.J.: Role of ascorbic acid in the formation and maintenance of dental structures. Ann. N.Y. Acad. Sci., *92*:286–294, 1961.

Gore, I., Fujinami, T., and Shirahama, T.: Endothelial changes produced by ascorbic acid deficiency in guinea pigs. Arch. Pathol., *80*:371–376, 1965.

Kim, J.C.S.: Ultrastructural studies of vascular and muscular changes in ascorbic acid deficient guinea-pigs. Lab. Anims., *11*:113–117, 1977.

LaDu, B.N., and Zannoni, V.C.: The role of ascorbic acid in tyrosine metabolism. Ann. N.Y. Acad. Sci., *92*:175–191, 1961.

Manning, P., Wagner, J., and Harkness, J.: Biology and diseases of guinea pigs. *In* Laboratory Animal Medicine. Edited by J. Fox, B. Cohen, and F. Loew. Orlando, Academic Press, 1984.

Navia, J.M., and Hunt, C.E.: Nutrition, nutritional diseases, and nutrition research applications. *In* The Biology of the Guinea Pig. Edited by J.E. Wagner and P.J. Manning. New York, Academic Press, 1976, pp. 235–267.

Norkus, E.P., Bassi, J., and Rosso, P.: Maternal-fetal transfer of ascorbic acid in the guinea pig. J. Nutr., *109*:2205–2212, 1979.

Nungester, W.J., and Ames, A.M.: The relationship between ascorbic acid and phagocytic activity. J. Infect. Dis., *83*:50–54, 1948.

Pirani, C.L., Bly, C.G., and Sutherland, K.: Scorbutic arthropathy in the guinea pig. Arch. Pathol., *49*:710–732, 1950.

Reid, M.E.: Guinea pig nutrition. Proc. Anim. Care Panel, *8*:23–33, 1957.

Sato, P., and Udenfriend, S.: Scurvy-prone animals, including man, monkey, and guinea pig, do not express the gene for gulonolactone oxidase. Arch. Biochem. Biophys., *187*:158–162, 1978.

Schwartz, E.R., Leveille, C., and Oh, W.H.: Experimentally-induced osteoarthritis in guinea pigs: effect of surgical procedure and dietary intake of vitamin C. Lab. Anim. Sci., *31*:683–687, 1981.

Smith, D.F., and Balagura, S.: Taste and physiological need in vitamin C intake by guinea pigs. Physiol. Behav., *14*:545–549, 1975.

LYMPHOCYTIC CHORIOMENINGITIS

Hosts. The natural reservoir host for the lymphocytic choriomeningitis (LCM) virus is the wild mouse population, where the prevalence of infection may approach 100%. The infection also occurs in mice, guinea pigs, rabbits, rats, canines, and primates and can be transmitted to hamsters, which have been the recent source of several hundred cases of LCM in humans. Infected pet hamsters were widely dispersed in the United States and all can be traced to one breeder and a nearby supplier of transplantable hamster tumors. Transplantable tumors of mice, hamsters, and guinea pigs, tissue culture cell lines, and virus and protozoan stocks may become persistently contaminated with the LCM virus and spread the virus to inoculated animals.

Etiology. The LCM virus is an arenavirus (RNA). On the basis of clinical signs, viscerotropic and neurotropic strains exist, although these effects may vary more quantitatively than qualitatively.

Transmission. In utero transmission is common within an infected mouse population, but bite wounds may be a more common route in hamsters. The virus is passed in the urine, saliva, milk, and feces and enters susceptible individuals via traumatized skin, the conjunctiva, or respiratory passages. Blood-sucking arthropod vectors, such as ticks, lice, and mosquitoes, as well as dust, may be transmission vehicles. Animals infected prenatally shed the virus for approximately 9 months, whereas animals infected postneonatally shed the virus for 2 to 3 months.

Predisposing Factors. Factors predisposing to the entry or persistence of LCM infection in a colony include wild rodent invasion, the introduction of carrier animals, tumors, cell lines, and other biologic products, decreased host resistance, and cutaneous traumatization.

Clinical Signs. The clinical signs of lymphocytic choriomeningitis depend on the host's resistance and age when infected, although the various categories of the disease are not always clearly delineated. Animals infected *in utero* or during the first 48 hours postpartum may develop a transient viremia but recover completely within a few weeks. Other animals similarly infected may develop a persistent tolerant infection (PTI) that continues asymptomatically for 6 or more months. Animals infected after the first few days, when the virus will be recognized as "foreign," often overcome the infection completely, but an acute, usually fatal syndrome can develop.

Signs of the acute infection in mice continue for 1 or 2 weeks and include decreased growth, rough hair coat, hunched posture, blepharitis, weakness, photophobia, tremors, and convulsions. The terminal stage of the PTI, which occurs over several weeks in 5- to 12-month-old mice, is characterized by weight loss, blepharitis, and impaired reproductive performance and runted litters.

Necropsy Signs. The important necropsy signs are microscopic. Visceral organs, including the liver, kidneys, lungs, pancreas, blood vessels, and meninges, are infiltrated by lymphocytes. A glomerulonephritis of probable immune complex origin is a characteristic feature of terminal PTI.

Diagnosis. Except for the PTI, infected animals either die or develop circulating antibodies, which can be detected serologically by neutralizing antibody, IFA, or ELISA techniques. PTI-infected animals may develop a glomerulonephrosis.

Mouse antibody production (MAP) tests may be useful in screening suspect animals, tumors, and other biologic materials for LCM virus. The MAP test is performed by inoculating known LCM-free mice with the suspect material and monitoring for serum antibody production.

Prevention. Scanning diagnostic tests are methods for monitoring and maintaining an LCM-free colony. Filter cage covers reduce aerosol transmission, and the exclusion of insect and wild rodent vectors from the colony prevents introduction of the virus. Vertical or transuterine passage of the LCM virus complicates eradication.

Public Health Significance. Lymphocytic choriomeningitis may be transmitted from infected animals to man through direct or indirect contact with feces or urine or with infected murine tissues. LCM is also transmitted via biting. Care must be taken in handling, bleeding, or processing tissues from suspect infected animals. The disease in man runs a 2- to 4-week course and in some aspects mimics influenza or mononucleosis. Signs, if any, may include persistent, intermittent fever, headache, adenopathy, pharyngitis, myalgia, fatigue, rash, arthritis, and a rarely fatal encephalomyelitis.

REFERENCES

Armstrong, D., et al.: Meningitis due to lymphocytic choriomeningitis virus endemic in a hamster colony. J.A.M.A., *209*:265–267, 1969.

Baum, S.C., et al.: Epidemic nomeningitic lymphocytic-chorio meningitis-virus infection. An outbreak in a population of laboratory personnel. N. Engl. J. Med., *274*:934–936, 1966.

Biggar, R.J., et al.: Lymphocytic choriomeningitis outbreak associated with pet hamsters. J.A.M.A., *232*:494–500, 1975.

Bowen, G.S., et al.: Laboratory studies of a lymphocytic choriomeningitis virus outbreak in man and laboratory animals. Am. J. Epidemiol., *102*:233–240, 1975.

Deibel, R., et al.: Lymphocytic choriomeningitis virus in man. Serologic evidence of association with pet hamsters. J.A.M.A., *232*:501–504, 1975.

Gregg, M.B.: Recent outbreaks of lymphocytic choriomeningitis in the United States. Bull. WHO, *52*:549–553, 1975.

Grimwood, B.G.: Viral contamination of a subline of

Toxoplasma gondii HR. Infect. Immun., *50*:917–918, 1985.

Hirsch, M.S., et al.: Lymphocytic choriomeningitis-virus infection traced to a pet hamster. N. Engl. J. Med., *291*:610–612, 1974.

Hotchin, J.: The contamination of laboratory animals with lymphocytic choriomeningitis virus. Am. J. Pathol., *64*:747–769, 1971.

Hotchin, J., et al.: Lymphocytic choriomeningitis in a hamster colony causes infection of hospital personnel. Science, *185*:1173–1174, 1974.

Hotchin, J., and Sikora, E.: Laboratory diagnosis of lymphocytic choriomeningitis. Bull. WHO, *52*:555–559, 1975.

Ivanov, A.P., Bashkirtsev, V.N., and Tkachenko, E.A.: Enzyme-linked immunosorbent assay for detection of arenaviruses. Arch. Virol., *67*:71–74, 1981.

Parker, J.C., et al.: Lymphocytic choriomeningitis virus infection in fetal, newborn, and young adult Syrian hamsters (*Mesocricetus auratus*). Infect. Immun., *13*:967–981, 1976.

Skinner, H.H., and Knight, E.H.: Natural routes for post-natal transmission of murine lymphocytic choriomeningitis. Lab. Anims., *7*:171–184, 1973.

Skinner, H.H., Knight, E.H., and Grove, R.: Murine lymphocytic choriomeningitis: the history of a natural cross-infection from wild to laboratory mice. Lab. Anims., *11*:219–222, 1977.

Smadel, J.E., and Wall, M.J.: Lymphocytic choriomeningitis in the Syrian hamster. J. Exp. Med., *75*:581–591, 1942.

Smith, A.L., et al.: Two epizootics of lymphocytic choriomeningitis virus occurring in laboratory mice despite intensive monitoring programs. Can. J. Comp. Med., *48*:335–337, 1984.

Thacker, W.L., et al.: Infection of Syrian hamsters with lymphocytic choriomeningitis virus: Comparison of detection methods. Am. J. Vet. Res., *43*:1500–1502, 1982.

van der Zeijst, B.A.M., et al.: Persistent infection of some standard cell lines by lymphocytic choriomeningitis virus; transmission of infection by an intracellular agent. J. Virol., *48*:249–261, 1983.

Wilsnack, R.E.: Lymphocytic choriomeningitis. Natl. Cancer Inst. Monogr., *20*:77–84, 1966.

MALOCCLUSION

Hosts. Malocclusion is most commonly seen in rabbits and guinea pigs, although small rodents may also be affected.

Etiology and Predisposing Factors. The incisors of rabbits and all rodents continue to grow throughout life. Malocclusion and tooth overgrowth result when, for genetic, dietary, infectious, or traumatic reasons, open-rooted teeth do not properly occlude and are therefore not eroded. Loss of all or part of an opposite tooth may also lead to tooth overgrowth.

In the rabbit, the teeth commonly involved are the incisors, which, if not constantly worn down, will grow approximately 10 cm (4 in.) per year. The cause of malocclusion in the rabbit is probably an autosomal recessive trait involving an abnormally short maxilla. Overgrowth of the open-rooted cheek teeth of the rabbit is uncommon.

Malocclusion occurs sporadically in guinea pigs and is difficult to detect except with an otoscope. It occurs in the open-rooted cheek teeth, usually the premolars. The condition may be linked with poor nutrition, although there is evidence to the contrary, or it may have a genetic basis and appear in the young. Other common laboratory rodents have multirooted molars that do not grow continuously, but their incisors will overgrow if an opposite member is broken, missing, or malaligned.

Clinical Signs.   Overgrown teeth result in trauma to the tongue and mouth and cause ptyalism (slobbers), anorexia, weight loss, starvation, and eventually death. The drooling around the mouth and onto the forequarters predisposes to moist dermatitis. Malocclusion is a common cause of chronic weight loss in rodents, especially guinea pigs and rabbits; however, inspection of the mouth is frequently neglected in the physical examination.

Treatment.   Treatment of tooth overgrowth involves filing or clipping (but not splitting or shattering) the teeth with a sharp clipper or dental bur. Use of a mouth speculum facilitates work on the teeth. A split tooth may become impacted and eventually abscessed.

REFERENCES

Fox, R.R., and Gary, D.D.: Mandibular prognathism in the rabbit: genetic studies. J. Hered., *62*:23–27, 1971.

Hard, G.C., and Atkinson, F.F.V.: "Slobbers" in laboratory guinea-pigs as a form of chronic fluorosis. J. Pathol. Bacteriol., *94*:95–102, 1967.

Hard, G.C., and Atkinson, F.F.V.: The aetiology of "slobbers" (chronic fluorosis) in the guinea-pig. J. Pathol. Bacteriol., *94*:103–112, 1967.

Harkness, J.E., et al.: Weight loss and impaired reproduction in the hamster attributable to an unsuitable feeding apparatus. Lab. Anim. Sci., *27*:117–118, 1977.

Huang, C.M., Mi, M.P., and Vogt, D.W.: Mandibular prognathism in the rabbit: discrimination between single-focus and multifactorial models of inheritance. J. Hered., *72*:296–298, 1981.

Lukefahr, S.D.: Malocclusion: a noxious genetic condition in rabbits. J. Appl. Rabbit Res., *4*:14–15, 1981.

Pollock, S.: Slobbers in the rabbit. J. Am. Vet. Med. Assoc., *119*:443–444, 1951.

Rest, J.R., Richards, T., and Ball, S.E.: Malocclusion in inbred strain-2 weanling guinea pigs. Lab. Anims., *16*:84–87, 1982.

Weisbroth, S.H., and Ehrman, L.: Malocclusion in the rabbit: a model for the study of the development, pathology and inheritance of malocclusion. I. Preliminary note. J. Hered., *58*:245–246, 1967.

Zeman, W.V., and Fielder, F.G.: Dental malocclusion and overgrowth in rabbits. J. Am. Vet. Med. Assoc., *155*:1115–1119, 1969.

MASTITIS

Etiology.   Acute inflammation of the mammary gland is most often encountered during lactation, when the milk provides an excellent medium for bacterial growth, the glands are pendulous, and the young traumatize the nipples. Mastitis is established when one or several genera of microorganisms (*Pasteurella, Klebsiella,* coliform bacilli, *Streptococcus, Staphylococcus, Pseudomonas*) enter the gland via the bloodstream, through a cutaneous lesion, or by the teat canal.

Predisposing Factors and Clinical Signs.   Unsanitary conditions, abrasive bedding or caging, biting young, and mammary impaction following early weaning predispose to mastitis. The affected gland becomes diffusely or focally enlarged, hyperemic, warm, and cyanotic. Depression and death from septicemia or toxemia often follow.

Among the pet rodents, guinea pigs are most often affected. Causative agents are usually α-hemolytic streptococci and *Escherichia coli,* although *Klebsiella* and *Staphylococcus* may also be involved. In acute cases, the affected gland is red-purple, and the milk appears bloody. A chronic, suppurative mastitis may also occur. In acute cases, maternal neglect and death of the sow occur within days.

Rabbit mastitis may be diffuse, when it causes a lesion known as "blue breasts" or "caked udder," or focal and suppurative. Affected rabbits are usually lactating and housed in a dirty environment. *Staphylococcus, Pasteurella,* and *Streptococcus* are usually the causative agents. Fever exceeding 104° F, anorexia, depression, death of neonates, and death of the doe from septicemia occur. In rats, mammary gland abscesses are more common than acute mastitis. *Pasteurella pneumotropica* is a frequent cause.

Treatment and Prevention.   Treatment of an acute infection in rabbits includes lancing the abscesses, if present, and administration of antibiotics. Chloramphenicol is the antibiotic of choice for treating mastitis in small rodents. The infected animal's environment should be disinfected with a sodium hypochlorite solution.

REFERENCES

Cheeke, P.R., et al.: Rabbit Production, 6th Edition. Danville, IL, Interstate Printers and Publishers, Inc., 1987.

Frisk, C.S., Wagner, J.E., and Owens, D.R.: Streptococcal mastitis in golden hamsters. Lab. Anim. Sci., *26*:97, 1976.

Gupta, B.N., Langham, R.F., and Conner, G.H.: Endotoxin-induced mastitis in the guinea pig. Am. J. Vet. Res., *32*:1785–1793, 1971.

Gupta, B.N., Langham, R.F., and Conner, G.H.: Mastitis in guinea pigs. Am. J. Vet. Res., *31*:1703–1707, 1970.

Hong, C.C., and Ediger, R.D.: Chronic necrotizing mastitis in rats caused by *Pasteurella pneumotropica*. Lab. Anim. Sci., *28*:317–320, 1978.

Kinkler, R.J., Jr., et al.: Bacterial mastitis in guinea pigs. Lab. Anim. Sci., *26*:214–217, 1976.

Williams, W.L., and Patnode, R.A.: Mastitis of the mouse as related to post-secretory mammary involution. Arch. Pathol., *45*:229–238, 1948.

METABOLIC TOXEMIAS OF PREGNANCY

Hosts.   This group of clinically similar diseases occurs in cattle, rabbits, guinea pigs, hamsters, sheep, goats, nonhuman primates, and man.

Etiology.   Probable primary causes or important contributing factors to toxemic states include fasting, especially during late pregnancy, low energy intake, stress, and endocrine imbalances. A possible contributing cause of the toxemias of late pregnancy is inadequate blood flow or nutrient supply to the gravid uteroplacental unit. Uterine ischemia may cause thromboplastin to be released from degenerating trophoblasts to initiate a disseminated intravascular coagulation. The renin-angiotensin system may also be involved.

Predisposing Factors.   Factors predisposing a guinea pig or rabbit to pregnancy toxemia are multiple and their roles uncertain. Specific influences include obesity, heredity, change in diet, increasing age, large fetal load (which may either compress or overextend the vascular supply to the uterus), anorexia, hypoplasia of vessels supplying the uterus or failure of those vessels to enlarge for the primigravid uterus, lack of exercise, and nonspecific environmental stressors. A tendency for more cases to occur during the winter has been noted. Obese boars may also develop the syndrome.

Obesity and fasting are also predisposing factors in ketosis in rabbits. The Dutch, Polish, and English breeds have an apparent predisposition to toxemia, as do pregnant, pseudopregnant, and postparturient does, although resting does and bucks may also succumb.

Clinical Signs.   Pregnancy toxemia, a metabolic condition of low morbidity and high mortality, usually occurs during the last week of gestation in rabbits and during the last 2 weeks of gestation or the first postpartum week in guinea pigs. The condition may be asymptomatic, rapidly fatal, or involve weakness, depression, reluctance to move, incoordination, anorexia, abortion, dyspnea, convulsions, coma, and death over a 1- to 5-day period. The urine becomes clear (acidic), and proteinuria, ketonuria, and hyperkalemia may develop. This condition is considered responsible for high fetal wastage in many guinea pig colonies.

General statements regarding serum calcium, phosphorus, and glucose are misleading because of considerable variation among species and stages of the disease within a single species.

Necropsy Signs.   Necropsy signs of pregnancy toxemia, when evident, include reduced gastric content, ample body fat, dead and decomposed fetuses, uterine and placental hemorrhage, adrenal enlargement and hemorrhage, and a tan liver that is greasy on cut surfaces. Glomerular and tubular cells are necrotic, and Bowman's capsule and tubular lumina contain protein. There may be marked fatty metamorphosis in the liver and disseminated intravascular coagulation. Diagnosis is based on clinical, necropsy, and microscopic signs.

Treatment.   If toxemia in progress is detected, several unproven and almost inevitably abortive treatments can be tried. The administration of lactated Ringer's solution, calcium gluconate, propylene glycol, 5% glucose, corticosteroids, and even magnesium sulfate solution to dilate arteries has been suggested and even attempted in small animals, but results are inconclusive. Guinea pigs under treatment for ketosis frequently die from an acute enteritis apparently due, in part, to stress and lack of ingesta in the intestinal tract.

Prevention.   A critical consideration in preventing toxemia is to avoid fasting and to supply fresh water and a nutritious diet. Animals should not become obese, and primiparous breeders should not be stressed. An energy-dense diet in late pregnancy may be helpful.

REFERENCES

Abitbol, M.M., et al.: Production of experimental toxemia in the pregnant rabbit. Am. J. Obstet. Gynecol., *124*:460–470, 1976.

Abitbol, M.M., Driscoll, S.G., and Ober, W.B.: Placental lesion in experimental toxemia in the rabbit. Am. J. Obstet. Gynecol., *125*:942–948, 1976.

Assali, N.S., Longo, L.D., and Holm, L.W.: Toxemia-like syndromes in animals, spontaneous and experimental. Obstet. Gynecol. Surv., *15*:151–181, 1960.

Bruce, N.W.: The effect of ligating a uterine artery on fetal and placental development in the rat. Biol. Reprod., *14*:246–247, 1976.

Foley, E.J.: Toxemia of pregnancy in the guinea pig. J. Exp. Med., *75*:539–547, 1942.

Ganaway, J.R., and Allen, A.M.: Obesity predisposes to pregnancy toxemia (ketosis) of guinea pigs. Lab. Anim. Sci., *21*:40–44, 1971.

Golden, J.G., Hughes, H.C., and Lang, C.M.: Experimental toxemia in the pregnant guinea pig (*Cavia porcellus*). Lab. Anim. Sci., *30*:174–179, 1980.

Greene, H.S.N.: Toxemia of pregnancy in the rabbit. I. Clinical manifestations and pathology. J. Exp. Med., *65*:809–832, 1937.

Greene, H.S.N.: Toxemia of pregnancy in the rabbit. II. Etiological considerations with especial reference to hereditary factors. J. Exp. Med., *67*:369–388, 1938.

Richter, A.G., Lausen, N.C., and Lage, A.L.: Pregnancy toxemia (eclampsia) in Syrian golden hamsters. J. Am. Vet. Med. Assoc., *185*:1357–1358, 1984.

Rogers, J.B., Vanloon, E.J., and Beattie, M.F.: Familial incidence of a toxemia of pregnancy in the guinea pig. J. Exp. Zool., *117*:247–258, 1951.

Seidl, D.C., et al.: True pregnancy toxemia (preeclampsia) in the guinea pig (*Cavia porcellus*). Lab. Anim. Sci., *29*:472–478, 1979.

Speroff, L.: Toxemia of pregnancy. Mechanism and therapeutic management. Am. J. Cardiol., *32*:582–591, 1973.

MOIST DERMATITIS

Hosts.   Moist dermatitis occurs sporadically in rabbits and rodents.

Etiology.   Moist dermatitis can be caused by a variety of traumatic and infectious agents, although these agents usually are not involved unless the skin is traumatized or continually wetted. Noninfectious factors include scratching at an allergic reaction, abrasions, cuts, puncture wounds, urine scald, and chemical or thermal burns. Bacteria that may be involved are *Staphylococcus, Treponema, Streptococcus, Fusobacterium, Corynebacterium,* and *Pseudomonas,* whose blue-green pigment causes a striking "blue-fur" disease. Specific or named conditions include "sore nose" in gerbils and "wet dewlap," "slobbers," and "hutch burn" in rabbits.

Sore nose or nasal dermatitis is a common condition of gerbils that is associated with excessive accumulation of porphyrins around the nasal area. Normally, porphyrin is continually removed from the nasal area by grooming with the forepaws. Porphyrin appears to be highly irritating if left to accumulate on the external nares. Cleaning of the nose, provision of absorbent bedding, such as clay products, and possibly a compatible cagemate are ways of reducing the severity of this condition. Harderianectomy will eliminate the source (harderian glands) of the irritating porphyrin and result in rapid remission of nasal lesions in chronic cases that do not respond to other kinds of treatment.

Predisposing Factors. Conjunctivitis, drooling from malocclusion, the use of water pans and crocks, chronic diarrhea, and damp cages predispose to moist dermatitis. Overcrowding, stress, fighting, and use of abrasive bedding appear to exacerbate or aggravate sore nose. Reducing or eliminating these factors may result in alleviation of symptoms.

Clinical Signs. Moist dermatopathies may range from superficial erythema to suppuration and deep ulceration. The hair surrounding the lesion is usually matted. Lesions occur on the face or ventrum. Sore nose starts around the external nares as an acute moist dermatitis characterized by erythema, alopecia, and serous exudate. Aside from the association with porphyrin, the primary or inciting cause of the condition is not known. Frequently, staphylococcal agents can be recovered from the inflamed or ulcerated nasal area. The condition can occur in epizootic proportions among weanling gerbils. Many cases persist for long periods, but others are transitory.

Prevention. Prevention involves maintaining nonabrasive, clean, dry bedding in a nonabrasive cage. Hopper feeders and sipper-tube waterers reduce throat abrasion. Balls or stones placed in water crocks reduce wetting while drinking.

Treatment. Moist dermatitis is treated by removing hair over and adjacent to the lesion, cleansing the area affected, administering topical or systemic antibiotics, and removing the cause.

REFERENCES

Andrews, E.J.: Muzzle trauma in the rat associated with the use of feeding cups. Lab. Anim. Sci., *27*:278, 1977.

Beattie, J.M., Yates, A.G., and Donaldson, R.: An epidemic disease in rabbits resembling that produced by *B. necrosis* (Schmorl), but caused by an aerobic bacillus. J. Pathol. Bacteriol., *18*:34–46, 1913.

Maronpot, R.R., and Chavannes, J-M.: Dacryoadenitis, conjunctivitis, and facial dermatitis of the mouse. Lab. Anim. Sci., *27*:277–278, 1977.

Peckham, J.C., et al.: Staphylococcal dermatitis in Mongolian gerbils (*Meriones unguiculatus*). Lab. Anim. Sci., *24*:43–47, 1974.

MOUSE HEPATITIS VIRUS INFECTION

Hosts. Mice are the hosts for the mouse hepatitis virus (MHV). Mouse hepatitis virus infection is widespread, often difficult to detect, highly contagious, and has manifold effects on the host's immune system.

Etiology. The mouse hepatitis virus, of which there are several strains within this hepatoencephalitis group, is a corona (RNA) virus. These viral strains, because of varied tissue affinities and host susceptibilities, induce a wide spectrum of diseases in mice.

Transmission. Mouse hepatitis virus infection is usually latent but always highly contagious. The virus is disseminated in the feces, by respiratory aerosol, and, according to experimental findings, through the placenta.

Predisposing Factors. Depending on the strain, mouse hepatitis viruses usually affect either the intestine or respiratory system, but if the host is debilitated, affected with certain lymphomas, injected with cortisone, urethane, cyclophosphamides, or antilymphocyte serum, X-irradiated, splenectomized, neonatally thymectomized, or infected with K virus, all conditions that alter function of reticuloendothelial cells, other tissues will be affected. Infection is suppressed by reticuloendothelial cell stimulators such as triolein or *Salmonella typhosa* endotoxin.

Genotype and other host factors highly influence the outcome of MHV infections. Adult mice are more resistant to clinical disease than are young mice.

Clinical Signs. Most MHV infections are latent, enzootic, and subclinical. Susceptible suckling mice, usually between 7 and 13 days of age, can develop an encephalitis with tremors and spasticity or a severe, epizootic, diarrheal disease with high mortality. This diarrhea is yellow and sticky, and in this aspect resembles that of rotavirus and reovirus infections of suckling mice. In older mice, loss of weight and breeding efficiency, jaundice, and death may result. Mouse hepatitis virus alters lymphocyte differentiation, immunoglobulin responses to sheep red blood cells, phagocytosis, tumor growth, antibody response, interferon production, and hepatic enzyme activity. Nude mice typically develop a chronic progressive emaciation (wasting syndrome) and an associated persistent infection of the intestinal or nasal mucosa.

Necropsy Signs.   A prominent lesion of MHV in the adult is a necrotizing hepatitis with gray or red foci on a tan liver. These foci may contain syncytial giant cells and other components of an inflammatory or granulomatous response.

The enteritis of suckling mice produces a flaccid gut distended with gas and yellow fluid. Histopathologic changes in the gut consist of villous shortening and vacuolation, desquamation, and syncytial cell formations in the absorptive epithelium.

Diagnosis.   An ELISA test is commonly used to detect serum antibody to MHV infection. Female mice usually have higher titers than males. ELISA tests are preferred to the CF test because the latter lacks sensitivity. Strains of mice vary markedly in their serologic responses to various mouse hepatitis viruses. C57BL/6 mice produce a relatively high antibody titer and are therefore a good sentinel strain. DBA/2 mice, on the other hand, are poor antibody responders and therefore should not be used to test for MHV if other strains are available.

Clinical history, necropsy, and histopathologic features are helpful in establishing a diagnosis of MHV infection. The finding of syncytial giant cells in the absorptive epithelium of the small intestine or cecum is highly indicative of MHV infection and is a valuable diagnostic procedure. Use of a "Swiss roll" technique greatly facilitates histopathologic examination of the gastrointestinal tract.

Treatment.   There is no treatment for MHV infection.

Prevention.   Because the source of MHV and other infections is usually other mice, feral or domestic, exclusion of infected animals is paramount to maintaining disease-free populations. The use of cesarean-derived mice in a barrier-sustained colony, repeated serologic testing of breeding females, histopathologic examinations of weanlings, filter top cages, gloves, disinfected forceps for handling mice, and laminar flow units are preventive measures. Unless one is dealing with strains of mice susceptible to chronic and latent infection, e.g., C3H and nude strains, it should be possible to eradicate MHV from enzootically infected closed colonies by breaking the infectious cycle by quarantine or cessation of breeding without introduction of new mice.

Because of the ubiquitous and highly contagious nature of MHV in wild rodents and resident mouse populations, the infection is very difficult to exclude from contemporary research animal facilities. The virus may persist for several days outside the host in feces and on fomites. Incoming animals and mouse biologic materials must be free of infection.

Public Health Significance.   The mouse hepatitis virus does not affect humans.

REFERENCES

Barthold, S.W., et al.: Epizootic coronaviral typhlocolitis in suckling mice. Lab. Anim. Sci., *32*:370–383, 1982.

Barthold, S.W., and Smith, A.L.: Mouse hepatitis virus strain-related patterns of tissue tropism in suckling mice. Arch. Virol., *81*:103–112, 1984.

Barthold, S.W.: Research complications and state of knowledge of rodent coronaviruses. *In* Complications of Viral and Mycoplasma Infections in Rodents to Toxicology Research and Testing. Edited by T.E. Hamm, Jr. Washington, DC, Hemisphere Press, 1986, pp. 53–90.

Barthold, S.W.: Host age and genotypic effects on enterotropic mouse hepatitis virus infection. Lab. Anim. Sci., *37*:36–40, 1987.

Boorman, G., et al.: Peritoneal and macrophage alterations caused by naturally occurring mouse hepatitis virus. Am. J. Pathol., *106*:110–117, 1982.

Broderson, J.R., Murphy, F.A., and Hierholzer, J.C.: Lethal enteritis in infant mice caused by mouse hepatitis virus. Lab. Anim. Sci., *26*:824, 1976.

Brownstein, D.G., and Barthold, S.W.: Mouse hepatitis virus immunofluorescence in formalin- or Bouin's-fixed tissues using trypsin digestion. Lab. Anim. Sci., *32*:37–39, 1982.

Carthew, P.: Lethal intestinal virus of infant mice is mouse hepatitis virus. Vet. Rec., *101*:465, 1977.

Cheever, F.S., et al.: A murine virus (JHM) causing disseminated encephalomyelitis with extensive destruction of myelin. I. Isolation and biological properties of the virus. J. Exp. Med., 90:181–194, 1949.

Garlinghouse, L.E., Jr., and Smith, A.L.: Responses of mice susceptible or resistant to lethal infection with mouse hepatitis virus, strain JHM, after exposure by a natural route. Lab. Anim. Sci., 35:469–472, 1985.

Goto, N., et al.: Giant cell formation in the brain of suckling mice infected with mouse hepatitis virus, JHM strain. Jpn. J. Exp. Med., 49:169–177, 1979.

Ishida, T., et al.: Isolation of mouse hepatitis virus from infant mice with fatal diarrhea. Lab. Anim. Sci., 28:269–276, 1978.

Ishida, T., and Fujiwara, K.: Pathology of diarrhea due to mouse hepatitis virus in the infant mouse. Jpn. J. Exp. Med., 49:33–41, 1979.

Katami, K., et al.: Vertical transmission of mouse hepatitis virus infection in mice. Jpn. J. Exp. Med., 48:481–490, 1978.

Lavi, E., et al.: Limbic encephalitis after inhalation of a murine coronavirus. Lab Investigation, 58:31–36, 1988.

Li, L.H., et al.: Effect of mouse hepatitis virus infection on combination therapy of P388 leukemia with cyclophosphamide and pyrimidinones. Lab. Anim. Sci., 37:41–44, 1987.

Lindsey, J.R.: Prevalence of viral and mycoplasmal infections in laboratory rodents. In Viral and Mycoplasmal Infections of Laboratory Rodents. Edited by P.N. Bhatt, et al. Orlando, Academic Press, 1986.

Peters, R.L., et al.: Enzyme-linked immunosorbent assay for detection of antibodies to murine hepatitis virus. J. Clin. Microbiol., 10:595–597, 1979.

Rowe, W.P., Hartley, J.W., and Capps, W.I.: Mouse hepatitis virus infection as a highly contagious, prevalent, enteric infection in mice. Proc. Soc. Exp. Biol. Med., 112:161–165, 1963.

Sebesteny, A.: Hepatitis and brain lesions due to mouse hepatitis virus accompanied by wasting in nude mice. Lab. Anims., 8:317–326, 1974.

Sugiyama, K., and Amano, Y.: Morphological and biological properties of a new coronavirus associated with diarrhea in infant mice. Arch. Virol., 67:241–251, 1981.

Viguera, C., et al.: Hepatic jaundice in a colony of nude mice. Lab. Anim. Sci., 28:714–719, 1978.

Ward, J.M., Collins, M.J., Jr., and Parker, J.C.: Naturally occurring mouse hepatitis virus infection in the nude mouse. Lab. Anim. Sci., 27:372–376, 1977.

Weir, E.C., et al.: Elimination of mouse hepatitis virus from a mouse breeding colony by temporary cessation of breeding. Lab. Anim. Sci., 35:524, 1985.

Williams, D.C., and Di Luzig, N.R.: Glucan-induced modification of murine viral hepatitis. Science, 208:67–69, 1980.

MOUSEPOX (ECTROMELIA)

Hosts.   The mouse is the only natural host of the ectromelia virus, which causes mousepox. (Ectromelia means absence or imperfection of limb.) Experimental infection of other laboratory rodents results in a brief noncontagious infection with antibody production. The highly infectious and potentially devastating disease appears to have become enzootic in the United States during 1979 and 1980, as has been the case in many areas of the world for a long time. Outbreaks in the United States have been among noncommercial inbred and congenic strains used in immunogenetic research. Clinically, the disease is severest among certain inbred strains, especially BALB/c, DBA, CBA, C3H, and A. C57BL strains are relatively resistant.

Etiology.   The ectromelia virus is a large (175 × 290 nm) oval- to brick-shaped, cytoplasmic, double-stranded DNA orthopoxvirus, with a characteristic dumbbell-shaped nucleoid, in the family Poxviridae, vaccinia subgroup.

Transmission.   The virus is relatively stable in dry environments. The probable transmission routes are respiratory aerosol, skin abrasions, contact with skin debris, and ingestion of contaminated feces. The virus is excreted from the intestinal tract for long periods following infection. Urine, ectoparasites, other animals, humans, fomites, and infected tissues may also disseminate the virus. About 10 days after infection, characteristic skin lesions develop, and more virus is shed into the environment. The virus is spread among institutions by investigators who exchange infected mice or mouse tissues, e.g., tumors, cell lines, hybridoma lines, and sera. The virus in these products may remain viable in ultracold freezers for years.

Predisposing Factors.   Different mouse strains vary in susceptibility to the ectromelia virus, although such differences have been determined on the basis of a few outbreaks. In recent outbreaks high mortality was observed among BALB/c, DBA, A, and CBA mice. Mice of these strains tended to die suddenly with few or no skin lesions.

C3H mice are susceptible, but death is delayed and skin lesions frequently develop. C57BL/6 and black congenic strains of mice are highly resistant.

Clinical Signs. Mousepox infections may be asymptomatic, latent, acute, subacute, or chronic. Otherwise, there is considerable variation in clinical signs. The nature of clinical disease expressed is largely dependent on mouse genotype.

The acute, systemic form of infectious mousepox with high morbidity and mortality occurs in epizootic outbreaks in susceptible strains. Clinical signs of acute mousepox include hunched posture, rough hair coat, conjunctivitis, swelling of the face or extremities, diarrhea, and high mortality. The cutaneous rash, an important source of virus dissemination, is seldom seen in acute outbreaks.

The subacute to chronic enzootic or cutaneous form involves a generalized papular rash with eventual swelling, ulceration, and amputation of appendages and variable mortality. The cutaneous lesions may resemble bite wounds, mite allergies, and *Corynebacterium* spp. arthritis.

Necropsy Signs. The gross lesions associated with acute mousepox are hyperemia and edema of the viscera, enlarged Peyer's patches, lymphoid hyperplasia, splenomegaly, a peritoneal exudate, and, as the disease progresses, hemorrhage into the intestinal lumen and focal necrosis of the spleen, liver, pancreas, lymph nodes, thymus, and other organs.

In the subacute to chronic forms, focal necrosis becomes more extensive; the vesicular, crusted cutaneous pox lesions develop, and swelling and necrosis of the extremities occur. At necropsy the observation of scars in the spleen is indicative of prior ectromelia virus infection.

Histopathologic changes are characterized by massive splenic necrosis that originates in the lymphoid follicles and spreads throughout the entire organ. Hepatic necrosis is often widespread and focal. Necrosis of Peyer's patches, lymph nodes, and thymus resembles microscopic changes seen in the spleen. Characteristic eosinophilic type A cytoplasmic inclusion bodies can be found in infected epithelial cells early in the rash at sites of focal epidermal hyperplasia and erosion. Infected hepatocytes adjacent to areas of hepatic necrosis contain basophilic type B cytoplasmic inclusion bodies.

Diagnosis. Mousepox is diagnosed by clinical and gross necropsy signs; by the demonstration of intracytoplasmic, eosinophilic inclusion bodies (Marchal bodies) in the epithelial cells of the skin, small intestine, and pancreas; by a fluorescent antibody test; and through the use of a hemagglutination inhibition test on sera from mice with subacute or chronic cases. When inapparent infections are suspected, known susceptible, disease-free mice may be introduced into the colony as sentinel animals.

Finding splenic and hepatic necrosis and other typical lesions allows a diagnosis of acute mousepox. Confirmation is through electron microscopy of affected organs, which reveals characteristic poxviruses. Caution is urged because vaccination with live vaccinia virus may introduce viruses with the same morphologic characteristics as the natural infection. The virus can be isolated from spleen and liver during early stages of disease by culture on mouse fibroblasts (L929), Vero, or HeLa cells. It grows well on the chorioallantoic membrane of hen's eggs. Virus can be demonstrated in tissues by immunocytochemical methods, e.g., fluorescent antibody.

Serologic testing is useful for screening suspect colonies and for detecting latent infections. The hemagglutination inhibition (HAI) test is commonly used for serologic screening. Possible prior vaccination is a consideration in interpretation of serologic results. The IHD-T strain of vaccinia virus is commonly used in vaccination; however, no hemagglutination inhibiting antibody is produced, and the HAI test can be used for screening vaccinated colonies. Vaccination interferes with the more sensitive ELISA and immunofluorescent antibody (IFA) tests. Although the HAI test is relatively insensitive and unexplained false-positive results occasionally appear in certain groups of mice, it is widely used for screening large numbers of sera. The ELISA test is more sensitive. The IFA test is highly sensitive and specific and is used as a confirmatory procedure.

Treatment. There is no treatment for infectious mousepox. Vaccination, using a live IHD-T strain of vaccinia virus cultured in an egg, may be used to limit outbreaks in small closed colonies; however, elimination of affected colonies is preferred to preclude the possibility of establishing the enzootic disease.

Prevention. Selection of ectromelia-free stock, careful husbandry and quarantine measures, serologic screening tests, and the use of susceptible or sentinel mice are measures used to exclude latent carriers from entering the colony. Particular care should be taken with inbred and congenic strains from sources where mousepox has been enzootic. Imported mice or mouse tissues, mice from unknown sources, or suspect mice should be strictly quarantined. Mouse tissue destined for transmission to susceptible mice should be checked for ectromelia virus by passage through susceptible hosts or tested for mouse antibody production (MAP). Infected colonies must be eliminated, and rooms and equipment should be thoroughly sanitized and disinfected because the virus remains viable in dry environments. Several disinfectants are effective: vapor phase formaldehyde, iodophores (150 to 300 ppm), and sodium hypochlorite (1000 ppm available chlorine). Dead animals, animal wastes, and contaminated bedding should be isolated and incinerated. Movement into and out of suspect colonies should be strictly limited. Intrauterine infections with ectromelia virus limit the effectiveness of hysterectomy derivation to eliminate the infection.

Susceptible mice may be vaccinated with the IHD-T strain, which does not produce HAI antibodies and interfere with subsequent serologic testing. The vaccine is given by scarification at the tail base. If a vacci-

nation "take" is not obvious, a latent carrier state should be suspected. Successive passage of tumor lines through at least two vaccinated mice results in loss of ectromelia virus from that tumor line.

Public Health Significance. Man is not susceptible to infection by the mousepox virus.

REFERENCES

Allen, A.M., et al.: Pathology and diagnosis of mousepox. Lab. Anim. Sci., *31*:599–608, 1981.

Auernhammer, H., et al.: Vaccination against mouse poxvirus infections. Z. Versuchstierkd., *20*:233–240, 1978.

Bhatt, P.N., et al.: Transmission of mousepox in genetically resistant or susceptible mice. Lab. Anim. Sci., *35*:523–524, 1985.

Bhatt, P.N., and Jacoby, R.O.: Mousepox: pathogenesis, diagnosis and rederivation. *In* Viral and Mycoplasmal Infections of Laboratory Rodents. Edited by P.N. Bhatt, et al. Orlando, Academic Press, 1986, pp. 557–570.

Bhatt, P.N., and Jacoby, R.O.: Mousepox in inbred mice innately resistant or susceptible to lethal infection with ectromelia virus. I. Clinical responses. Lab. Anim. Sci., *37*:11–15, 1987.

Bhatt, P.N., and Jacoby, R.O.: Mousepox in inbred mice innately resistant or susceptible to lethal infection with ectromelia virus. III. Experimental transmission of infection and derivation of virus-free progeny from previously infected dams. Lab. Anim. Sci., *37*:23–27, 1987.

Bhatt, P.N., and Jacoby, R.O.: Stability of ectromelia virus strain NIH-79 under various laboratory conditions. Lab. Anim. Sci., *37*:33–35, 1987.

Bhatt, P.N., and Jacoby, R.O.: Effect of vaccination on the clinical response, pathogenesis and transmission of mousepox. Lab. Anim. Sci., *37*:610–614, 1987.

Briody, B.A.: The natural history of mouse pox. Natl. Cancer Inst. Monogr., *20*:105–116, 1966.

Briody, B.A.: Response of mice to ectromelia and vaccinia viruses. Bacteriol. Rev., *23*:61–95, 1959.

Buller, R.M.L., and Wallace, G.D.: Reexamination of the efficacy of vaccination against mousepox. Lab. Anim. Sci., *35*:473–476, 1985.

Buller, R.M.L., et al.: Observations on the replication of ectromelia virus in mouse-dervied cell lines: Implications for epidemiology of mousepox. Lab. Anim. Sci., *37*:28–32, 1987.

Christensen, L.R., Weisbroth, S., and Mantanic, B.: Detection of ectromelia virus and ectromelia antibodies by immunofluorescence. Lab. Anim. Care, *16*:129–141, 1966.

Collins, M.J., Jr., Peters, R.L., and Parker, J.C.: Serological detection of ectromelia virus antibody. Lab. Anim. Sci., *31*:595–598, 1981.

Fenner, F.: The clinical features and pathogenesis of mousepox (infectious ectromelia of mice). J. Pathol. Bacteriol., *60*:529–552, 1948.

Fenner, F.: Mouse-pox (infectious ectromelia of mice): a review. J. Immunol., *63*:341–373, 1949.

Fenner, F.: Mousepox (infectious ectromelia). *In* The Laboratory Mouse. Vol. II. Edited by H. Foster, D. Small, and J. Fox. New York, Academic Press, 1982.

Flynn, R.J.: The diagnosis and control of ectromelia infection of mice. Lab. Anim. Care, *13*:130–136, 1963.

Jacoby, R.O., and Bhatt, P.N.: Mousepox in inbred mice innately resistant or susceptible to lethal infection with ectromelia virus. II. Pathogenesis. Lab. Anim. Sci., *37*:15–22, 1987.

Roberts, J.A.: Histopathogenesis of mouse pox. I. Respiratory infection. Br. J. Exp. Pathol., *43*:451–461, 1962.

Roberts, J.A.: Histopathogenesis of mouse pox. II. Cutaneous infection. Br. J. Exp. Pathol., *43*:462–468, 1962.

Small, J.D., and New, A.E.: Prevention and control of mousepox. Lab. Anim. Sci., *31*:616–629, 1981.

Tantawi, H.H., Zaghloul, T.M., and Zakaria, M.: Poxvirus infection in a rat (*Rattus Norvegicus*) in Kuwait. Int. J. Zoon., *10*:28–32, 1983.

Trentin, J.J., and Ferrigno, M.A.: Control of mouse pox (infectious ectromelia) by immunization with vaccinia virus. J. Natl. Cancer Inst., *18*:757–767, 1957.

Voller, A., Bidwell, D.E., and Bartlett, A.: Microplate enzyme immunoassays for the immunodiagnosis of virus infections. *In* Manual of Clinical Immunology. Edited by N.R. Rose and H. Friedman. Washington, DC, American Society for Microbiology, 1976, pp. 506–512.

Wagner, J.E., and Daynes, R.A.: Observations of an outbreak of mousepox in laboratory mice in 1979 at the University of Utah Medical Center, USA. Lab. Anim. Sci., *31*:565–569, 1981.

Wallace, G.D., and Buller, R.M.L.: Kinetics of ectromelia virus (mousepox) transmission and clinical response in C57BL/6J, BALB/cByJ and AKR/J inbred mice. Lab. Anim. Sci., *35*:41–46, 1985.

Wallace, G.D., and Buller, R.M.L.: Ectromelia virus (mousepox): Biology, epizootiology, prevention and control. *In* Viral and Mycoplasmal Infections of Laboratory Rodents. Edited by P.N. Bhatt, et al. Orlando, Academic Press, 1986, pp. 539–556.

Whitney, R.A., Jr.: Ectromelia in U.S. mouse colonies (letter). Science, *184*:609, 1974.

MUCOID ENTEROPATHY

Hosts.   Mucoid enteropathy of rabbits is a distinct disease entity with unusual characteristics and, in this description, is not a catarrhal diarrhea. The disease occurs in adults but weanling rabbits 7 to 10 weeks of age are most susceptible.

Etiology.   Uncomplicated mucoid enteropathy has no known cause, although bacterial toxins and numerous metabolic, pathogenic, and nutritional factors have been mentioned. Ingesta blockage in the ileum, cecum, or colon may be involved in the causation through augmentation of goblet cell function. Mucoid enteropathy may be complicated by inflammatory conditions, but when this happens, the classic mucoid enteropathy becomes difficult to separate from the usual catarrhal reaction encountered in other enteric diseases of rabbits.

Transmission.   If an infectious organism is involved, transmission is by the fecal-oral route. The spread of this disease through a colony would indicate such an etiology and route.

Predisposing Factors.   Dietary changes, dietary fiber under 6% (Cheeke and Patton 1978) or over 20% (Patton and Cheeke 1981), antibiotic influences, environmental stress, and encounters with bacteria have been mentioned as factors predisposing to mucoid enteropathy. The disease is more common in the spring, in certain rabbit families, in young does and in does with large litters, and in young fryers with high food intake.

Clinical Signs.   Disease outbreaks diagnosed as mucoid enteropathy have occurred in both weanling and adult animals; in fact, a distinguishing feature of this disease is the involvement of adult rabbits. Abdominal distention, slushing of intestinal contents, hunched posture, tooth grinding, rapid weight loss, dehydration, clear mucous diarrhea, subnormal temperature, and death in many cases are clinical signs, although many of these signs are common to other enteropathies. Distinguishing signs of the condition are the subnormal temperature, clear mucous diarrhea or constipation, occurrence in adults, and necropsy signs.

Necropsy Signs.   Uncomplicated mucoid enteropathy is noninflammatory; there is little discoloration of the intestinal wall. The stomach and small intestine contain gas and watery fluid, and the large intestine has a thick, nearly clear, gelatinous mucous plug. There may be an ingesta block at the sacculus rotundus. The only discoloration of the fluid and mucus is from bile. On opening the distended colon, the plug can be removed as a single mass. The cecum contains either mucus or fluid. An inspissated, dried mass of ingesta may be present, perhaps blocking the ileum, cecum, or colon. Histologically, the primary change is goblet cell hyperplasia of the colonic wall.

Diagnosis. Diagnosis of mucoid enteropathy is based on clinical and necropsy signs. Culture of the gut or a fecal examination may be indicated to detect colibacillosis, salmonellosis, or coccidiosis.

Treatment. The only treatment of a rabbit constipated with mucus is an enema. Treatment is essentially useless, but chloramphenicol, electrolyte solutions, vitamins, and analgesics can be tried.

Prevention. Prevention of mucoid enteropathy is discussed under "Enteropathy, Nonspecific," in this chapter. Provision of 16% to 20% of adequate large-particle dietary fiber may be important in prevention.

Public Health Significance. There is no known public health hazard with mucoid enteropathy.

REFERENCES

Cheeke, P., and Patton, N.: Effect of alfalfa and dietary fiber on the growth performance of weanling rabbits. Lab. Anim. Sci., *28*:167–172, 1978.

Greenham, L.W.: Some preliminary observations on rabbit mucoid enteritis. Vet. Rec., *74*:79–85, 1962.

McCuiston, W.R.: Rabbit mucoid enteritis. Vet. Med. Small Anim. Clin., *59*:815–818, 1964.

Lelkes, L., and Chang, C.-L.: Microbial dysbiosis in rabbit mucoid enteropathy. Lab. Anim. Sci., *37*:757–764, 1987.

Meshorer, A.: Histological findings in rabbits which died with symptoms of mucoid enteritis. Lab. Anims., *10*:199–202, 1976.

Patton, N., and Cheeke, P.: A precautionary note on high fiber levels and mucoid enteritis. J. Appl. Rabbit Res., *4*:56, 1981.

Pout, D.: Mucoid enteritis in rabbits. Vet. Rec., *89*:214–216, 1971.

Sinkovics, G.: Intestinal flora studies in rabbit mucoid enteritis. Vet. Rec., *98*:151–152, 1976.

Toofanian, F., and Targouski, S.: Experimental production of rabbit mucoid enteritis. Am. J. Vet. Res., *44*:705–708, 1983.

Toofanian, F., and Hamar, D.W.: Cecal short-chain fatty acids in experimental rabbit mucoid enteropathy. Lab. Anim. Sci., *87*:2423–2425, 1986.

van Kruiningen, H.J., and Williams, C.B.: Mucoid enteritis of rabbits. Comparison to cholera and cystic fibrosis. Vet. Pathol., *9*:53–77, 1972.

Vetesi, F., and Kutas, F.: Mucoid enteritis in the rabbit associated with *E. coli* changes in water, electrolyte and acid base balance. Acta Vet. Acad. Sci. Hung., *23*:381–388, 1973.

MURINE ENCEPHALOMYELITIS (MOUSE POLIO)

Hosts. Laboratory mice are the natural hosts of the usually asymptomatic infections with Theiler's murine encephalomyelitis viruses (TMEV), which are also referred to as the mouse encephalomyelitis viruses. Wild mice may also be a source of infection, but rats are resistant.

Etiology. Strains of the RNA *Picornadviridae* viruses vary greatly in their virulence for mice. GD VII and FA (both highly neurovirulent) and Theiler's Original (TO) are among the more widely studied and reported strains. The TMEV viruses should not be confused with the encephalomyocarditis viruses, which are picornaviruses with a predilection to cause heart damage.

Transmission. Spread of the infection is by the fecal-oral route because most strains of the virus that infect mice naturally replicate in the intestine. In colonies enzootically infected, mice become infected around the time of weaning, at approximately 3 weeks of age. Whereas these agents usually cause enteric infections, some strains also infect the central nervous system and, like polio in man, produce a demyelinating disease. Unlike the transient intestinal infection, the CNS form of the disease may persist for the life of the mouse.

Clinical Signs. Infections with wild-type strains of TMEV are usually subclinical; however, small numbers of mice may develop an extended viremia with dissemination of the virus to the spinal cord and brain, causing paralysis of the rear legs, circling, rolling, and other signs of CNS disease. There is little or no mortality. Signs of CNS disease in mice are by no means pathognomonic of mouse polio, and CNS signs in mice may be associated with disease processes other than mouse polio.

Whether or not symptoms are seen, the

severity of signs depends on the strain of infecting virus and the strain of mouse. BALB/c and C57B1/6 mice tend to be resistant, whereas DBA/2, SJL, SWR, and CD-1 outbred mice are more susceptible.

Diagnosis.   Diagnosis is usually based on serologic testing using either ELISA or IFA tests. Viral isolation, the presence of microscopic immune-related lesions of poliomyelitis, and demyelination in the CNS are also useful in establishing a diagnosis. Infection resembles polio of man and is sometimes referred to as mouse polio. Infected mice are an important model of polio in man; therefore, this disease of mice has been studied extensively. Much of the credit for our current success in controlling human polio worldwide can be attributed to continuing research using laboratory mice and other animals, especially monkeys, in polio research. Millions of cases of crippling disease and death in humans have been prevented through use of these animals.

Control and Prevention.   Use of barrier conditions and mice proved free of mouse polio by serologic testing are the most commonly used methods of preventing introduction of this disease into viral-free mouse colonies.

REFERENCES

Descoteaux, J.P., Grignon-Archambault, D., and Lussier, G.: Serologic study of the prevalence of murine viruses in five Canadian mouse colonies. Lab. Anim. Sci., *27*:621–626, 1977.

Downs, W.G.: Mouse encephalomyelitis virus. *In* The Mouse in Biomedical Research, Vol. II. Edited by H.L. Foster, J.D. Small, and J.G. Fox. New York, Academic Press, 1982, pp. 341–352.

Lipton, H.L., and Dal Canto, M.C.: Susceptibility of inbred mice to chronic central nervous system infection by Theiler's murine encephalomyelitis virus. Infect. Immun., *26*:369–374, 1979b.

Lipton, H.L., and Melvold, R.M.: Genetic analysis of susceptibility to Theiler's virus-induced demyelinating disease in mice. J. Immunol., *132*:1821–1825, 1984.

Lipton, H.L., and Rozhon, E.J.: The Theiler's murine encephalomyelitis viruses. *In* Viral and Mycoplasmal Infections of Laboratory Rodents. Edited by P.N. Bhatt, et al. New York, Academic Press, 1986, pp. 253–275.

Nitayaphan, S., Toth, M.M., and Roos, R.P.: Neutralizing monoclonal antibodies to Theiler's murine encephalomyelitis viruses. J. Virol., *53*:651–657, 1985.

Stroop, W.G., Baringer, J.R., and Brahic, M.: Detection of Theiler's virus RNA in mouse central nervous system by *in situ* hybridization. Lab. Invest., *45*:504–509, 1981.

Theiler, M.: Spontaneous encephalomyelitis of mice— A new virus disease. Science, *80*:122, 1934.

MURINE MYCOPLASMOSIS

Hosts.   Rats and mice are the principal natural hosts of *Mycoplasma pulmonis* respiratory and genital infection. Infection and disease are common in pets and non-barrier-housed rats and mice. Rabbits, guinea pigs, hamsters, and other rodents may on occasion carry the organism, but they are not clinically affected.

Etiology.   Mycoplasmas are very small pleomorphic organisms with no distinctive cell wall and a diameter between 0.2 and 1.0 μm. *Mycoplasma pulmonis,* which may accompany Sendai virus or bacterial infections, is an extracellular mucosal pathogen that colonizes mucosal surfaces and is responsible for the clinical signs and lesions of murine mycoplasmosis. Bacteria that may accompany *M. pulmonis* in respiratory disease include *Pasteurella pneumotropica, Actinobacillus* spp., *Streptococcus pneumoniae, Bordetella bronchi-* septica, cilia-associated respiratory bacillus (CAR), and *Corynebacterium kutscheri. Mycoplasma arthritidis,* though usually causing inapparent infections of rats and mice, may cause polyarthritis in rats.

Transmission.   *Mycoplasma pulmonis* is carried in the upper respiratory system. Transmission of the extremely contagious mycoplasmal infections is by direct contact between mother and young, respiratory aerosol over short distances, sexual transfer, animal carriers, and *in utero* passage.

Predisposing Factors.   Agents that damage the protective capacity of the respiratory epithelium predispose to *M. pulmonis* infection. Such agents include ammonia, sulfur dioxide, Sendai virus, and bacterial infections.

Ammonia gas in an animal colony is generated from urine and feces by urease-positive bacteria. Factors involved in ammonia accumulation include poor ventilation and sanitation, cage crowding, bacterial growth, and excessive populations. The metaplastic and ciliary inhibiting effects of ammonia can extend an innocuous upper respiratory infection into a bronchopneumonia.

Clinical Signs.   Although *M. pulmonis* infections are usually subclinical, at least to the casual observer, clinical signs occur and represent the three major foci of infection: upper respiratory, bronchopulmonary, and genital. In all areas the clinical onset is usually slow and progressive, but acute episodes may occur in young and susceptible animals.

The upper respiratory disease, involving the nasal passages and middle ears, is signaled by sniffling, occasional squinting, rough hair coat, and sneezing. If the inner ear becomes involved, torticollis or head tilt may occur.

The bronchopulmonary syndrome, initiated or exacerbated by ammonia, bacterial infections, or Sendai virus infection, is characterized by lethargy, rough hair coat, hunched posture, chattering, weight loss, labored breathing, and eventually death. Porphyrin may accumulate around the eyes and external nares. Unless the respiratory infections are complicated by bacteria, the terminal clinical stages of mycoplasmosis may last weeks or months.

In the ascending genital infection, which is common, infertility, embryonic resorptions, and small litters occur.

Necropsy Signs.   The upper respiratory infection is characterized by serous to purulent inflammation in affected tissues. In murine mycoplasmosis, unilateral or bilateral otitis media is a common finding, often the only gross abnormality. The pulmonary lesions in the early stages of the disease are well-demarcated foci of firm red or gray atelectasis and consolidation or a "cobblestone appearance." As the disease progresses, inflammatory debris accumulates in the air passages, resulting in bulging, mucopurulent areas of bronchiectasis. The content of these lumps is viscid to caseous and yellow-gray.

The genital infection, which may exist independently of the respiratory infection, is an ascending process that can involve the entire reproductive tract. Older females are more often affected. Metritis, pyometra, and purulent oophoritis and salpingitis characterize the serious genital infection.

Microscopic lesions may be acute or chronic and begin with neutrophils in airways, hyperplasia of mucosal epithelium, and lymphoid proliferation in the submucosa of the bronchi and progress to bronchiolar invasion, consolidation, bronchiectasis, and pulmonary abscesses.

Diagnosis.   Diagnosis of murine mycoplasmosis is based on gross and microscopic lesions and on the cultural isolation of *M. pulmonis* from the nasal pharynx, tympanic bullae, trachea, uterus, or lungs. The organism may be carried in the upper respiratory passages in the absence of clinical disease. Culture of *M. pulmonis* requires special media (Hayflick 1965) enriched with yeast extract and 10% swine or horse serum. The plates are incubated at 37° C in an atmosphere of normal or reduced oxygen and increased humidity.

The recent development of ELISA tests for detecting mycoplasmal infections has eliminated some of the uncertainty associated with the diagnosis of mycoplasmosis by cultural means, e.g., culture negatives, but false positives and cross reactions with other *Mycoplasma* occur.

The cilia-associated respiratory (CAR) bacillus may produce lesions resembling those associated with respiratory mycoplasmosis. In silver stained sections, large numbers of the bacilli can be found among cilia in the ciliated respiratory epithelium of the nasal passages, trachea, and other airways. An ELISA test can be used to detect serum antibody to the *CAR bacillus.*

Treatment.   Elimination of a mycoplasmal infection from large populations of affected rats and mice is, for all practical purposes, impossible. Antimicrobials placed in the drinking water, however, may suppress infection and clinical signs. Tetracycline hydrochloride at 5 mg/ml given fresh daily for 5 days or longer in deionized, sweetened drinking water (5% sucrose) often suppresses clinical signs if rats drink the concentrated solution (Stunkard, Schmidt and Cordano 1971). Some tetracycline solutions at this concentration in tap water form a scale that blocks the sipper tube. Use of distilled water reduces this effect. Lower levels of tetracycline may have an effect on secondary bacterial complications.

Valuable strains of rodents have been rendered free of mycoplasma by treatment of pregnant dams with oxytetracycline, hysterectomy, and foster nursing on mycoplasma-free dams. *In vitro,* tetracycline and tylosin have been shown more effective against *M. pulmonis* than are tiamulin, spectinomycin, lincomycin, and gentamycin.

Sulfamerazine at 0.02% in the drinking water or 1 mg/4 g feed, tylosin at 66 mg/L (2.5 g/10 gal) for 21 days, and chloramphenicol at 30 mg/kg body weight for 5 days (Habermann, et al. 1963) are other treatment suggestions, but the prognosis for recovery remains poor, and treatment should never be advocated as a method to eliminate *Mycoplasma* from a colony.

Prevention. Prevention of murine mycoplasmosis involves placing rodents that are free of *Mycoplasma pulmonis* into a barrier-sustained facility. Strict husbandry standards, exclusion of wild rodents, serologic and postmortem monitoring, good ventilation, and low population densities in cage and room help to maintain a colony free of *Mycoplasma pulmonis.* The experimental development of intranasal and other vaccines against *M. pulmonis* is a significant advance in prophylaxis. Use of inactivated or viable *M. pulmonis* vaccines decreased the severity of lower respiratory tract lesions in rats.

Public Health Significance. *Mycoplasma pulmonis* does not affect man, although the organism may be carried in the human nasal passage.

REFERENCES

Atobe, H., and Ogata, M.: Protective effect of killed *Mycoplasma pulmonis* vaccine against experimental infection in mice. Jpn. J. Vet. Sci., *39*:39–46, 1977.

Banerjee, A.K., et al.: An alternative method for the decontamination of rats carrying *Mycoplasma pulmonis* without the use of germfree isolators. Lab. Anims., *21*:138–142, 1987.

Banerjee, A.K., et al.: Naturally occurring genital mycoplasmosis in mice. Lab. Anims., *19*:275–276, 1985.

Broderson, J.R., Lindsey, J.R., and Crawford, J.E.: The role of environmental ammonia in respiratory mycoplasmosis of rats. Am. J. Pathol., *85*:115–130, 1976.

Carter, K.K., et al.: Tylosin concentrations in rat serum and lung tissue after administration in drinking water. Lab. Anim. Sci., *37*:468–470, 1987.

Cassell, G.H., Carter, P.B., and Silvers, S.H.: Genital disease in rats due to *Mycoplasma pulmonis*: development of an experimental model. Proc. Soc. Gen. Microbiol., *3*:150, 1976.

Cassell, G.H., and Davis, J.K.: Protective effect of vaccination against *Mycoplasma pulmonis* respiratory disease in rats. Infect. Immun., *21*:69–75, 1978.

Cassell, G.H., et al.: Mycoplasmal and rickettsial diseases. *In* The Laboratory Rat, Vol. I. Edited by H.J. Baker, J.R. Lindsey, and S.H. Weisborth. Orlando, Academic Press, 1979, pp. 243–269.

Cassell, G.H., Lindsey, J.R., and Davis, J.K.: Respiratory and genital mycoplasmosis of laboratory rodents: Implications for biomedical research. Isr. J. Med. Sci., *17*:548–554, 1981.

Cassell, G.H., Davis, J.K., and Lindsey, J.R.: Control of *Mycoplasma pulmonis* infection in rats and mice: Detection and elimination vs. vaccination. Isr. J. Med. Sci., *17*:674–677, 1981.

Cassell, G.H., and Brown, M.B.: Enzyme-linked immunosorbent assay (ELISA) for detection of antimycoplasmal antibody. *In* Methods in Mycoplasmology, Vol. I. Edited by S. Razin and J.G. Tully. Orlando, Academic Press, 1983, pp. 457–469.

Cassell, G.H., et al.: Recovery and identification of murine mycoplasmas. *In* Methods in Mycoplasmology, Vol. II. Edited by G. Tully and S. Razin. Orlando, Academic Press, 1983, pp. 129–142.

Cassell, G.H., et al.: Mycoplasmal infections: Disease pathogenesis, implications for biomedical research, and control. *In* Viral and Mycoplasmal Infections of Laboratory Rodents. Edited by P.N. Bhatt, et al. Orlando, Academic Press, 1986, pp. 87–130.

Cassell, G.H., et al.: State-of-the-art detection methods for rodent mycoplasmas. *In* Complications of Viral and Mycoplasma Infections in Rodents to Toxicology Research and Testing. Edited by T.E. Hamm, Jr. Washington, DC, Hemisphere Press, 1986, pp. 143–160.

Cole, B.C., and Cassell, G.H.: Mycoplasma infections as models of chronic joint inflammation. Arthritis Rheum., *22*:1375–1381, 1979.

Cole, B.C., and Ward, J.R.: Fate of intravenously injected *Mycoplasma arthritidis* in rodents and effect of vaccines. Infect. Immun., *7*:416–425, 1973.

Davis, J.K., and Cassell, G.H.: Murine respiratory mycoplasmosis in LEW and F344 rats: strain differences in lesion severity. Vet. Pathol., *19*:280–293, 1982.

Fortney, R.E., et al.: A transport medium for murine mycoplasmas. Lab. Anim. Sci., *30*:652–656, 1980.

Ganaway, J.R., and Allen, A.M.: Chronic murine pneumonia of laboratory rats. Production and description of pulmonary-disease-free rats. Lab. Anim. Care, *19*:71–79, 1969.

Ganaway, J.R., et al.: Isolation, propagation, and characterization of a newly recognized pathogen, cilia-associated respiratory bacillus of rats: An etiological agent of chronic respiratory disease. Infect. Immun., 47:472–479, 1983.

Gardner, M.C., Owens, D.R., and Wagner, J.E.: *In vitro* activity of select antibiotics against *Mycoplasma pulmonis* from rats and mice. Lab. Anim. Sci., 31:143–145, 1981.

Giddens, W.E., Jr., Whitehair, C.K., and Carter, G.R.: Morphologic and microbiologic features of the nasal cavity and middle ear in germfree, defined-flora, conventional and chronic respiratory disease-affected rats. Am. J. Vet. Res., 32:99–114, 1971.

Giddens, W.E., Jr., Whitehair, C.K., and Carter, G.R.: Morphologic and microbiologic features of trachea and lungs in germfree, defined-flora, conventional, and chronic respiratory disease-affected rats. Am. J. Vet. Res., 32:115–129, 1971.

Griffith, J.W., et al.: Cilia-associated respiratory (CAR) bacillus infection of obese mice. Vet. Pathol., 25:72–76, 1988.

Habermann, R.T., et al.: The effect of orally administered sulfamerazine and chlortetracycline on chronic respiratory disease in rats. Lab. Anim. Care, 13:28–40, 1963.

Halliwell, W.H., McCune, E.L., and Olson, L.D.: *Mycoplasma pulmonis*-induced otitis media in gnotobiotic mice. Lab. Anim. Sci., 24:57–61, 1974.

Hannan, P.C.T., and Hughes, B.O.: Reproducible polyarthritis in rats caused by *Mycoplasma arthritidis*. Ann. Rheum. Dis., 30:316–321, 1971.

Harwick, H.J., et al.: Arthritis in mice due to infection with *Mycoplasma pulmonis*. I. Clinical and microbiologic features. J. Infect. Dis., 128:533–540, 1973.

Hatsushita, S., Koshima, M., and Joshima, H.: Serodiagnosis of cilia-associated respiratory bacillus infection by the indirect immunofluorescence assay technique. Lab. Anims., 21:356–359, 1987.

Hayflick, L.: Tissue cultures and mycoplasmas. Tex. Rep. Biol. Med., 23:285–303, 1965.

Hill, A.C.: *Mycoplasma colis*, a new species isolated from rats and mice. Int. J. Syst. Bacteriol., 33:847–851, 1983.

Horowitz, S.A., and Cassell, G.H.: Detection of antibodies to *Mycoplasma pulmonis* by an enzyme-linked immunosorbent assay. Infect. Immun., 22:161–170, 1978.

Howard, C.J., Stott, E.J., and Taylor, G: The effect of pneumonia induced in mice with *Mycoplasma pulmonis* on resistance to subsequent bacterial infection and the effect of a respiratory infection with Sendai virus on the resistance of mice to *Mycoplasma pulmonis*. J. Gen. Microbiol., 109:79–87, 1978.

Jakab, G.J.: Interactions between Sendai virus and bacterial pathogens in the murine lung: a review. Lab. Anim. Sci., 31:170–177, 1981.

Jopp, A.J.: Myxomatosis—A personal view. N. Z. Vet. J., 34:52–54, 1986.

Kappel, H.K., Nelson, J.B., and Weisbroth, S.H.: Development for screening technic to monitor a *Mycoplasma*-free Blu(LE) Long-Evans rat colony. Lab. Anim. Sci., 24:768–772, 1974.

Kirchoff, H., et al.: Studies of polyarthritis caused by *Mycoplasma arthritidis* in rats. I. Detection of the persisting mycoplasma antigen by the enzyme-immune assay (EIA) and conventional culture technique. Zbl. Bakt. Hyg. I. Abt. Orig. A., 254:129–138, 1983.

Kohn, D.F.: Sequential pathogenicity of *Mycoplasma pulmonis* in laboratory rats. Lab. Anim. Sci., 21:849–855, 1971.

Kohn, D.F., Magill, L.S., and Chinookoswong, N.: Localization of *Mycoplasma pulmonis* in cartilage. Infect. Immun., 35:730–733, 1982.

LaRegina, M., Lonigro, J., and Steffen, E.: A comparison of three ELISA systems for the detection of *Mycoplasma pulmonis* antibody in rats. Lab. Anim. Sci., 37:331–334, 1987.

Lindsey, J.R., and Cassell, G.H.: Experimental *Mycoplasma pulmonis* infection in pathogen-free mice. Models of studying mycoplasmosis of the respiratory tract. Am. J. Pathol., 72:63–90, 1973.

Lindsey, J.R., et al.: Murine chronic respiratory disease. Significance as a research complication and experimental production with *Mycoplasma pulmonis*. Am. J. Pathol., 64:675–716, 1971.

McGarrity, G.J., et al.: *Mycoplasma muris*, a new species from laboratory mice. Int. J. Syst. Bacteriol., 33:350–355, 1983.

Mia, A.S., Kravcak, D.M., and Cassell, G.H.: Detection of *Mycoplasma pulmonis* antibody in rats and mice by a rapid micro enzyme-linked immunosorbent assay. Lab. Anim. Sci., 31:356–359, 1981.

Minion, F.C., Brown, M.B., and Cassell, G.H.: Identification of cross-reactive antigens between *Mycoplasma pulmonis* and *Mycoplasma arthritidis*. Infect. Immun., 43:115–121, 1984.

Organick, A.B., and Lutsky, I.I.: *Mycoplasma pulmonis* infection in gnotobiotic and conventional mice: aspects of pathogenicity including microbial enumeration and studies of tracheal involvement. Lab. Anim. Sci., 26:419–429, 1976.

Saito, M., et al.: Effects of gaseous ammonia on *Mycoplasma pulmonis* infection in mice and rats. Exp. Anim., 31:203–206, 1982.

Saito, M., et al.: Strain differences of mouse in susceptibility to *Mycoplasma pulmonis* infection [sic]. Jpn. J. Vet. Sci., 40:697–705, 1978.

Schoeb, T.R., Davidson, M.K., and Lindsey, J.R.: Intracage ammonia promotes growth of *Mycoplasma pulmonis* in respiratory tracts of rats. Infect. Immun., 38:212–217, 1982.

Stewart, D.D., and Buck, G.E.: The occurrence of *Mycoplasma arthritidis* in the throat and middle ear of rats with chronic respiratory disease. Lab. Anim. Sci., 25:769–773, 1975.

Stunkard, J.A., Schmidt, J.P., and Cordano, J.T.: Consumption of oxytetracycline in drinking water by healthy mice. Lab. Anim. Sci., 21:121–122, 1971.

Tanaka, H.: *Mycoplasma pulmonis* arthritis in congenitally athymic (nude) mice. Clinical and biological features. Microbiol. Immunol., 23:1055–1065, 1979.

Taylor-Robinson, D., and Furr, P.M.: Observations on the occurrence of mycoplasmas in the central nervous system of some laboratory animals. Lab. Anims., 15:223–227, 1981.

Thirkill, C.E., and Gregerson, D.S.: *Mycoplasma arthritidis*-induced ocular inflammatory disease. Infect. Immun., 36:775–781, 1982.

van Zwieten, M.J., et al.: Respiratory disease in rats associated with a filamentous bacterium. Lab. Anim. Sci., 30:215–221, 1980.

MYXOMATOSIS

Hosts. Myxomatosis is a viral disease of domestic rabbits. Wild rabbits (*Sylvilagus*) act as the natural or reservoir host. The virus causes local skin tumors in wild lagomorphs.

Etiology. The disease is caused by several strains of poxviruses on the more virulent end of the myxoma-fibroma spectrum of viruses. The highly virulent variant occurring in western California and Oregon is, appropriately, the California strain; other strains exist in South America, Australia, and Europe.

Transmission. The highly virulent virus is transmitted from the wild reservoir primarily by arthropod vectors. The dimensions of this transmission vary with the populations of mosquitoes, mites, and fleas feeding on rabbits. An increase in myxomatosis is seen from August to November. As these populations vary with seasons and years, so does the incidence of myxomatosis. Arthropods act as mechanical vectors, as might birds, plants, and fomites.

Clinical Signs. The first signs following infection with the California strain produce a "sleepy-eyed" rabbit: mild lethargy, red eyes and swollen lids, fever, and a watery ocular discharge. If the rabbit survives the acute stage, the reddening and swelling extend to the lips, face, ears, and anogenital areas. Death follows in a large percentage of rabbits. Those surviving develop cutaneous hemorrhages. Affected rabbits are anorectic and dehydrated, but many do survive and the lesions regress over 1 to 3 months. In chronic cases, the disease is often complicated by pasteurellosis, which causes the deaths.

Lesions in the European and South American disease forms include the development of skin tumors, which may eventually rupture and ooze.

Necropsy. Gross necropsy signs following infection with the California strain are subcutaneous edema and widespread visceral hemorrhage. Microscopically, there is extensive epithelial proliferation with ballooning of cells in the stratum granulosum and hyperkeratinization. Large, eosino-philic, intracytoplasmic inclusion bodies are present in the stratum germinativum. There is lymphocytic depletion in the spleen and necrosis of lymphatic tissue in several organs. European rabbits develop mucinous skin tumors and endothelial proliferation in capillaries and small venules.

Diagnosis. Diagnosis is based on clinical signs, necropsy findings, and the characteristic histopathologic appearance of the lesions.

Treatment. There is no treatment for myxomatosis.

Prevention. Vector control through spraying and screening, avoidance of wild rabbits, quarantine of new arrivals, and vaccination with attenuated vaccines prepared in the face of an outbreak are methods of preventing myxomatosis. The vaccine used to combat myxomatosis in Europe is not approved for use in the United States.

Public Health Significance. The myxomatosis virus does not affect man.

REFERENCES

Cheeke, P.R., Patton, N.M., and Templeton, G.S.: Rabbit Production. Danville, IL, Interstate Printers and Publishers, Inc., 1987.

Fenner, F., and Marshall, I.D.: A comparison of the virulence for European rabbits (*Oryctolagus cuniculus*) of strains of myxoma virus recovered in the field in Australia, Europe and America. J. Hyg. (Camb.), *55*:149–191, 1957.

Fenner, F., and Woodroofe, G.M.: The pathogenesis of infectious myxomatosis: the mechanism of infection and the immunological response in the European rabbit (*Oryctolagus cuniculus*). Br. J. Exp. Pathol., *34*:400–411, 1953.

Grodhaus, G., Regnery, D.C., and Marshall, I.D.: Studies in the epidemiology of myxomatosis in California. II. The experimental transmission of myxomatosis in brush rabbits (*Sylvilagus bachmani*) by several species of mosquitoes. Am. J. Hyg., *77*:205–212, 1963.

Gumbrell, R.C.: Myxomatosis and rabbit control in New Zealand. N. Z. Vet. J., *34*:54–55, 1986.

Hurst, E.W.: Myxoma and the Shope fibroma. I: The histology of myxoma. Br. J. Exp. Pathol., *18*:1–14, 1937.

Kessel, J.F., Prouty, C.C., and Meyer, J.W.: Occurrence of infectious myxomatosis in Southern California. Proc. Soc. Exp. Biol. Med., *28*:413–414, 1931.

Marshall, I.D., and Regnery, D.C.: Myxomatosis in a California brush rabbit (*Sylvilagus bachmani*). Nature, *188*:73–74, 1960.

McKercher, D.G., and Saito, J.K.: An attenuated live virus vaccine for myxomatosis. Nature, 202:933–934, 1964.

Patton, N.M., and Holmes, H.T.: Myxomatosis in domestic rabbits in Oregon. J. Am. Vet. Med. Assoc., 171:560–562, 1977.

Rivers, T.M.: Infectious myxomatosis of rabbits. Observations on the pathological changes induced by virus myxomatosum (Sanarelli). J. Exp. Med., 51:965–976, 1930.

Rosamond, C.H., Shepherd, C.H., and Edmonds, J.W.: Myxomatosis: changes in the epidemiology of myxomatosis coincident with the establishment of the Eu-

ropean rabbit flea Spilopsyllus cuniculi (Dale) in the Mallee region of Victoria. J. Hyg. (Camb.), 81:399–403, 1978.

Ross, J., and Sanders, M.F.: The development of genetic resistance to myxomatosis in wild rabbits in Britain. J. Hyg. (Camb.), 92:255–261, 1984.

Saito, J.K., McKercher, D.G., and Castrucci, G.: Attenuation of the myxoma virus and use of the living attenuated virus as an immunizing agent for myxomatosis. J. Infect. Dis., 114:417–428, 1964.

Stewart, F.W.: The fundamental pathology of infectious myxomatosis. Am. J. Cancer, 15:2010–2028, 1931.

NEOPLASIA

Although literature surveys of the incidence of neoplasia in rabbits and rodents contain lengthy lists of tumor types, each species, if not each strain, actually possesses a limited number of "common" tumors. Therefore, even though this section describes only certain tumors, it is important to note that individual animals may develop one or more of a wide variety of neoplasms.

REFERENCES

Fenner, F.: Classification of myxoma and fibroma viruses. Nature (London), 171:562–563, 1963.

Jones, S.R., et al.: Naturally occurring neoplastic diseases. In Handbook of Laboratory Animal Science, Vol. III. Edited by E.C. Melby, Jr., and N.H. Altman. Cleveland, OH, CRC Press, Inc., 1976, pp. 221–381.

Pour, P., Ii, Y., and Althoff, J.: Comparative studies on spontaneous tumor incidence based on systematic histologic examination of rat and hamster strains of the same colony. Prog. Exp. Tumor Res., 24:199–206, 1979.

Prejean, J.D., et al.: Spontaneous tumours in Sprague-Dawley rats and Swiss mice. Cancer Res., 33:2768–2773, 1973.

Roe, F.J.C.: Spontaneous tumours in rats and mice. Food Cosmet. Toxicol., 3:707–720, 1965.

Squire, R.A., et al.: Tumors. In Pathology of Laboratory Animals, Vol. II. Edited by K. Benirschke, F.M. Garner, and T.C. Jones. New York, Springer-Verlag, 1978, pp. 1051–1283.

NEOPLASIA IN THE RABBIT

The adenocarcinoma of the uterine endometrium is the most common tumor of Oryctolagus, but its occurrence may be influenced by genetic background, age, and probably endocrinologic factors. This tumor has been reported more often in the Tan and Dutch breeds and less often in the Polish and Rex breeds. Rabbits of the higher-incidence breeds under 3 years of age have an incidence of approximately 4%, whereas rabbits over 3 years old have an incidence of uterine adenocarcinoma approaching 50% to 80%. There may also be higher incidences among hybrid than inbred strains. The association of uterine adenocarcinoma with pregnancy toxemia, pseudopregnancy, and hyperestrogenism remains controversial. The dose of estrogen may be the important factor influencing the carcinogenic stimulus or inhibition.

If there is an immediate, predisposing influence to tumor development, then senile atrophy of the endometrium probably assumes that role. With increasing age, the endometrial cells, beginning deep in the glandular crypts, become less specialized, and the connective tissue stroma becomes less cellular and more collagenous. These progressive, senile changes may underlie the transition to adenomatous and cystic hyperplasia and the in situ carcinoma. The position of cystic hyperplasia in the transition is uncertain.

The clinical signs associated with uterine neoplasia in the rabbit include an altered reproductive performance and eventually death over a period of 5 to 20 months. During the subclinical, in situ period, affected does become less fertile, have smaller litters, abort, resorb fetuses, deliver dead young, retain fetuses past term, and may have a bloody vulvar discharge. The multiple neoplastic nodules, which may be from 1 to 5 cm in diameter by 6 months, can be palpated through the abdominal and uterine walls.

The hyperplastic stage lasts approximately 3 months, the carcinoma *in situ* until 7 months, and the metastatic stage from 10 to 12 months. The tumors are usually ovoid, firm, and hemorrhagic and are regularly spaced along the mesometrial junction. The neoplasm may invade the myometrium and peritoneal cavity before hemotogenous metastasis occurs. Multiple tumors may be of different sizes, but they are usually at the same stage of differentiation.

The myxofibroma group of poxvirus-induced tumors occurs as fibromas in the *Sylvilagus* (cottontail) in both the Eastern and Western United States and in Europe, South America, and Australia. These antigenically related viruses are transmitted mechanically by mosquitoes and biting insects from cottontails to domestic rabbits, in which disease processes range from peracute, fatal myxomatosis to transitory and single fibromas.

Myxomatosis, described elsewhere in this chapter, induces a generalized proliferation of reticuloendothelial cells and their mucin, which with some strains of virus results in the formation of irregular, subcutaneous, gelatinous tumors. These tumors, however, are not characteristic of the California disease.

A related poxvirus, endemic in Eastern cottontails (*S. floridanus*), will cause self-limiting, subcutaneous fibromas in adult, domestic rabbits. In young *Oryctolagus,* the disease may become a disseminated fibromatosis. Prevention of the myxofibromatous diseases involves the exclusion of biting insects from the rabbitry and vaccination during an epidemic.

Other neoplasms of the rabbit include lymphosarcoma, papilloma, embryonal nephroma, squamous cell carcinoma, bile duct tumor, and osteogenic sarcoma. Lymphosarcoma, a tumor of juvenile and young adult rabbits, is probably the second most common tumor of rabbits. Lymphosarcomas are characterized by pale, enlarged kidneys, hepato- and splenomegaly, and lymphad-

enopathy. The homozygous state of an autosomal recessive gene may be necessary for development of lymphosarcoma.

REFERENCES

Baba, N., and von Haam, E.: Animal model: spontaneous adenocarcinoma in aged rabbits. Am. J. Pathol., *68*:653–656, 1972.

Burrow, H.: Spontaneous uterine and mammary tumours in the rabbit. J. Pathol. Bacteriol., *51*:385–390, 1940.

Cloyd, G.G., and Johnson, G.R.: Lymphosarcoma with lymphoblastic leukemia in a New Zealand white rabbit. Lab. Anim. Sci., *28*:66–69, 1978.

Dominguez, J.A., Corella, E.L., and Auró, A.: Oral papillomatosis in two laboratory rabbits in Mexico. Lab. Anim. Sci., *31*:71–73, 1981.

Finnie, J.W., Bustock, D.E., and Walden, N.B.: Lymphoblastic leukaemia in a rabbit: a case report. Lab. Anims., *14*:49–51, 1980.

Flatt, R.E.: Pyometra and uterine adenocarcinoma in a rabbit. Lab. Anim. Care, *19*:398–401, 1969.

Flatt, R.E., and Weisbroth, S.H.: Interstitial cell tumor of the testicle in rabbits: a report of two cases. Lab. Anim. Sci., *24*:682–685, 1974.

Fox, R.R., et al.: Lymphosarcoma in the rabbit: genetics and pathology. J. Natl. Cancer Inst., *45*:719–729, 1970.

Ginder, D.R.: Rabbit papillomas and the rabbit papilloma virus. A review. N.Y. Acad. Sci., *54*:1120–1125, 1952.

Green, H.S.N., and Strauss, J.S.: Multiple primary tumors in the rabbit. Cancer, *2*:673–691, 1949.

Gupta, B.N.: Lymphosarcoma in a rabbit. Am. J. Vet. Res., *37*:841–843, 1976.

Hagen, K.W.: Spontaneous papillomatosis in domestic rabbits. Bull. Wildl. Dis. Assoc., *2*:108–110, 1966.

Heiman, J.: Spontaneous mammary carcinoma in a rabbit. Am. J. Cancer, *29*:93–101, 1937.

Hinton, M., and Regan, M.: Cutaneous lymphosarcoma in a rabbit. Vet. Rec., *103*:140–141, 1978.

Joiner, G.N., Jardine, J.H., and Gleiser, C.A.: An epizootic of Shope fibromatosis in a commercial rabbitry. J. Am. Vet. Med. Assoc., *159*:1583–1587, 1971.

Kaufmann, A.F., and Quist, K.D.: Spontaneous renal carcinoma in a New Zealand white rabbit. Lab. Anim. Care, *20*:530–532, 1970.

Kinkler, R.J., and Jepsen, P.L.: Ependymoma in a rabbit. Lab. Anim. Sci., *29*:255–256, 1979.

Mews, A.R., et al.: Detection of oral papillomatosis in a British rabbit colony. Lab. Anims., *6*:141–145, 1972.

Port, C.D., and Sidor, M.A.: A sebaceous gland carcinoma in a rabbit. Lab. Anim. Sci., *28*:215, 1978.

Pulley, L.T., and Shively, J.N.: Naturally occurring infectious fibroma in the domestic rabbit. Vet. Pathol., *10*:509–519, 1973.

Ruflo, C.P., et al.: Characterization of a fibroma virus isolated from naturally-occurring skin tumors in domestic rabbits. Lab. Anim. Sci., *23*:525–532, 1973.

Szczech, G.M., et al.: Fibroma in Indiana cottontail rabbits. J. Am. Vet. Med. Assoc., *165*:846–849, 1974.

Walberg, J.A.: Osteogenic sarcoma with metastasis in a rabbit (*Oryctolagus cuniculus*). Lab. Anim. Sci., *31*:407–408, 1981.

Weisbroth, S.H., and Scher, S.: Spontaneous oral papillomatosis in rabbits. J. Am. Vet. Med. Assoc.,
157:1940–1944, 1970.

NEOPLASIA IN THE GUINEA PIG

Rhabdomyomatosis, an accumulation of glycogen-bearing myocardial cells, and not a neoplasm, is visible grossly as pale foci on the mural and valvular endocardial surfaces of the atria and ventricles. Another non-neoplastic proliferation in the guinea pig is the embryonic placentoma, a multiple-layered transitory growth of parthogenic origin occurring within the ovary of the young female. The placentoma is resolved by fibrosis and may be related to the neoplastic ovarian teratoma.

True neoplastic processes are rare in guinea pigs, but an age-related increase in incidence has been noted. Estimates of the incidence of neoplasia in guinea pigs over 3 years of age range up to 30%. Pulmonary neoplasia, usually the bronchogenic papillary adenoma, is the most common category of tumors in the guinea pig, comprising 35% of the total. The second category (15%) is tumors of the skin and subcutis, although a single report of several trichofolliculomas biases the incidence. Tumors of the reproductive tract, the mammary glands, and the hematopoietic system comprise the remainder of the guinea pig tumors. Spontaneous lymphocytic leukemia is an acute, virally induced (c-type RNA virus) lymphoblastic leukemia often fatal within 5 days of onset. The hair coat of the affected animal becomes rough, the mucous membranes appear pale, and the liver and lymphatic tissues are greatly enlarged. The white blood cell count may reach 250,000 mm^3.

REFERENCES

Congdon, C.C., and Lorenz, E.: Leukemia in guinea-
pigs. Am. J. Pathol., *30*:337–359, 1954.
Ediger, R.D., and Rabstein, M.M.: Spontaneous leukemia in a Hartley strain guinea pig. J. Am. Vet.
Med. Assoc., *153*:954–956, 1968.
Frisk, C.S., Wagner, J.E., and Doyle, R.E.: An ovarian
teratoma in a guinea pig. Lab. Anim. Sci., *28*:199–
201, 1978.
Hong, C.C.: Spontaneous papillary cystadenocarci-
noma of the ovary in Duncan-Hartley guinea pigs.
Lab. Anims., *14*:39–40, 1980.

Hong, C.C., and Liu, P.I.: Osteogenic sarcoma in 2
guinea pigs. Lab. Anims., *15*:49–51, 1981.
Hong, C.C., Liu, P.I., and Poon, K.C.: Naturally occurring lymphoblastic leukemia in guinea pigs. Lab.
Anim. Sci., *30*:222–226, 1980.
Kitchen, D.N., Carlton, W.W., and Bickford, A.A.: A
report of fourteen spontaneous tumors of the guinea
pig. Lab. Anim. Sci., *25*:92–102, 1975.
Vink, H.H.: Rhabdomyomatosis (nodular glycogenic
infiltration) of the heart in guinea-pigs. J. Pathol.,
97:331–334, 1969.
Wolff, A., et al.: Cervical lymphoblastic lymphoma in
an aged guinea pig. Lab. Anim. Sci., *38*:83–84, 1988.
Zarrin, K.: Thyroid carcinoma of a guinea pig: A case
report. Lab. Anims., *8*:145–148, 1974.
Zwart, P., et al.: Cutaneous tumours in the guinea pig.
Lab. Anim., *15*:375–377, 1981.

NEOPLASIA IN THE HAMSTER

The incidence of spontaneous neoplasia in the golden hamster is low, but how low depends on the source of the survey and the ages of the animals described. Reports of incidences vary from 4% as an overall population incidence to 50% or more in hamsters over 2 years of age. Although spontaneous neoplasia is uncommon in hamsters, these animals are remarkably susceptible to a wide range of experimentally induced tumors. The hamster's eversible cheek pouch is an easily seen, relatively immunologically protected site for tumor transplantation.

Reports of incidences vary from colony to colony, but tumors of the adrenal cortex comprise the largest reported group. Next in incidence are tumors of the gastrointestinal tract (polyps, papillomas, and adenocarcinomas), tumors of the lymphoreticular system, and tumors of the skin and subcutis.

REFERENCES

Ambrose, K.R., and Coggin, J.H., Jr.: An epizootic in
hamsters of lymphomas of undetermined origin and
mode of transmission. J. Natl. Cancer Inst., *54*:335–
336, 1969.
Banfield, W.G.: Hamster lymphomas, TM. Natl. Cancer Inst. Monogr., *32*:335–336, 1969.
Barthold, S.W., Bhatt, P.N., and Johnson, E.A.: Further
evidence for papovavirus as the probable etiology of
transmissible lymphoma of Syrian hamsters. Lab.
Anim. Sci., *37*:283–288, 1987.
Fortner, J.G.: Spontaneous tumors, including gastrointestinal neoplasms and malignant melanomas in the
Syrian hamster. Cancer, *10*:1153–1156, 1957.
Homburger, F.: Background data for tumor incidence
in control animals (Syrian hamsters). Prog. Exp. Tumor Res., *26*:259–265, 1983.

Kesterson, J.W., and Carlton, W.W.: Multiple malignant neoplasms in a golden hamster. A case report and literature survey. Lab. Anim. Care, *20*:220–225, 1970.

Mangkoewidjojo, S., and Kim, J.C.S.: Malignant melanoma metastatic to the lung in a pet hamster. Lab. Anims., *11*:125–127, 1977.

Pour, P., et al.: Spontaneous tumors and common diseases in two colonies of Syrian hamsters. I. Incidence and sites. J. Natl. Cancer Inst., *56*:937–948, 1976.

Pour, P., et al.: Spontaneous tumors and common diseases in two colonies of Syrian hamsters. II. Respiratory tract and digestive system. J. Natl. Cancer Inst., *56*:937–948, 1976.

Pour, P., et al.: Spontaneous tumors and common diseases in two colonies of Syrian hamsters. III. Urogenital system and endocrine glands. J. Natl. Cancer Inst., *56*:949–961, 1976.

Pour, P., et al.: Spontaneous tumors and common disease in three types of hamsters. J. Natl. Cancer Inst., *63*:797–811, 1979.

Trahan, C.J., and Mitchell, W.C.: Spontaneous transitional cell carcinoma in the urinary bladder of a Strain 13 guinea pig. Lab. Anim. Sci., *36*:691–693, 1986.

Yabe, Y., et al.: Spontaneous tumors in hamsters: incidence, morphology, transplantation, and virus studies. Gan., *63*:329–336, 1972.

NEOPLASIA IN THE GERBIL

The incidence of spontaneous neoplasia in gerbils over 2 years of age has been reported to be approximately 24%, with a higher incidence in aged animals. Pseudoadenomatous structures of the skin, cystic ovaries, and periovarian cysts, which can cause infertility, are common, non-neoplastic processes occurring in the gerbil. Neoplasms in gerbils cover a wide range of types, with tumors of the female reproductive system perhaps most common. Granulosa cell tumors, uterine adenocarcinomas, lutein cell tumors, and dysgerminomas have been reported. Adrenal adenomas and adenocarcinomas are also common tumors of gerbils. Neoplasms of the skin, often found in association with the ventral scent gland, include basal cell carcinomas, melanomas, sebaceous adenomas, and squamous cell carcinomas. Neoplasms have been reported in several other tissues, most being of mesenchymal origin.

REFERENCES

Benitz, K.-F., and Kramer, A.W., Jr.: Spontaneous tumors in the Mongolian gerbil. Lab. Anim. Care, *15*:281–294, 1965.

Cramlet, S.H., Toft, J.D., II, and Olsen, N.W.: Malignant melanoma in a black gerbil (*Meriones unguiculatus*). Lab. Anim. Sci., *24*:545–547, 1974.

Meckley, P.E., and Zwicker, G.M.: Naturally-occurring neoplasms in the Mongolian gerbil, *Meriones unguiculatus*. Lab. Anims., *13*:203–206, 1979.

Ringler, D.H., Lay, D.M., and Abrams, G.D.: Spontaneous neoplasms in aging Gerbillinae. Lab. Anim. Sci., *22*:407–414, 1972.

Shumaker, R.C., Paik, S.K., and Houser, W.D.: Tumors in Gerbillinae: A literature review and a report of a case. Lab. Anim. Sci., *24*:688–690, 1974.

Vincent, A.L., and Ash, L.R.: Further observations on spontaneous neoplasms in the Mongolian gerbil, *Meriones unguiculatus*. Lab. Anim. Sci., *28*:297–300, 1978.

Vincent, A.L., Rodrick, G.E., and Sodeman, W.A., Jr.: The pathology of the Mongolian gerbil (*Meriones unguiculatus*): A review. Lab. Anim. Sci., *29*:645–651, 1979.

NEOPLASIA IN THE MOUSE

Neoplasia in mice is one of the most extensively investigated disease processes in rodents; volumes have been written about the causation, pathogenesis, structure, and resolution of murine tumors. With the development of inbred strains and selection for tumor susceptibility and resistance, the pattern of tumor incidence has become quite different from the pattern in the wild house mouse or random-bred white mouse. Mice develop a great variety of tumors, and as in other animals, neoplasia may be manifested clinically as weight gain or loss, cutaneous, subcutaneous, or abdominal swellings, infertility, external or internal bleeding, increased susceptibility to infection, and death.

Adult wild or random-bred female breeder mice have the following approximate tumor incidences: pulmonary tumors, 28%; hemangioendotheliomas, 8%; ovarian tumors, 6%; mammary tumors, 6%; hepatomas, 4%; leukemias, 2%; reticulum cell sarcomas, 2%; and subcutaneous sarcomas, 2%. Representative inbred strains and their approximate tumor incidences are described in the *Handbook on the Laboratory Mouse* by Crispens (1975). In general, mammary tumors and lymphosarcomas are among the most common tumors of inbred strains.

Some representative inbred strains and their associated neoplasms are leukemias in AKR and C3H mice, mammary tumors in the C3H strain, plasma cell tumors in BALB/c mice, and type B reticulum cell sarcomas in SJL mice. Tumor incidence varies with sex, age, parity, and substrain. Oncogenic murine viruses (oncornaviruses) include the mouse mammary tumor virus (MTV), the mouse leukemia virus (MLV), and the mouse sarcoma virus (MSV). The mouse mammary tumor virus, or the Bittner agent, is an RNA virus, which, when present in a susceptible strain, predisposes to mammary adenocarcinoma. This virus is widely distributed in the host and is passed to the fetus and, via the milk, to the neonate. The virus may be detected by electron microscopy or by the oncogenic consequences of injection into BALB/c mice. Unlike the rat mammary tumor, which is easily removed, the mouse mammary adenocarcinoma is soft, fleshy, highly vascularized, and infiltrative. It cannot be removed without extensive tissue damage and hemorrhage. Although polyoma virus may infect cell lines and produce tumors in inoculated mice, hamsters, and guinea pigs, its occurrence as a natural infection is unlikely.

The mouse leukemia viruses are transmitted through the placenta or passed in the milk. C58 and AKR strains are highly susceptible to leukemia. Other predisposing factors include radiation exposure and chemical carcinogens.

REFERENCES

Abbott, D.P., Gregson, R.L., and Imm, S.: Spontaneous ovarian teratomas in laboratory mice. J. Comp. Pathol., 93:109–114, 1983.

Artzt, K., and Damjanov, I.: Spontaneous extragonadal teratocarcinoma in a mouse. Lab. Anim. Sci., 28:584–586, 1978.

Cavaliere, A., Bacci, M., and Fratini, D.: Spontaneous pancreatic adenocarcinoma in a mouse (*Mus musculus*). Lab. Anim. Sci., 31:502–503, 1981.

Charles, R.T., and Turusov, V.S.: Bone tumors in CF-1 mice. Lab. Anims., 8:137–144, 1974.

Crispens, C.G.: Handbook on the Laboratory Mouse. Springfield, IL, Charles C Thomas, 1975.

Eaton, G.J., et al.: The Icr:Ha(ICR) mouse: a current account of breeding, mutations, diseases and mortality. Lab. Anims., 14:17–24, 1980.

Frith, C.H., Johnson, B.P., and Highman, B.: Osteosarcomas in BALB/c female mice. Lab. Anim. Sci., 32:60–63, 1982.

Goodall, C.M., Bielschowsky, M., and Forster, D.R.: Incidence and metastatic pattern of lymphoreticular neoplasms in untreated NZO/B1 mice. Lab. Anim., 6:85–94, 1972.

Horn, H.A., and Stewart, H.L.: A review of some spontaneous tumors in noninbred mice. J. Natl. Cancer Inst., 13:591–603, 1952.

Sass, B.: Mixed mesenchymal tumours of the mouse uterus. Lab. Anims., 15:365–369, 1981.

Sass, B.: The occurrence of a bilateral mandibular mast cell neoplasm in a mouse with lymphocytic leukemia. Lab. Anim. Sci., 29:492–494, 1979.

Sass, B., Peters, R.L., and Kelloff, G.J.: Differences in tumor incidence in two substrains of Claude BALB/c (BALB/cfCd) mice, emphasizing renal, mammary, pancreatic, and synovial tumors. Lab. Anim. Sci., 26:736–741, 1976.

Sheldon, W.G., et al.: Distribution of mammary gland neoplasms and factors influencing metastases in hybrid mice. Lab. Anim. Sci., 32:166–168, 1982.

Skinner, H.H., Knight, E.H., and Lancaster, M.C.: Lymphomas associated with a tolerant lymphocytic choriomeningitis virus infection in mice. Lab. Anims., 14:117–121, 1980.

Turnsov, V.S. (Ed.): Pathology of Tumours in Laboratory Animals. Vol. II. Tumours of the Mouse. Lyon, International Agency for Research on Cancer, 1979.

NEOPLASIA IN THE RAT

As with the mouse and other rodents, the reported incidence of spontaneous neoplasia in the rat varies with the age, sex, strain, and environmental circumstances of the colony surveyed. Neoplasia in rats has a reported incidence of up to 87% in rats over 2 years of age. The most common tumor in the rat is the mammary fibroadenoma, followed (in order of incidence) by testicular interstitial cell tumors, pheochromocytomas, pituitary adenomas, uterine endometrial polyps, malignant lymphomas, mononuclear cell leukemia, and thyroid adenomas. Polyoma virus infections do not occur naturally in rats.

Mammary neoplasms are most often benign fibroadenomas; adenocarcinomas comprise less than 10% of mammary tumors in the rat. The benign tumor, which may occur in males, is usually single and may grow to 8 or 10 cm in diameter. The fibroadenomas are well demarcated, ovoid or discoid, firm, and nodular, although considerable variation exists in size, growth rate, color, and consistency. Mammary tumors of rats may be ulcerated, hemorrhagic, and necrotic. The tumor is well tolerated by the host until the mass hinders locomotion or a septicemia or toxemia results from the ulceration and necrosis. Rats with large tumors lose weight and die.

The encapsulated tumors may be surgically removed, although care should be taken to ligate the large vessels entering the mass. Because mammary tissue is widely distributed in the subcutis of murine rodents, mammary tumors may be found behind the shoulders, on the ventral abdomen and flank, or around the tail base.

Testicular tumors of old rats most frequently occur in the Leydig or interstitial cells. These benign neoplasms are usually multiple, bilateral, soft, and yellow to brown.

Pituitary gland neoplasms occur frequently in older female rats on diets high in protein or calories. Chromophobe adenomas, the most common type of pituitary tumors, are soft, nonsecreting, and well circumscribed, with irregular surfaces. Brain tissue compression from chromophore adenomas may cause hydrocephalus and result in head tilt or depression.

REFERENCES

Adams, S.W., and Crowley, A.M.: Posterior paralysis due to spontaneous oligodendroglioma in the spinal cord of a rat. Lab. Anim. Sci., 37:345–347, 1987.

Altman, N.H., and Goodman, D.G.: Neoplastic diseases. In The Laboratory Rat. Vol. I. Edited by H.J. Baker, J.R. Lindsey, and S.H. Weisbroth. New York, Academic Press, 1979, pp. 333–376.

Benitz, K.F., and Roth, R.N.: A spontaneous metastasizing exocrine adenocarcinoma of the pancreas in the rat. Lab. Anim. Sci., 30:64–66, 1980.

Bullock, F.D., and Curtis, M.R.: Spontaneous tumors of the rat. J. Cancer Res., 14:1–115, 1930.

Coleman, G.L., et al.: Pathological changes during aging in barrier-reared Fischer 344 male rats. J. Gerontol., 32:258–278, 1977.

Crain, R.C.: Spontaneous tumors in the Rochester strain of the Wistar rat. Am. J. Pathol., 34:311–335, 1958.

Deerberg, F., and Rehm, S.: Tumours of the external auditory canal and the auditory sebaceous glands in Han:WIST rats. Short Communication. Z. Versuchstierkd., 23:134–137, 1981.

Deerberg, F., Rehm, S., and Pittermann, W.: Uncommon frequency of adenocarcinomas of the uterus in virgin Han:Wistar rats. Vet. Pathol., 18:707–713, 1981.

Fitzgerald, J.E., Schardein, J.L., and Kaump, D.H.: Several uncommon pituitary tumors in the rat. Lab. Anim. Sci., 21:581–584, 1971.

Glaister, J.R., Samuels, D.M., and Tucker, M.J.: Ganglioneuroma-containing tumours of the adrenal medulla in Alderly Park rats. Lab. Anims., 11:35–37, 1977.

Goodman, D.G., et al.: Neoplastic and nonneoplastic lesions in aging Osborne-Mendel rats. Toxicol. Appl. Pharmacol., 55:433–447, 1980.

Heath, J.E.: Granulocytic leukemia in rats: a report of two cases. Lab. Anim. Sci., 31:504–506, 1981.

Heslop, B.F.: Cystic adenocarcinoma of the ascending colon in rats occurring as a self-limiting outbreak. Lab. Anims., 3:185–195, 1969.

MacKenzie, W.F., and Garner, F.M.: Comparison of neoplasms in six sources of rats. J. Natl. Cancer Inst., 50:1243–1257, 1973.

Magnusson, G., Majeed, S., and Gopinath, C.: Infiltrating pituitary neoplasms in the rat. Lab. Anims., 13:111–113, 1979.

Magnusson, G., Majeed, S., and Offer, J.M.: Intraocular melanoma in the rat. Lab. Anim., 12:249–252, 1978.

Newman, A.J., and Mawdesley-Thomas, L.E.: Spontaneous tumours of the central nervous system of laboratory rats. J. Comp. Pathol., 84:39–50, 1974.

Noble, R.L., and Cutts, J.H.: Mammary tumors of the rat: a review. Cancer Res., 19:1125–1139, 1959.

Pollard, M., and Luckert, P.H.: Spontaneous liver tumors in aged germfree Wistar rats. Lab. Anim. Sci., 29:74–77, 1979.

Reznik, G., and Ward, J.M.: Morphology of hyperplastic and neoplastic lesions in the clitoral and preputial gland of the F344 rat. Vet. Pathol., 18:228–238, 1981.

Robertson, J.L., Garman, R.H., and Fowler, E.H.: Spontaneous cardiac tumors in eight rats. Vet. Pathol., 19:30–37, 1982.

Schardein, J.L., and Fitzgerald, J.E.: Teratoma in a Wistar rat. Lab. Anim. Sci., 27:114, 1977.

Stefanski, S.A., Elwell, M.R., and Yoshitomi, K.: Malignant hybridoma in a Fischer 344 rat. Lab. Anim. Sci., 37:347–350, 1987.

Turnsov, V.S. (Ed.): Pathology of Tumours in Laboratory Animals. Vol. I. Parts 1 & 2. Tumours of the Rat. Lyon, International Agency for Research on Cancer, 1973 and 1976.

NEPHROSIS

Hosts.　　Chronic, progressive renal disease occurs in several species, including rats, mice, guinea pigs, and hamsters.

Etiology.　　The specific etiologies of chronic renal disease in rats and hamsters have not been fully determined, but amino acid toxicity, excessive protein in diet, mineral inadequacies, viral infection, reaction against normal intestinal flora, and autoimmune phenomena have been suggested. Tubular occlusion by proteinaceous casts, periarteritis, and tubular cell hyperplasia and necrosis have been suggested as the immediate causes of renal impairment in rats. This very common condition is found in laboratory rats on commercial laboratory diets fed *ad libitum.*

A variable but large percentage of hamsters over 12 months of age develop an interstitial or glomerular deposition of hyaline or amyloid. The lesion in aged guinea pigs is nephrosclerosis; both affected hamsters and guinea pigs may be otherwise normal, or the guinea pigs may be afflicted with a chronic staphylococcal infection. Lesions in mice are frequently caused by autoimmune disease, particularly in the MRL and New Zealand black and white strains.

Transmission.   These diseases are not thought to be contagious.

Predisposing Factors.   Laboratory rats on conventional laboratory diets have a high incidence of chronic renal disease. Advancing age, the male sex, prolactin secretion or administration, the strain of rat, and diets high in protein, carbohydrates, and calories or low in potassium may all predispose to chronic renal disease in rats. Nephrosis in rats is already evident at 3 to 6 months of age, although age-related renal changes are delayed and reduced in germfree rats.

Renal amyloidosis in hamsters and nephrosclerosis and amyloidosis in guinea pigs occur in otherwise healthy animals. Chronic bacterial infections, such as staphylococcal pododermatitis in guinea pigs, have been related to an accelerated deposition of amyloid. Demodecosis may become manifest in hamsters with advanced chronic renal disease.

Clinical Signs.   Chronic renal disease is usually a subclinical condition detected only on necropsy. In severe cases affected animals progressively lose weight, become inactive, and die. A pronounced proteinuria (mostly albumin) appears in affected yearling rats. Polyuria may appear in hamsters and guinea pigs, and mice with autoimmune nephrosis may also have anemia and generalized edema.

Necropsy Signs.   Chronic renal disease in rats is a bilateral nephrosis in which the kidneys may be enlarged two or three times their normal size, discolored tan to yellow, appear nodular, granular, or pitted, and have radial, pale striations on section. Cortical cysts may be up to 3 mm in diameter. Affected kidneys in other animals are enlarged, have surface irregularities, and are pale.

Histologic changes include albuminous casts within the lumen of the tubules and Bowman's capsule, tubular atrophy, thickened basement membranes within the capsule, glomerular adhesions, and lymphocytic infiltration.

Diagnosis.   Diagnosis of chronic nephrosis in rats is based on gross and microscopic lesions. Histologic abnormalities include dilation of the tubules (loop of Henle and distal convoluted tubules) with proteinaceous material, atrophy of the tubular epithelium, and fibrosis and mild lymphocytic infiltration of the renal interstitium. These lesions may be focal at the corticomedullary junction, or they may be wedge-shaped, with the broad portion at the capsule. The glomeruli and capsule, if affected, are thickened. Amyloid may be demonstrated either with a differential stain (Congo red) or with polarized light.

Treatment.   There is no treatment for this progressive disease of conventional rats. If a staphylococcal infection underlies a renal disorder in a guinea pig, or is a potential cause of a disorder, the infection may be treated with antibiotics. Such treatment, however, is difficult in guinea pigs.

Prevention.   As the specific causative factors of chronic nephrosis are unknown, preventive measures are uncertain. Reducing food intake may slow the course of the disease in some animals.

Public Health Significance.   Nephrosis in rodents has no known public health significance.

REFERENCES

Alt, J.M., et al.: Proteinuria in rats in relation to age-dependent renal changes. Lab. Anims., *14:*95–101, 1980.

Andrew, W., and Pruett, D.: Senile changes in the kidneys of Wistar Institute rats. Am. J. Anat., *100*:51–80, 1957.

Berg, B.N.: Spontaneous nephrosis with proteinuria, hyperglobulinemia and hypercholesterolemia in the rat. Proc. Soc. Exp. Biol. Med., *119*:417–420, 1965.

Blatherwick, N.R., and Medlar, E.M.: Chronic nephritis in rats fed high protein diets. Arch. Intern. Med., *59*:572–596, 1937.

Bolton, W.K., and Sturgill, B.C.: Spontaneous glomerular sclerosis in aging Sprague-Dawley rats. II. Ultrastructural studies. Am. J. Pathol., *98*:339–356, 1980.

Bras, G.: Age-associated kidney lesions in the rat. J. Infect. Dis., *120*:131–135, 1969.

Bras, G., and Ross, M.H.: Kidney disease and nutrition in the rat. Toxicol. Appl. Pharmacol., *6*:247–262, 1964.

Gleiser, C.A., et al.: Amyloidosis and renal paramyloid in a closed hamster colony. Lab. Anim. Sci., *21*:197–202, 1971.

Gray, J.E.: Chronic progressive nephrosis in the albino rat. *In* Critical Reviews in Toxicology. West Palm Beach, Florida, CRC Press Inc., 1977, pp. 115–144.

Gray, J.E., Weaver, R.N., and Purmalis, A.: Ultrastructural observations of chronic progressive nephrosis in the Sprague-Dawley rat. Vet. Pathol., *11*:153–164, 1974.

Haensly, W.E., et al.: Proximal-tubule-like epithelium in Bowman's capsule in spontaneously hypertensive rats: changes with age. Am. J. Pathol., *107*:92–97, 1982.

Heymann, W., and Lund, H.Z.: Nephrotic syndrome in rats. Pediat., *7*:691–706, 1951.

Hinton, M.: Kidney disease in the rabbit: a histological survey. Lab. Anims., *15*:263–265, 1981.

Klassen, J.: Immunologic renal disease in laboratory animals. Lab. Anim. Sci., *23*:86–91, 1973.

Murphy, J.C., Fox, J.G., and Niemi, S.H.: Nephrotic syndrome associated with renal amyloidosis in a colony of Syrian hamsters. J. Am. Vet. Med. Assoc., *185*:1359–1362, 1984.

Queisser, G., and Drommer, W.: Mesangium in rats after *E. coli* neurotoxin shock: A morphometric and light microscopic analysis. Vet. Pathol., *19*:294–304, 1982.

Snell, K.C.: Renal disease of the rat. *In* The Pathology of Laboratory Rats and Mice. Edited by E. Cotchin and F.J.C. Roe. Oxford and Edinburgh, Blackwell Scientific Publications, 1967, pp. 105–147.

Van Marck, E.A.E., et al.: Spontaneous glomerular basement membrane changes in the golden Syrian hamster (*Mesocricetus auratus*): A light and electron microscope study. Lab. Anims., *12*:207–211, 1978.

Weaver, R.N., Gray, J.E., and Schultz, J.R.: Urinary proteins in Sprague-Dawley rats with chronic progressive nephrosis. Lab. Anim. Sci., *25*:705–710, 1975.

OXYURIASIS (PINWORMS)

Hosts. Pinworms, oxyurid nematodes inhabiting the intestinal tract, are common and widespread among vertebrate hosts, but only those affecting laboratory rodents and rabbits are discussed here.

Etiology. *Passulurus ambiguus* occurs in rabbits, cottontails, and hares.

Syphacia obvelata occurs in mice and other rodents.

Syphacia muris occurs in rats and other rodents.

Aspiculuris tetraptera occurs in mice and other rodents.

Pinworms are commensal, essentially nonpathogenic, ubiquitous, bacteria-feeding roundworms often present in the hundreds in the intestinal tracts of clinically normal animals. *Syphacia* has a direct life cycle and a prepatent period of 8 to 15 days, whereas *Aspiculuris* requires 23 days and *Passalurus* 55 to 65 days from ingestion to the appearance of ova in the feces.

Transmission. Hosts are infected by ingesting embryonating eggs in feces or fecal-contaminated feed, water, or debris or through contact with eggs adhering to the perianal skin (*Syphacia* only). *Aspiculuris* and *Passulurus* ova are passed in the feces; embryonation of *Aspiculuris* eggs requires 6 days. *Syphacia* ova on the perianal skin embryonate within a few hours of deposition, and retrograde infection may occur. Pinworms are highly infectious.

Predisposing Factors. Animals with diminished resistance are more susceptible to pinworm infections. Intestinal pinworm populations vary with the host's age: In enzootically infected colonies, *Syphacia* numbers plateau at 5 to 6 weeks and then diminish with age of the host, but *Aspiculuris* populations rise to a plateau when the host reaches 10 weeks. In addition, the presence of one pinworm species may inhibit habitation by the other. Male hosts are more heavily parasitized than females, and worm populations vary with the composition of the intestinal microflora and among inbred strains. Heavy pinworm infections may predispose to a variety of other diseases. DBA/2 and RF strains are highly susceptible; C3H strain mice are more resistant.

Clinical Signs. Clinical signs related directly to oxyuriasis are uncommonly observed, even in hosts harboring hundreds of worms, but reported signs involve diminished weight gains, decreased activity, rectal prolapse or constipation from rectal irritation associated with severe infestations, impaction and intussusception, increased stickiness of feces, self-mutilation of the tail base, and reproductive depression.

Necropsy Signs. Necropsy signs of pinworm infection are rare and are limited to intussusception and catarrhal enteritis of otherwise uncertain etiology. The small, adult worms are visible in the ingesta or feces, *Syphacia* in the cecum and less so in the colon, and *Aspiculuris* and *Passalurus* in the anterior colon and cecum. Pinworms are confined to the intestinal lumen except in rare cases of penetration through the intestinal epithelium by *Aspicularis* larvae. Histologically there may be slight colonic inflammation but generally there are no lesions. Profiles of the characteristic nematodes are usually seen in the lumen.

Diagnosis. Diagnosis of oxyuriasis depends on finding the adult nematodes in the cecum or colon or the ova in the feces or on the perianal skin. *Aspiculuris* and *Passalurus,* however, do not deposit eggs around the anus.

Adult worms, with females longer than males, range in length from 1 mm for *Syphacia* males to 11 mm for *Passalurus* females, but most are between 2 and 4 mm in length. Ova are ellipsoidal, approximately 90 to 120 μm $\times$ 30 to 40 μm, and, with the exception of the symmetrical *Aspiculuris* ovum, somewhat flattened on one side. Adult worms and ova are seen in fecal smears; ova are seen on fecal flotation, or, with *Syphacia,* on a (clear) cellophane tape impression of the anus.

Treatment. Although numerous treatment regimens have been reported, treatment procedures alone usually result in only a transient elimination of the intestinal ne-matodes. Ova on the perianal skin or in the environment must be removed with a detergent wash if eradication is desired, and repeated treatment of both animals and environment is necessary. A consideration in the treatment of *Aspiculuris* is the 6-day period required for embryonation and the 23-day prepatent period. Prior to treating any parasitism, the prepatent period of the parasite must be considered.

Some anthelmintics are heat sensitive and will not withstand autoclaving; therefore, they cannot be mixed in feed that will be autoclaved before being passed into a barrier. Thiabendazole (Merck, Rahway, NJ) is heat stable and recommended for in-barrier use.

Food intake per animal is estimated to determine the appropriate anthelmintic dose to be added to the diet. Addition of food dye to colorless piperazine compounds aids in determining which mice are receiving treatment. Determination of the volume of water consumed aids in the calculation of dose levels to be added to water.

Some treatments that have been used with various degrees of success, mostly in small rodents, are:

For mice piperazine citrate at 200 mg/kg body weight in drinking water 1 week on, 1 week off, 1 week on (Hoag 1961).

For rabbits piperazine adipate 0.5 g/kg/day for 2 days for adult rabbits and 0.75 g/kg/day for 2 days in young rabbits with *P. ambiguus* (Kraus et al. 1984).

Pyrvinium pamoate at concentrations of 0.0016% in feed or 0.008% in drinking water for 1 month for mice (Blair, Thompson and Vandenbelt 1968). Pyrvinium pamoate at 12 mg/kg per rat in feed (Veletzky, Bubna-Littitz, and Prosl 1985).

Trichlorfon (1.75 g/L), often combined with atropine and sucrose for palatability, given in the drinking water for 14 days (Simmons, Williams and Wright 1965).

Dichlorvos at 0.5 mg/g feed for 1 day (Wagner 1970).

Uredofos at 125 ppm in feed for 6 days; 25 mg/kg body weight by gavage, or in drinking water at 25 mg/kg for 24 hours for mice (Tetzlaff and Weir 1978).

Mebendazole at 40 to 500 mg/kg in single oral dose or 40 mg/kg in 2 oral doses for mice (Sharp and Wescott 1976).

Thiabendazole at 0.1% in feed, alternating 3 days on and 4 days off medicated feed for 3 or more weeks for *Syphacia* spp.

Thiabendazole at 0.1% in feed for 3 days on and 11 days off medicated feed for 3 or more treatments for *Aspiculuris* spp.

Thiabendazole at 400 mg/kg in a single dose given by gastric gavage for *Passalurus ambiguus* in rabbits (Barth 1974).

Thiabendazole at 0.3% in feed for 7 to 10 days in hamsters (Sebesteny 1979).

Thiabendazole at 0.1% in feed for 3 months for pinworms of rats and mice (Owen and Turton 1979).

Fenbendazole at 50 ppm in feed for 5 days for *P. ambiguus* in rabbits (Düwel and Williams 1981).

Prevention.   Prevention of pinworm infection in rodents and rabbits is extremely difficult and usually ineffective. Bedding dusts and wild rodents should be reduced or eliminated, filter tops should be used, and cages, rooms, and air ducts should be cleaned frequently. Higher fiber diets reduce worm populations. Two percent peracetic acid or formaldehyde vapor treatments may not kill all pinworm eggs in the environment. Super-heating of barrier facilities (115° F for 48 hours) following decontamination procedures may be worth considering as a means of killing residual pinworm ova.

Public Health Significance.   *Syphacia* infections have been reported in nonhuman primates and in humans, but the effects were negligible, and the reports have not been substantiated by further reports.

REFERENCES

Barth, D.: Efficacy of thiabendazole versus *Passalurus ambiguus* (Rudolphi 1819) in domestic rabbits. Dtsch. Tieraerztl. Wochenschr., *81*:477–800, 1974.

Battles, A.H., et al.: Efficacy of ivermectin against natural infection of *Syphacia muris* in rats. Lab. Anim. Sci., *37*:791–792, 1987.

Blair, L.S., Thompson, P.E., and Vandenbelt, J.M.: Effects of pyrvinium pamoate in the ration of drinking water of mice against pinworms *Syphacia obvelata* and *Aspiculuris tetraptera*. Lab. Anim. Care, *18*:314–327, 1968.

Brody, G., and Elward, T.E.: Comparative activity of 29 known anthelmintics under standardized drug-diet and gavage medication regimens against four helminth species in mice. J. Parasitol., *57*:1068–1077, 1971.

Düwel, D., and Brech, K.: Control of oxyuriasis in

rabbits by fenbendazole. Lab. Anims., *15*:101–105, 1981.

Habermann, R.T., and Williams, F.P., Jr.: Treatment of female mice and their litters with piperazine adipate in the drinking water. Lab. Anim. Care, *13*:41–45, 1963.

Harwell, J.F., and Boyd, D.D.: Naturally occurring oxyuriasis in mice. J. Am. Vet. Med. Assoc., *153*:950–953, 1968.

Herrlein, H.G.: Elimination of oxyurids from the laboratory mouse. Proc. Anim. Care Panel, *9*:165–166, 1959.

Heyneman, D.: Nematodes. *In* Parasites of Laboratory Animals. Edited by R.J. Flynn. Ames, The Iowa State University Press, 1973.

Hoag, W.G.: Oxyuriasis in laboratory mouse colonies. Am. J. Vet. Res., *22*:150–153, 1961.

Hussey, K.L.: *Syphacia muris* vs. *S. obvelata* in laboratory rats and mice. J. Parasitol., *43*:555–559, 1957.

King, M., and Cosgrove, G.E.: Intestinal helminths in various strains of laboratory mice. Lab. Anim. Sci., *13*:46–48, 1963.

Kraus, A.L., et al.: Biology and diseases of rabbits. *In* Laboratory Animal Medicine. Edited by J. Fox, B. Cohen, and F. Loew. Orlando, Academic Press, Inc., 1984.

MacArthur, J.A., and Wood, M.: Control of oxyurids in mice using thiabendazole. Lab. Anims., *12*:141–143, 1978.

McNair, D.M., and Timmons, E.H.: Effects of *Aspiculuris tetraptera* and *Syphacia obvelata* on exploratory behavior of an inbred mouse strain. Lab. Anim. Sci., *27*:38–42, 1977.

Mohn, G., and Phillip, E-M.: Effects of *Syphacia muris* and the anthelmintic fenbendazole on the microsomal monooxygenase system in mouse liver. Lab. Anim., *15*:89–95, 1981.

Mullink, J.W.M.A.: Pathological effects of oxyuriasis in the laboratory mouse. Lab. Anims., *4*:197–201, 1970.

Owen, D., and Turton, J.A.: Eradication of the pinworm *Syphacia obvelata* from an animal unit by anthelmintic therapy. Lab. Anim., *13*:115–118, 1979.

Ross, C.R., et al.: Experimental transmission of *Syphacia muris* among rats, mice, hamsters, and gerbils. Lab. Anim. Sci., *30*:35–37, 1980.

Sebesteny, A.: Syrian hamsters. *In* Handbook of Diseases of Laboratory Animals. Edited by J. Hime and P. O'Donoghue. London, Heinemann, 1979.

Sharp, J.W., and Wescott, R.B.: Anthelmintic efficacy of mebendazole for pinworm infection of mice. Lab. Anim. Sci., *26*:222–223, 1976.

Simmons, M.L., Williams, H.E., and Wright, E.G.: Therapeutic value of the organic phosphate, trichlorfon, against *Syphacia obvelata* in inbred mice. Lab. Anim. Care, *15*:382–385, 1965.

Stone, W.B., and Manwell, R.D.: Potential helminth infections in humans from pet or laboratory mice and hamsters. Pub. Health Rep., *81*:647–653, 1966.

Taffs, L.F.: Pinworm infections in laboratory rodents: a review. Lab. Anims., *10*:1–13, 1976.

Tetzlaff, R.D., and Weir, W.D.: Anthelmintic control of concurrent *Hymenolepis nana* and *Syphacia obvelata* infections in the mouse with uredofos. Lab. Anim. Sci., *28*:287–289, 1978.

Veletzky, V.S., Bubna-Littitz, H., and Prosl, H.: Vorschlag zur Behandlung einer *Syphacia muris*—Infektion mit dem Wirkstoff Pyrviniumpamoate bei der Laborratte. Zbl. Vet. Med. B., *32*:149–153, 1985.

Wagner, J.E.: Control of mouse pinworms, *Syphacia obvelata*, utilizing dichlorvos. Lab. Anim. Care, *20*:39–44, 1970.

Wescott, R.B., Malczewski, A., and Van Hoosier, G.L.: The influence of filter top caging on the transmission of pinworm infections in mice. Lab. Anim. Sci., *26*:742–745, 1976.

Wightman, S.R., Wagner, J.E., and Corwin, R.M.: *Syphacia obvelata* in the Mongolian gerbil (*Meriones unguiculatus*): natural occurrence and experimental transmission. Lab. Anim. Sci., *28*:51–54, 1978.

PASTEURELLA MULTOCIDA INFECTION

Hosts.   *Pasteurella multocida* causes the major respiratory disease of rabbits, but rodents, birds, and farm animals can also be affected by this common animal pathogen.

Etiology.   *Pasteurella multocida* is a small, gram-negative, bipolar-staining, ovoid rod. Colonies on blood agar vary considerably and are designated mucoid, smooth, and rough. Several serotypes occur in rabbits, with types 12:A, 12:D, and the highly virulent 3:A (somatic:capsular) among the more common. An endotoxin is produced.

Transmission.   The bacterium is transmitted by direct contact between a chronically infected doe and her litter or between breeding pairs. Neonates may be infected in vaginal passage, during nursing, and from contaminated sipper tubes. Transmission by respiratory aerosol occurs over short distances, but noninfected rabbits housed in cages adjacent to *Pasteurella* shedders require weeks to months to demonstrate nasal infection.

Asymptomatic carriers of *P. multocida* are common. The organism resides in the nares, tympanic bullae, conjunctivae, vagina, and lungs, where only a minimal immune response is elicited.

Predisposing Factors.   The ubiquity of asymptomatic carriers or chronically diseased adult rabbits, the ease of contact transmission to the young, and the virulence of the organism make any listing of predisposing factors almost academic. Environmental factors such as increased ammonia level, temperature changes and drafts, reproduction, older age, existence of carriers, and poor sanitation all contribute to the development of clinical pasteurellosis.

Clinical Signs.   Clinical signs of *P. multocida* infection range from inapparent in carrier animals to peracute, septicemic death to chronic abscessation and suppuration in adult rabbits. The primary locus of the infection can occur in almost any tissue and spread to any other tissue, although the nasal passage is the most common primary site. Signs exhibited depend on the site of infection and may include nasal discharge, snuffling, runny eyes, torticollis, cutaneous ulceration, subcutaneous swelling, enlarged testes, vaginal discharge, infertility, weight loss, and sudden or lingering death. Pneumonia has few clinical signs in rabbits.

Acute enzootic pneumonia is the common clinical syndrome in young rabbits, whereas abscesses and suppuration occur in older animals. Neonates may die from a septicemia. Pulmonary impairment may cause cyanosis of the iris and a short, terminal period of dyspnea.

Necropsy Signs.   Gross lesions of *P. multocida* infection may involve one or more of the following: 1. generalized visceral congestion, sanguineous nasal discharge, and focal hemorrhages resulting from a septicemia; 2. well-demarcated reddish-gray foci of bronchopneumonia; and 3. fibrinopurulent or mucopurulent reactions in the meninges and brain, middle and inner ear, thoracic and abdominal viscera, nasal passages, subcutaneous tissues, bones, or the reproductive organs. The pus in a *P. multocida* abscess is creamy-white and usually thick.

Diagnosis.   Diagnosis of a *P. multocida* infection is based on the isolation of the causative organism from blood or affected tissues. Nasal swab cultures are unreliable indicators of infection because some carrier animals may be culture negative. Culture of larger numbers of animals and culture of other sites in addition to the nasal cavities will help to confirm the carrier state. Purulent processes in rabbits are usually due to *P. multocida* or *Staphylococcus aureus.* Both grow on blood agar media. *Pasteurella multocida* is nonhemolytic and produces large, translucent, mucoid colonies up to 4 mm in diameter. *Staphylococcus* colonies are smaller, dry, opaque, and often hemolytic. Gram staining will provide a final diagnosis. *Pasteurella multocida* must also be differentiated from *Pasteurella pneumotropica* and *Bordetella bronchiseptica* through the use of indole, glucose, and urea cultures.

Treatment.   Concerns in the treatment of pasteurellosis in rabbits include the advisability of treating an essentially incurable infection, the number and location of purulent foci and fistulous tracts, the potential for septicemia if a focus is entered, the antibiotic sensitivity of the bacterium, and the ability of the antibiotic to penetrate the purulent accumulation.

If surgical drainage is indicated, as with an abscess, the rabbit can be restrained with ketamine and xylazine, and the abscess can be opened, drained, and flushed or packed with an antimicrobial wash or ointment. A systemic antibiotic is also usually indicated to prevent hematogenous spread of the bacterium and to act on the abscess itself. Even extensive, deep abscesses, if properly attended, heal rapidly and well in rabbits.

Rhinitis or snuffles is easily treated and the clinical signs suppressed, but a cure is difficult, if not impossible. The bacteria lie sequestered within pus in the labyrinthine nasal passages and are isolated from antibiotics. Procaine penicillin G, or other suitable antibiotic, will suppress the nasal discharge, but bacteria will remain and repopulate the nasal passage.

Conjunctivitis is treated with an appropriate ophthalmic ointment and by flushing the nasolacrimal duct with injectable kanamycin or gentamicin.

Other internal *Pasteurella multocida* infections, including pneumonia, metritis, and otitis interna, are usually only detected after considerable tissue damage has occurred and antibiotics have little effect, although

it is often possible to stop the progression of a head tilt. Does with pyometra can be spayed, and bucks with orchitis can be castrated.

In vitro, P. multocida is sensitive to a wide range of antibiotics, although sensitivities vary from culture to culture. Procaine penicillin (60,000 IU/kg body weight for 10 days) can be used for individual rabbits (Kraus et al. 1984). Penicillin G at 60,000 IU/kg has also been shown to effectively reduce the numbers of *P. multocida* organisms (Welch, Lu and Bawdon 1987). Whereas penicillin is often the antibiotic of choice, chloramphenicol, erythromycin, gentamicin, tylosin, kanamycin, polymyxin B, triple sulfa, and neomycin have been used to treat pasteurellosis in rabbits, but recurrence of the disease is common. There are penicillin-resistant *Pasteurella.* Withdrawal times must be observed in animals used for human consumption. Antibiotics may have adverse effects on experimental protocols using treated rabbits and may cause enteritis.

Prevention.   Rabbit pasteurellosis is endemic in most rabbitries and difficult to eradicate. Selection of *Pasteurella*-free or *Pasteurella*-resistant stock, repeated culture surveys of rabbits in the colony with culling of affected individuals, and provision for a 3-week entry quarantine with cultural or serologic screening are methods of preventing and eliminating the organism from a colony.

High husbandry standards, elimination of ammonia, and stabilization of the environment reduce the development of the more severe forms of the disease in adults. As the young are susceptible to pulmonary infection, chronically infected does should be culled. Weaning the young at 4 or 5 weeks reduces exposure to the infected mother. Manure removal and good ventilation reduce accumulation of ammonia.

The prophylactic use of sulfaquinoxaline-supplemented water (0.05%) may reduce the incidence of enzootic pneumonia in exposed young until they can be isolated at weaning. Antibiotic eradication systems have many shortcomings, including the generation of diarrhea, and are no substitute for culling, cleaning, and repopulation with *Pasteurella multocida*-free or *Pasteurella multocida*-resistant rabbits.

Bacterins to prevent pasteurellosis in rabbits have been discussed and investigated, but no commercial product is available. Antibodies elicited have been either nonprotective or limited to a certain capsular type. Natural antibodies to the agent may be directed against the endotoxin. An attenuated avian bacterin, bacterins of temperature-sensitive mutants, and a streptomycin-dependent live vaccine for type A *P. multocida* given intranasally or subcutaneously show promise.

Public Health Significance. Although transmission of a rabbit infection to man is extremely unlikely, *P. multocida* can cause skin infections, arthritis, meningitis, peritonitis, pneumonia, and septicemia in susceptible humans.

REFERENCES

Anderson, L.C., Rush, H.G., and Glorioso, J.C.: Strain differences in the susceptibility and resistance of *Pasteurella multocida* to phagocytosis and killing by rabbit polymorphonuclear neutrophils. Am. J. Vet. Res., 45:1193–1198, 1984.

Belin, R.P., and Banta, R.G.: Successful control of snuffles in a rabbit colony. J. Am. Vet. Med. Assoc., 159:622–623, 1971.

Boisvert, P.L., and Fousek, M.D.: Human infection with *Pasteurella lepiseptica* following a rabbit bite. J.A.M.A., 116:1902–1903, 1941.

Brogden, K.A.: Physiological and serological characteristics of 48 *Pasteurella multocida* cultures from rabbits. J. Clin. Microbiol., 11:646–649, 1980.

Chengappa, M.M., Myers, R.C., and Carter, G.R.: A streptomycin dependent live *Pasteurella multocida* vaccine for the prevention of rabbit pasteurellosis. Lab. Anim. Sci., 30:515–518, 1980.

DiGiacomo, R.F., Garlinghouse, L.E., Jr., and Van Hoosier, G.L., Jr.: Natural history of infection with *Pasteurella multocida* in rabbits. J. Am. Vet. Med. Assoc., 183:1172–1175, 1983.

DiGiacomo, R.F., et al.: Safety and efficacy of a streptomycin dependent live *Pasteurella multocida* vaccine in rabbits. Lab. Anim. Sci., 37:187–190, 1987.

DiGiacomo, R.F., Jones, C.D.R., and Wathes, C.M.: Transmission of *Pasteurella multocida* in rabbits. Lab. Anim. Sci., 37:621–623, 1987.

Flatt, R.E., DeYoung, D.W., and Hogle, R.M.: Suppurative otitis media in the rabbit: prevalence, pathology, and microbiology. Lab. Anim. Sci., 27:343–347, 1977.

Flatt, R.E., and Dungworth, D.L.: Enzootic pneumonia in rabbits: naturally occurring lesions in lungs of apparently healthy young rabbits. Am. J. Vet. Res., 32:621–626, 1971.

Fox, R.R., Norberg, R.F., and Myers, D.D.: The relationship of *Pasteurella multocida* to otitis media in the domestic rabbit (*Oryctolagus cuniculus*). Lab. Anim. Sci., 21:45–48, 1971.

Furie, R.A., et al.: *Pasteurella multocida* infection: report in urban setting and review of spectrum of human disease. N.Y. State J. Med., 80:1597–1602, 1980.

Garlinghouse, L.E., Jr., et al.: Selective media for *Pasteurella multocida* and *Bordetella bronchiseptica*. Lab. Anim. Sci., 31:39–42, 1981.

Hagen, K.W., Jr.: Enzootic pasteurellosis in domestic rabbits. I. Pathology and bacteriology. J. Am. Vet. Med. Assoc., 133:77–80, 1958.

Holmes, H.T., et al.: Serologic methods for detection of *Pasteurella multocida* infections in nasal culture negative rabbits. Lab. Anim. Sci., 36:650–654, 1986.

Holmes, H.T., et al.: A method for culturing the nasopharyngeal area of rabbits. Lab. Anims., 21:353–355, 1987.

Hwang, E.J., et al.: Characterization of antigen purified from Type 3 strains of *Pasteurella multocida* and its use for an enzyme-linked immunosorbent assay. Lab. Anim. Sci., 36:633–639, 1986.

Jaslow, B.W., et al.: *Pasteurella* associated rhinitis of rabbits: efficacy of penicillin therapy. Lab. Anim. Sci., 31:382–385, 1981.

Kluger, M.J., and Vaughn, L.K.: Fever and survival in rabbits infected with *Pasteurella multocida*. J. Physiol., 282:243–251, 1978.

Kraus, A.L., et al.: Biology and diseases of rabbits. *In* Laboratory Animal Medicine. Edited by J.G. Fox, B.J. Cohen, and F.M. Loew. Orlando, Academic Press, 1984, pp. 207–240.

Kunstyr, I., and Naumann, S.: Head tilt in rabbits caused by pasteurellosis and encephalitozoonosis. Lab. Anim., 19:208–213, 1985.

Lu, Y.S., Ringler, D.H., and Park, J.S.: Characterization of *Pasteurella multocida* isolates from the nares of healthy rabbits and rabbits with pneumonia. Lab. Anim. Sci., 28:691–697, 1978.

Lukas, V.S., et al.: An enzyme-linked immunosorbent assay to detect IgG to *Pasteurella multocida* in naturally and experimentally infected rats. Lab. Anim. Sci., 37:60–64, 1987.

Manning, P.J.: Serology in *Pasteurella multocida* in laboratory rabbits: A review. Lab. Anim. Sci., 32:666–671, 1982.

Manning, P.J., et al.: A dot-immuno-binding assay for the serodiagnosis of *Pasteurella multocida* infection in laboratory rabbits. Lab. Anim. Sci., 37:615–620, 1987.

Percy, D.H., and Black, W.D.: Pharmokinetics of tetracycline in the domestic rabbit following intravenous or oral administration. Can. J. Vet. Res., 52:5–11, 1988.

Petkus, A.R., et al.: Experimental chronic *Pasteurella multocida* infection of subcutaneous chambers in the rabbit. Lab. Anim. Sci., 29:749–754, 1979.

Plant, J.W.: Control of *Pasteurella multocida* infections in a small rabbit colony. Lab. Anims., 8:39–40, 1974.

Rush, H.G., et al.: Resistance of *Pasteurella multocida* to rabbit neutrophil phagocytosis and killing. Am. J. Vet. Res., 42:1760–1768, 1981.

Savage, N.L., and Sheldon, W.G.: Torticollis in mice due to *Pasteurella multocida* infection. Can. J. Comp. Med., 35:267–268, 1971.

Scharf, R.A., Monteleone, S.A., and Stark, D.M.: A modified barrier system for maintenance of *Pasteurella*-free rabbits. Lab. Anim. Sci., *31*:513–515, 1981.

Ward, G.M.: Development of a *Pasteurella*-free rabbit colony. Lab. Anim. Sci., *23*:671–674, 1973.

Watson, W.T., et al.: Experimental respiratory infection with *Pasteurella multocida* and *Bordetella bronchiseptica* in rabbits. Lab. Anim. Sci., *25*:459–464, 1975.

Webster, L.T.: The epidemiology of a rabbit respiratory infection. I. Introduction. J. Exp. Med., *39*:837–841, 1924.

Welch, W.D., Lu, Y.-S., and Bawdon, R.E.: Pharmacokinetics of penicillin-G in serum and nasal washings of *Pasteurella multocida* free and infected rabbits. Lab. Anim. Sci., *37*:65–68, 1987.

Wright, J.: An epidemic of *Pasteurella* infection in a guinea pig stock. J. Pathol. Bacteriol., *42*:209–212, 1936.

Yarnoff, S.R.: What is your diagnosis? J. Am. Vet. Med. Assoc., *183*:1347–1348, 1983.

PASTEURELLA PNEUMOTROPICA INFECTION

Hosts. Mice, rats, and occasionally hamsters are susceptible to disease caused by *Pasteurella pneumotropica.*

Etiology. *Pasteurella pneumotropica* is a gram-negative, nonmotile, pleomorphic coccobacillus. The ability of the organism to cause disease is increased as the number of infecting organisms is increased. Some of the reports of rodent disease implicating *P. pneumotropica* as the causative agent were apparently caused instead by an *Actinobacillus.*

Transmission. *Pasteurella pneumotropica* often exists in a latent, carrier state in the upper respiratory or gastrointestinal tracts, and may be disseminated by respiratory aerosol or fecal contamination, biting, licking, and intrauterine contamination.

Predisposing Factors. Since *P. pneumotropica* is an opportunistic pathogen, circumstances that lower a host's resistance, particularly other infections, may precipitate the clinical disease. *Pasteurella pneumotropica* in conjunction with Sendai virus infection of mice frequently results in fatal pneumonia, and *P. pneumotropica* may complicate *Mycoplasma pulmonis* infection in rats and mice.

Clinical Signs. *Pasteurella pneumotropica* is widespread as a latent infection, but the bacterium causes clinical disease only sporadically. Signs associated with *P. pneumotropica* infection include chattering, labored respiration, weight loss, skin abscesses, harderian gland abscesses, conjunctivitis, panophthalmitis, mastitis, infertility, bulbourethral gland infections, abortion, and internal and subcutaneous abscesses.

Necropsy Signs. The pulmonary lesions seen in the early stages of *P. pneumotropica* pneumonia resemble the well-demarcated, red foci of consolidation seen with a *Mycoplasma pulmonis* infection; however, the bacterial infection may produce scattered abscessation, especially in athymic, nude mice. Suppurative reactions may also occur in the middle ears, orbital glands, uterus, skin, mammary gland, lymph nodes, accessory sex glands, and the urinary system.

Diagnosis. A definitive diagnosis of *Pasteurella pneumotropica* infection is established through recovery of the organism on culture. On 24-hour incubation on blood agar, the colonies are small (1 mm), circular, convex, smooth, and surrounded by a zone of slight greenish discoloration.

Treatment. Treatment of a bacterial disease in rodents with an antibiotic effective *in vitro* is relatively easy; the outcome of such therapy, however, is rarely the elimination of the organism or the infection from the colony. *Pasteurella pneumotropica* is sensitive to several antibiotics. Among specific antibiotic regimens used with this organism are chloramphenicol in the drinking water (0.25 mg/ml for 2 weeks), ampicillin administered subcutaneously (mouse—9 mg daily for 5 days); and oxytetracycline administered intramuscularly (mouse—3.5 mg per day) (Moore and Aldred 1978), although some isolants are resistant to ampicillin and oxytetracycline, and streptomycin can be toxic for rodents.

Prevention. Elimination of murine respiratory infection from a colony requires a known disease-free stock placed into a clean and barrier-sustained colony. Newly arrived animals should be quarantined until their microbial status has been determined. Caesarean derivation should be done with the knowledge that *P. pneumotropica* is a relatively common uterine inhabitant, as is *Mycoplasma pulmonis.*

Public Health Significance.   A strain of *Pasteurella pneumotropica* can infect man, but the possibility of rodent-to-man transmission is unlikely.

REFERENCES

Blackmore, D.K., and Casillo, S.: Experimental investigation of uterine infections of mice due to *Pasteurella pneumotropica.* J. Comp. Pathol., *82*:471–475, 1972.

Brennan, P.C., Fritz, T.E., and Flynn, R.J.: *Pasteurella pneumotropica:* cultural and biochemical characteristics, and its association with disease in laboratory animals. Lab. Anim. Care, *15*:307–312, 1965.

Brennan, P.C., Fritz, T.E., and Flynn, R.J.: Role of *Pasteurella pneumotropica* and *Mycoplasma pulmonis* in murine pneumonia. J. Bacteriol., *97*:337–349, 1969.

Carthew, P., and Gannon, J.: Secondary infection of rat lungs with *Pasteurella pneumotropica* after Kilham rat virus infection. Lab. Anims., *15*:219–221, 1981.

Carthew, P., and Aldred, P.: Embryonic death in pregnant rats owing to intercurrent infection with Sendai virus and *Pasteurella pneumotropica.* Lab. Anims., *22*:92–97, 1988.

Gray, D.F., and Campbell, A.L.: The use of chloramphenicol and foster mothers in the control of natural pasteurellosis in experimental mice. Aust. J. Exp. Biol. Med. Sci., *31*:161–166, 1953.

Hoag, W.G., et al.: A study of latent *Pasteurella* infection in a mouse colony. J. Infect. Dis., *111*:135–140, 1962.

Hong, C.C., and Ediger, R.D.: Chronic necrotizing mastitis in rats caused by *Pasteurella pneumotropica.* Lab. Anim. Sci., *28*:317–320, 1978.

Jawetz, E.: A latent pneumotropic *Pasteurella* of laboratory animals. Proc. Soc. Exp. Biol. Med., *68*:46–48, 1948.

Jawetz, E., and Baker, W.H.: A pneumotropic Pasteurella of laboratory animals. II. Pathological and im-munological studies with the organism. J. Infect. Dis., *86*:184–196, 1950.

Lentsch, R.H., and Wagner, J.E.: Isolation of *Actinobacillus lignieresii* and *Actinobacillus equuli* from laboratory rodents. J. Clin. Microbiol., *12*:351–354, 1980.

Lescher, R.J., Jeszenka, E.V., and Swan, M.E.: Enteritis caused by *Pasteurella pneumotropica* infection in hamsters. J. Clin. Microbiol., *23*:448, 1985.

Moore, G.J., and Aldred, P.: Treatment of *Pasteurella pneumotropica* abscesses in nude mice (*nu/nu*). Lab. Anims., *12*:227–228, 1978.

Moore, T.D., Allen, A.M., and Ganaway, J.R.: Latent *Pasteurella pneumotropica* infection of the gnotobiotic and barrier-held rats. Lab. Anim. Sci., *23*:657–661, 1973.

Needham, J.R., and Cooper, J.E.: An eye infection in laboratory mice associated with *Pasteurella pneumotropica.* Lab. Anims., *9*:197–200, 1975.

Sebesteny, A.: Abscesses of the bulbourethral glands of mice due to *Pasteurella pneumotropica.* Lab. Anims., *7*:315–317, 1973.

Simpson, W., and Simmons, D.J.C.: Two *Actinobacillus* species isolated from laboratory rodents. Lab. Anims., *14*:15–16, 1980.

Van der Schaff, A., et al.: *Pasteurella pneumotropica* as a causal microorganism of multiple subcutaneous abscesses in a colony of Wistar rats. Z. Versuchstierkd., *12*:356–362, 1970.

Wagner, J.E., et al.: Spontaneous conjunctivitis and dacryoadenitis of mice. J. Am. Vet. Med. Assoc., *155*:1211–1217, 1969.

Ward, G.E., Moffatt, R., and Olfert, E.: Abortion in mice associated with *Pasteurella pneumotropica.* J. Clin. Microbiol., *8*:177–180, 1978.

Weisbroth, S.H., Scher, S., and Boman, I.: *Pasteurella pneumotropica* abscess syndrome in a mouse colony. J. Am. Vet. Med. Assoc., *155*:1206–1210, 1969.

Wilson, P.: *Pasteurella pneumotropica* as the causal organism of abscesses in the masseter muscle of mice. Lab. Anims., *10*:171–172, 1976.

Wullenweber-Schmidt, M., et al.: An enzyme-linked immunosorbent assay (ELISA) for the detection of antibodies to *Pasteurella pneumotropica* in murine colonies. Lab. Anim. Sci., *38*:37–41, 1988.

PEDICULOSIS

Hosts.   Lice as a group have a wide host range, but individual species are host specific. Mites and lice sometimes occur together on a host.

Etiology.   Lice are flattened insects, with six legs and no wings. *Polyplax* and *Haemodipsus* are Anoplura, or sucking lice, and *Gliricola* and *Gyropus* are mallophagans, or biting lice.

Haemodipsus ventricosis, the rabbit louse, has a life cycle of 30 days and its eggs hatch in 7 days. *Gliricola porcelli,* the slender guinea pig louse, and *Gyropus ovalis,* the oval guinea pig louse, abrade the skin to obtain fluid.

Polyplax serrata, the mouse louse, is a sucking louse and has a life cycle of 13 days. *Polyplax spinulosa,* the spined rat louse, is common in wild rats and has a life cycle of about 26 days. Its eggs hatch after 6 days. Of the 5 species listed, *Gliricola* and *Polyplax* are fairly common, and *Gyropus* and *Haemodipsus* are uncommon.

Transmission.   Transmission of lice is by direct contact with an infected host or bedding. Lice seldom leave the host.

Predisposing Factors.   Young animals or animals with decreased resistance that are housed under unsanitary conditions may experience more severe infections and clinical consequences. Grooming by the rodent hosts controls the louse population.

Clinical Signs. Infestation by the rabbit louse, *Haemodipsus ventricosis,* may cause pruritus, alopecia, weakness, and anemia. The rabbit louse more often affects dorsolateral areas of the trunk than other areas of the body.

Gliricola and *Gyropus* are usually benign, but in heavy infestations they can cause scratching, partial alopecia, and scabs, usually around the ears.

Polyplax spp. are blood-sucking lice found on the neck and body. They may cause debilitation, anemia, scratching, small scabbed wounds over the side and back, and death. *P. serrata* transmits the agent of murine eperythrozoonosis, and *P. spinulosa* transmits the agent of murine haemobartonellosis.

Diagnosis. A hand lens or dissecting microscope is used to observe the pelt, especially the margin of lesions and the nape of the neck, for adult or immature ectoparasites. If the animal is dead, direct observation of lice is easier if the pelt is cooled in the refrigerator for 30 minutes, removed for 10 minutes, and then examined with a lens. The parasites migrate from the cool skin toward the warmer hair tips. Placing the suspect pelt on a black paper and within a frame of double-gummed cellophane tape or petroleum jelly will facilitate detection of the lice.

Haemodipsus is 1.2- to 2.5-mm long and has an oval abdomen. *Polyplax* spp. are slender and from 0.6- to 1.5-mm long. *Gliricola* has a narrow head and body approximately 1.0- to 1.5-mm long. *Gyropus* has a wide head and oval abdomen and is 1.0- to 1.2-mm long.

Treatment. Lice may be treated with the dusts and dips described in the section on acariasis. Elimination of ectoparasites from premises and equipment requires treatment, removal of the animals from the room, and thorough mechanical scrubbing and formaldehyde fumigation of equipment and premises.

Prevention. Pediculosis is prevented by installing clean stock in a clean facility. Examination for lice during quarantine and afterward and barrier housing will protect a colony. If lice are a problem elsewhere in a facility, the animals should be treated, and the equipment should be cleaned and disinfected. Wild rodents and lagomorphs frequently have pediculosis and should be excluded from colonies of domestic animals.

Public Health Significance. *Haemodipsus ventricosis* is a vector for the transmission of *Francisella tularensis* from rabbits to man. *Gliricola* and *Gyropus* are not known to affect man. *Polyplax* may serve as vectors for rickettsial organisms.

REFERENCES

Bell, J.F., Jellison, W.L., and Owen, C.R.: Effects of limb disability on lousiness in mice. I. Preliminary studies. Exp. Parasitol., *12*:176–183, 1962.

Eliot, C.P.: The insect vector for the natural transmission of *Eperythrozoon coccoides* in mice. Science, *84*:397, 1936.

Flynn, R.J.: Ectoparasites of mice. Proc. Anim. Care Panel, *6*:75–91, 1955.

Flynn, R.J.,: The diagnosis of some forms of ectoparasitism of mice. Lab. Anim. Care, *13*:111–125, 1963.

Kim, K.C.: Lice. *In* Parasites of Laboratory Animals. Edited by R.J. Flynn. Ames, The Iowa State University Press, 1973, pp. 376–397.

Murray, M.D.: The ecology of the louse *Polyplax serrata* (Burm.) on the mouse *Mus musculus* L. Aust. J. Zool., *9*:1–13, 1961.

Olewine, D.A.: An effective control of Polyplax infestation without the use of insecticide. Lab. Anim. Care, *13*:750–751, 1963.

Pratt, H.D., and Karp, H.: Notes on the rat lice *Polyplax spinulosa* (Burmeister) and *Hoplopleura oenomydis* Ferris. J. Parasitol., *39*:495–504, 1953.

PNEUMOCYSTOSIS

Hosts. Pneumocystosis is usually a latent or subclinical pulmonary disease that occurs widely in rodents, and is also found in domestic rabbits, primates, ungulates, carnivores and other animal species.

Etiology. *Pneumocystis carinii* is a unicellular protozoan or fungal extracellular parasite. Precysts, cysts, intracystic bodies, and trophozoites of the organism inhabit the alveoli of infected animals.

Transmission. Transmission of infection is probably by respiratory aerosol or contact with respiratory secretions. The cyst is antigenic and is probably the transmitted form.

Predisposing Factors. Pneumocystosis is a latent infection that in some species may cause clinical disease if the host is immunosuppressed by concurrent disease or chemotherapy. Thymectomy is not a predisposing factor.

Clinical Signs. Clinical signs include weight loss, cyanosis, and dyspnea.

Necropsy Signs. Signs of an active infection seen on gross necropsy include distended lungs with areas of consolidation and air passage rupture and ballooning. On histopathologic examination, alveoli contain a foam-like material (extracellular matrix and organisms). There may also be a pneumonitis and thickening of the alveolar lining.

Diagnosis. Infected animals have forms of the organism in the alveoli whether or not the alveolar protein-rich foam and interstitial reaction are present. The organisms may be seen with phase contrast microscopy. Wright, Giemsa, and Gram-Weigert stains are suitable for demonstration of sporozoites, whereas methenamine silver nitrate is the preferred stain for cyst forms. Organisms are difficult to see in Giemsa stained sections. The organism can be seen on lung impression prints.

Special concentration techniques may be necessary to detect the agent in the lungs of carrier rats. Cortisone acetate treatment (25 mg) and use of an 8% protein diet may increase organism populations in the lungs of suspect carriers. Complement fixation, fluorescent antibody, ELISA, counter immunoelectrophoresis, and the latex particle agglutination test of Pifer (1979) may be used for serologic screening of rodent colonies.

The cyst forms are 4 to 12 μm in diameter and contain 8 of the crescent-shaped sporozoites. Trophozoites (polymorphic with a single nucleus) are 1 to 5 μm in diameter; they adhere tightly to the alveolar wall.

Treatment. Sulfadiazine plus pyrinrethamine compounds are suggested treatments (Hsu 1979). Trimethoprim/sulfamethoxazole is a treatment commonly used in humans.

Prevention. Prevention of pneumocystosis is achieved by purchasing stock known to be free of *Pneumocystis* and by constant serologic and histopathologic surveillance.

Public Health Significance. The agent affects immunosuppressed humans. Isolants from different animal species tend to be antigenetically different. Subclinical infections are common.

REFERENCES

Barton, E.G., Jr., and Campbell, W.G., Jr.: *Pneumocystis carinii* in lungs of rats treated with cortisone acetate. Am. J. Pathol., *54*:209–236, 1969.

Cushion, M.T., Ruffolo, J.J., and Walzer, P.D.: Analysis of the developmental stages of *Pneumocystis carinii, in vitro.* Lab. Invest., *58*:324–331, 1988.

Frenkel, J.K., Good, J.T., and Shultz, J.A.: Latent pneumocystis infection of rats; relapse, and chemotherapy. Lab. Invest., *15*:1559–1577, 1966.

Furuta, R., et al.: Cellular and humoral immune responses of mice subclinically infected with *Pneumocystis carinii.* Infect. Immun., *47*:544–548, 1985.

Furuta, R., Fujiwara, K., and Yamanouchi, K.: Detection of antibodies to *Pneumocystis carinii* by enzyme-linked immunosorbent assay in experimentally infected mice. J. Parasitol., *41*:522–523, 1985.

Furuta, T., and Ueda, K.: Intra- and inter-species transmission and antigenic difference of *Pneumocystis carinii* derived from rat and mouse. Japan J. Exp. Med., *57*:11–17, 1987.

Hendley, J.O., and Weller, T.H.: Activation and transmission in rats of infection with *Pneumocystis.* Proc. Soc. Exp. Biol. Med., *137*:1401–1404, 1971.

Hsu, C-K.: Parasitic diseases. *In* The Laboratory Rat. Vol. I. Edited by H.J. Baker, J.R. Lindsey, and S.H. Weisbroth. New York, Academic Press, 1979.

Pifer, L.L.: Rapid diagnosis of parasitic infections. *In* Rapid Diagnosis in Infectious Diseases. Edited by M.J. Rytel. Boca Raton, FL, CRC Press, Inc., 1979, pp. 177–197.

Pifer, L.L., et al.: *Pneumocystis carinii* infection and antigenemia in immunocompromised patients: studies in a murine model. Clin. Res., *27*:811A, 1979.

Pifer, L.L., et al.: *Pneumocystis carinii* infection in germ-free rats: Implications for human patients. Diag. Micro. Infect. Dis., *2*:23–36, 1984.

Pintozzi, R.L., Blecka, L.J., and Nanos, S.: The morphologic identification of *Pneumocystis carinii.* Acta Cytol., *23*:35–39, 1979.

Ruffolo, J.J., Cushion, M.T., and Walzer, P.D.: Techniques for examining *Pneumocystis carinii* in fresh specimens. J. Clin. Microbiol., *23*:17–21, 1986.

Walzer, P.D., et al.: *Pneumocystis carinii*: Immunoblotting and immunofluorescent analyses of serum antibodies during experimental rat infection and recovery. Exper. Parasitol., *63*:319–328, 1987.

Walzer, P.D., and Linke, M.J.: A comparison of the antigenic characteristics of rat and human *Pneumocystis carinii* by immunoblotting. J. Immunol., *138*:2257–2265, 1987.

Weir, E.C., Brownstein, D.G., and Barthold, S.W.: Spontaneous wasting disease in nude mice associated with *Pneumocystis carinii* infection. Lab. Anim. Sci., *36*:140–144, 1986.

PROLIFERATIVE ILEITIS

Hosts.   The recently weaned golden hamster (3 to 8 weeks of age) is highly susceptible to proliferative ileitis.

Etiology.   Proliferative ileitis (PI) is a convenient name for the disease because both hyperplastic and inflammatory components are included. Alternative names are "wet tail" and "transmissible ileal hyperplasia." The specific etiology is unknown, but a bacterium or bacteria is probably involved because the disease can be experimentally transmitted through oral inoculation of intestinal tissue and content from affected hamsters. The abundance of intracellular bacteria observed electron microscopically by Wagner et al. (1973) in hyperplastic absorptive epithelial cells of the ileum were believed to have been etiologically significant, although unculturable. The enteritis component, an acute enteritis or a pyogranulomatous inflammation, may be caused by *Escherichia coli,* and the hyperplastic lesion may be caused by a rod-shaped, intracellular bacterium described as *Campylobacter*-like (Frisk, Wagner and Owens 1978). Several investigators have since isolated *Campylobacter* spp. from hamsters with PI.

Transmission.   The infectious agent or agents are passed from weanling to weanling, or from adult to weanling, by the fecal-oral route. The recent, but unconfirmed, suggestions that *Campylobacter* spp. may be involved in the etiology of this interesting disease raise speculation about swine, dogs, primates, and a wide variety of other animal species as sources of infection for hamsters, because so many species of animals may carry *Campylobacter* agents.

Predisposing Factors.   Recently weaned hamsters (3 to 8 weeks of age) are most often affected. Improper diet, exposure to infected animals, familial predisposition, primiparous litters, crowding, surgical treatment, dietary changes, and shipping predispose to the development of the disease. Long-haired and teddy bear hamsters seem to be highly susceptible.

Clinical Signs.   Proliferative ileitis, an epizootic enteropathy, has acute, subacute, and clinical consequences. The acute condition involves lethargy, matted hair coat, hunched stance, anorexia, irritability, watery diarrhea, emaciation, and death within 48 hours. In less acute cases, survivors fail to gain and may die at any time from ileal obstruction, intussusception, peritonitis, or impaction. Animals with intussusceptions may have a bloody diarrhea or prolapsed rectum and colon.

Necropsy Signs.   Lesions of proliferative ileitis are usually most obvious in the ileum. The ileal mucosa is thickened, often dramatically, and the gut is enlarged three to four times. In some outbreaks the cecum or colon may be involved instead of the ileum. The intestinal wall is hyperemic, often ulcerated, and contains a yellow fluid often including mucus or blood. A pseudomembranous lining may be present. Mesenteric lymph nodes are enlarged. Intussusceptions may occur during the acute or chronic stages, as can perforation of the gut wall and peritonitis. Surviving hamsters can develop strictures, diverticula, multifocal hepatic abscesses, and adhesions.

Microscopically, there is hyperplasia of the villous epithelium with resulting widening and elongation of the villi followed by strangulation and necrosis of villous tips. The inflammatory process can extend into all levels of the gut wall.

Diagnosis.   Diagnosis of proliferative ileitis is based on the clinical observation of diarrhea and on the enlargement of the distal small intestine in weanling hamsters. Recent developments in culture methods for recovery of *Campylobacter* isolants from the gastrointestinal tracts of a variety of animals and man pave the way for future surveys and studies of these agents in laboratory animals. Silver and other special stains are helpful in demonstrating the *Campylobacter*-like organisms in the apical cytoplasm of the hyperplastic, ideal epithelium.

Treatment.   Young hamsters often have enteritis without ileal proliferation; however, because this difference may not be apparent on abdominal palpation, symptomatic treatment with a few drops of kaolin and pectin suspension can be attempted, but always with a guarded prognosis.

Erythromycin at 20 mg/kg body weight is the drug of choice in *Campylobacter* infections but may cause fatal alterations of gut flora in the hamster. During an outbreak of proliferative ileitis, treatment with tetracycline hydrochloride added to the drinking water of hamsters at 400 mg/L for 10 days reduced mortality (La Regina, Fales and Wagner 1980).

Prevention.   Proliferative ileitis is best prevented through the selection of hamsters with no familial history of the disease. Otherwise, high husbandry standards, frequent bedding changes, absence of environmental stressors, use of cage filter covers, and, when all else fails, erythromycin in the drinking water may reduce mortality. In our experience, use of 0.5 g erythromycin phosphate (Gallimycin-poultry formula—Abbott, Chicago, IL) per gallon of drinking water provided continuously brought an outbreak of PI in a commercial breeding colony under control. Prior to treatment, 90% of weanlings from 2500 breeder females were dying. During treatment, mortality was sporadic. The colony was maintained on the drug until its phase out and replacement several years later. Success in treating the clinical outbreak with erythromycin is further suggestive evidence of the causative role of *Campylobacter* spp.

Public Health Significance.   Proliferative ileitis is not known to be transmissible to man, but *Campylobacter* spp. can be highly pathogenic for man and other animals.

REFERENCES

Amend, N.K., et al.: Transmission of enteritis in the Syrian hamster. Lab. Anim. Sci., 26:566–572, 1976.

Fernie, D.S., and Park, R.W.A.: The isolation and nature of campylobacters (microaerophilic vibrios) from laboratory and wild rodents. J. Med. Microbiol., 10:325–329, 1977.

Frisk, C.S., and Wagner, J.E.: Hamster enteritis: a review. Lab. Anims., 11:79–85, 1977.

Frisk, C.S., and Wagner, J.E.: Experimental hamster enteritis: An electron microscopic study. Am. J. Vet. Res., 38:1861–1868, 1977.

Frisk, C.S., Wagner, J.E., and Owens, D.R.: Enteropathogenicity of *Escherichia coli* isolated from hamsters (*Mesocricetus auratus*) with hamster enteritis. Infect. Immun., 20:319–320, 1978.

Jacoby, R.O.: Transmissible ileal hyperplasia of hamsters. I. Histogenesis and immunocytochemistry. Am. J. Pathol., 91:433–450, 1978.

Jacoby, R.O., Osbaldiston, G.W., and Jonas, A.M.: Experimental transmission of atypical ileal hyperplasia of hamsters. Lab. Anim. Sci., 25:465–473, 1975.

Jelinek, F., and Aldova, E.: Campylobacteriosis in golden hamsters (*Mesocricetus auratus*). Z. Versuchstierkd., 28:167–171, 1986.

Johnson, E.A., and Jacoby, R.O.: Transmissible ileal hyperplasia of hamsters. II. Ultrastructure. Am. J. Pathol., 91:451–468, 1978.

Kim, J.C.S., and Jourden, M.: Ultrastructure of proliferative ileitis in the hamster (*Mesocricetus auratus*). Lab. Anims., 11:171–174, 1977.

La Regina, M., Fales, W.H., and Wagner, J.E.: Effects of antibiotic treatment on the occurrence of experimentally induced proliferative ileitis of hamsters. Lab. Anim. Sci., 30:38–41, 1980.

La Regina, M., and Lonigro, J.: Isolation of *Campylobacter fetus* subspecies *jejuni* from hamsters with proliferative ileitis. Lab. Anim. Sci., 32:660–662, 1982.

Lentsch, R.H., et al.: *Campylobacter fetus* subspecies *jejuni* isolated from Syrian hamsters with proliferative ileitis. Lab. Anim. Sci., 32:511–514, 1982.

McNeil, P.E., et al.: Control of an outbreak of wet-tail in a closed colony of hamsters (*Mesocricetus auratus*). Vet. Rec., 119:272–273, 1986.

Sheffield, F.W., Beveridge, E., and Sebesteny, A.: Syrian hamsters. In Diseases of Laboratory Animals. Edited by J. Hime and P. O'Donoghue. London, Heineman Veterinary Books, 1979.

Wagner, J.E., Owens, D.R., and Troutt, H.F.: Proliferative ileitis of hamsters: electron microscopy of bacteria in cells. Am. J. Vet. Res., 34:249–252, 1973.

REOVIRUS TYPE 3 INFECTION OF MICE

Hosts.   Reoviruses have a wide host range. Naturally occurring disease is observed primarily in suckling mice from both conventional and cesarean-derived colonies. Hamsters, rabbits, rats, humans, nonhuman primates, and many other species, including mammals, marsupials, birds, reptiles, and fish, may develop titers to reoviruses. The importance of reoviruses in contemporary stocks of research animals is uncertain. They are not commonly seen as entities of clinical disease.

Etiology. Reovirus (*R*espiratory *E*nteric *O*rphan or hepatoencephalomyelitis virus) are double-stranded RNA viruses types 1, 2, and 3.

Transmission. Transmission is by the fecal-oral route, although passage through the placenta, urine, transplanted tissues, and respiratory secretions may occur.

Clinical Signs. Signs of infection in suckling mice include diarrhea, oily hair caused by excess unabsorbed fat in the feces, conjunctivitis, growth retardation, emaciation, tremors, and jaundice. Survivors past weaning age may develop hair loss and jaundice.

Necropsy Signs. At necropsy mice may have yellow, necrotic areas in the liver. Lesions may also be present in the brain, heart, pancreas, and skeletal muscle. Infant's intestines may be distended and red with a yellow content. The agent multiplies in many organs and exists as a viremia.

Diagnosis. Diagnosis is usually by ELISA serologic test or IFA.

Prevention and Treatment. Prevention is attained by purchasing and maintaining stock free of reoviruses. Because the reoviruses may be carried by man, people are a potential source of breaches in barrier containment efforts. Therefore it may be advisable to use people with antibody titers to reoviruses as in-barrier animal technicians. There is no treatment for the disease.

Public Health Significance. The agent has zoonotic potential and may cause enteritis in susceptible persons, usually the young, in contact with infected mice.

REFERENCES

Andrews, E.J.: Spontaneous viral infections of laboratory animals. *In* Handbook of Laboratory Animal Science, Vol. III. Edited by E.C. Melby, Jr. and N.H. Altman. Cleveland, OH, CRC Press, Inc., 1976, pp. 119–178.

Cook, I.: Reovirus type 3 infection in laboratory mice. Aust. J. Exp. Biol. Med. Sci., *41*:651–659, 1963.

Jenson, A.B., et al.: Reovirus encephalitis in newborn mice. An electron microscopic and virus assay study. Am. J. Pathol., *47*:223–239, 1965.

Papadimitriou, J.M.: Electron micrographic features of acute murine reovirus hepatitis. Am. J. Pathol., *47*:565–585, 1965.

Parker, J.C., Tennant, R.W., and Ward, T.C.: Prevalence of viruses in mouse colonies. Natl. Cancer Inst. Monogr., *20*:25–36, 1966.

Parker, J.C., et al.: Virus studies with germfree mice. I. Preparation of serologic diagnostic reagents and survey of germfree and monocontaminated mice for indigenous murine viruses. J. Natl. Cancer Inst., *34*:371–380, 1965.

Rosen, L.: Reoviruses. *In* Diagnostic Procedures for Viral and Rickettsial Infections. Edited by E.H. Lennette and N.J. Schmidt. New York, American Public Health Association, 1969, pp. 354–363.

Schmidt, N.J., Lennette, E.H., and Hanahoe, M.F.: Microneutralization test for the reoviruses. Application to detection and assay of antibodies in sera of laboratory animals. Proc. Soc. Exp. Biol. Med., *121*:1268–1275, 1966.

Stanley, N.F., Dorman, D.C., and Pensford, J.: Studies on the pathogenesis of a hitherto undescribed virus (hepato-encephalomyelitis) producing unusual symptoms in suckling mice. Aust. J. Exp. Biol. Med. Sci., *31*:147–159, 1953.

Stanley, N.F., et al.: Murine infection with reovirus: II. The chronic disease following reovirus type 3 infection. Br. J. Exp. Med., *45*:142–149, 1964.

Tyler, K.L., and Fields, B.N.: Reovirus infection in laboratory rodents. *In* Viral and Mycoplasmal Infections of Laboratory Rodents. Edited by P.N. Bhatt, et al. Orlando, Academic Press, 1986, pp. 277–303.

Walters, M.N.-I., et al.: Murine infection with reovirus: I. Pathology of the acute phase. Br. J. Exp. Pathol., *44*:427–436, 1963.

ROTAVIRUS INFECTION

Hosts. Rotaviruses have been isolated from normal and diarrheal mice, rats, rabbits, humans, and several farm and wild animals. The disease in mice has long been known as epizootic diarrhea of infant mice (EDIM). EDIM is widespread in conventional and barrier-sustained mouse colonies worldwide. Infectious diarrhea of infant rats (IDIR) is similar to EDIM of mice.

Etiology. Rotaviruses are 60 to 80 nm, contain double-stranded RNA, and are structurally (capsomers on surface and nonenveloped) and chemically related to the reoviruses. There are, however, serologic and biophysical differences from the reoviruses, which merit the rotaviruses a distinct classification. The virus causes an inhibition of the intestinal absorptive mechanism through epithelial destruction and transport system paralysis. Concurrent infection with the EDIM rotavirus plus the MHV coronavirus produces a disease that resembles what was once described as lethal intestinal viral infection of mice (LIVIM).

Transmission.   The viruses of EDIM and IDIR are transmitted perorally by direct contact, ingestion of contaminated feces, or contaminated dusts and aerosols. The disease is highly contagious. Animals surviving the clinical disease or exposed as adults may become chronic shedders of the virus for weeks. The filter cover concept was initially introduced to prevent transmission of the EDIM agent among susceptible mice. Little is known about rotavirus infection in rabbits.

Predisposing Factors.   The clinical diarrheal disease is seen in mice and rats up to 12 or 13 days of age, and it is more severe in mice born to primiparous females.

Clinical Signs.   EDIM and IDIR are diarrheal diseases of high morbidity and low mortality. Affected suckling mice and rats have a soft, yellow feces that wets and stains the perineum. In rats, diarrhea is usually seen in 8- to 12-day-old pups. The diarrhea persists for 5 to 6 days and is associated with erythema and cracking and bleeding of the perianal skin. Although affected rodents continue to eat, they fail to grow normally, and stunted growth is a common sequela. Weaning percentage (percentage of young weaned compared with litter size at birth) may be slightly reduced. Those that are weaned are somewhat underweight for age. Affected rabbits have a mild to severe diarrhea that, according to the few reports available, has caused deaths.

Necropsy Signs.   The gastrointestinal tract of the affected mouse, rabbit, or rat may appear normal or be distended with gas and a watery, yellow feces. Histologically there may be increased extrusion of necrotic epithelial cells from villous tips in the small intestine, thus leaving more cuboidal-shaped cells than normal columnar-shaped cells. In severe cases villi may slough their epithelial covering and there is villous atrophy. The vacuolar change seen in villous epithelial cells of mice with EDIM is quite characteristic but must be distinguished from "absorption" vacuoles found in normal suckling mice. Rats, like pigs with rotaviral infections, develop villous epithelial syncytial cell formations, particularly in the distal one third of the small intestine. Numerous viral particles can be seen in syncytial cells electron microscopically.

Diagnosis.   A diagnosis of rotaviral infection is usually made on the basis of clinical signs, case history, and histologic findings. Electron microscopy of negatively stained specimens prepared from diarrheic feces has been used to diagnose rotaviral infections. An ELISA test has been developed, and an immunofluorescent test for viral antigen in the cytoplasm of intestinal cells has been described.

Treatment.   Usually little can be done to treat EDIM or IDIR; however, supportive treatment of rabbits may be indicated.

Prevention.   Selection of disease-free stock and the use of cage filter covers will reduce or prevent outbreaks of EDIM. Barrier housing will markedly reduce chances of new infections.

Public Health Significance.   Animal rotaviruses are not known to affect man.

REFERENCES

Banfield, W.G., Kasnic, G., and Blackwell, J.H.: Further observations on the virus of epizootic diarrhea of infant mice: an electron microscope study. Virology, 36:411–421, 1968.

Bryden, A.S., Thouless, M.E., and Flewett, T.H.: [Rotavirus in calves] . . . and rabbits. Vet. Rec., 99:323, 1976.

Cheever, F.S.: Epidemic diarrheal disease of suckling mice. Ann. N.Y. Acad. Sci., 66:196–203, 1956.

Coelho, K.I.R., et al.: Pathology of rotavirus infection in suckling mice: A study by conventional biology, immunofluorescence, ultra thin sections, and scanning electron microscopy. Ultrastruct. Pathol., 2:59–80, 1981.

DiGiacomo, R.F., and Thouless, M.E.: Age-related antibodies to rotavirus in New Zealand rabbits. J. Clin. Microbiol., 19:710–711, 1984.

Eiden, J., Vonderfecht, S., and Yolken, R.H.: Evidence that a novel rotavirus-like agent of rats can cause gastroenteritis in man. Lancet, ii:8–11, 1985.

Foster, H.L.: Comparison of epizootic diarrhea of suckling rats and a similar condition in mice. J. Am. Vet. Med. Assoc., 133:198–201, 1958.

Gastrucci, G.: Isolation and characterization of cytopathic strains of rotavirus from rabbits. Arch. Virol., 83:99–104, 1985.

Jennings, L.F., and Rumpf, R.M.: Control of epizootic diarrhea in infant mice. Lab. Anim. Care, 15:386–391, 1965.

Kraft, L.M.: Studies on the etiology and transmission of epidemic diarrhea of infant mice. J. Exp. Med., 106:743–755, 1957.

Kraft, L.M., et al.: Practical control of diarrheal disease in a commercial mouse colony. Lab. Anim. Care, 14:16–19, 1964.

Kudron, E., et al.: Occurrence of rotavirus infection of rabbits in Hungary. Magyar Allatorvosok Lapja, 37:248–254, 1982.

Little, L.M., and Shadduck, J.A.: Pathogenesis of rotavirus infection in mice. Infect. Immun., 38:755–763, 1982.

Moon, H.W.: Mechanisms in the pathogenesis of diarrhea: a review. J. Am. Vet. Med. Assoc., 172:443–448, 1978.

Much, D.H., and Zajac, I.: Purification and characterization of epizootic diarrhea of infant mice virus. Infect. Immun., 6:1019–1024, 1972.

Panel report on the colloquium on selected diarrheal diseases of the young. J. Am. Vet. Med. Assoc., 173:315–318, 1978.

Petric, M., et al.: Lapine rotavirus: preliminary studies on epizoology and transmission. Can. J. Comp. Med., 42:143–147, 1978.

Poiley, S.M.: The development of an effective method for control of epizootic diarrhea in infant mice. Lab. Anim. Care, 17:501–510, 1967.

Schneider, H.A., and Collins, G.R.: Successful prevention of infantile diarrhea of mice during an epizootic by means of a new filter cage unopened from birth to weaning. Lab. Anim. Care, 16:60–71, 1966.

Sheridan, J.F., et al.: Prevention of rotavirus-induced diarrhea in neonatal mice born to dams immunized with empty capsids of simian rotavirus SA-11. J. Infect. Dis., 149:434–438, 1984.

Sheridan, J.F.: Mouse rotavirus. In Viral and Mycoplasmal Infections of Laboratory Rodents. Edited by P.N. Bhatt, et al. Orlando, Academic Press, 1986, pp. 217–243.

Smith, A.L., et al.: Detection of antibody to epizootic diarrhea of infant mice (EDIM) virus. Lab. Anim. Sci., 33:442–445, 1983.

Vonderfecht, S.L., et al.: Infectious diarrhea of infant rats produced by a rotavirus-like agent. J. Virol., 52:94–98, 1984.

Vonderfecht, S.L.: Infectious diarrhea of infant rats (IDIR) induced by an antigenically distinct rotavirus. In Viral and Mycoplasmal Infections of Laboratory Rodents. Edited by P.N. Bhatt, et al. Orlando, Academic Press, 1986, pp. 245–252.

SALMONELLOSIS

Hosts. Salmonellae are widespread in nature and affect a wide range of vertebrates. Guinea pigs are highly susceptible and develop severe clinical disease; mice and rats are also very susceptible and may carry subclinical infections for long periods. Rabbits, hamsters, and gerbils are less often infected, but severe outbreaks have occurred in these species.

Etiology. Salmonella typhimurium and S. enteritidis are the species most often isolated from laboratory animal species.

Transmission. Transmission is by the fecal-oral route through ingestion of feces or fecal-contaminated feed or bedding. The organism can exist in the carrier state in the intestinal tract and be continually shed into the environment.

Predisposing Factors. Among the factors predisposing to salmonellosis are youth or old age, nutritional deficiencies, concomitant diseases, genetic predisposition, serotype of organism involved, and environmental and experimental stresses.

Guinea pigs at parturition and at weaning, especially in winter, are highly susceptible. C57BL, DBA/2, C3H/He, and BALB/c mice are considered among the more susceptible strains.

Clinical Signs. Salmonellosis in laboratory animals is an enteric and systemic infection that may be enzootic or epizootic. In guinea pig colonies, sporadic outbreaks with high mortality are the rule, but in mice and rats, clinical signs are usually inapparent.

Specific signs, when present, include anorexia, rough hair coat, weight loss, light soft feces, ocular discharge, small litters, dyspnea, and abortions. A colony of affected rats had conjunctivitis, anorexia, weight loss, and sporadic deaths. Chronic carriers exist and make elimination of the infection difficult.

Necropsy Signs. Lesions of acute salmonellosis include enlargement, congestion, and focal necrosis of the liver, spleen, lymphoid tissues, and intestine. The gut may contain increased amounts of gas and fluid.

In subacute or chronic cases the spleen may be twice enlarged, the liver and gut may be hyperemic, and yellow necrotic foci may be prominent in the viscera. Enlarged and prominent Peyer's patches may be seen throughout the ileal serosa.

Microscopically, embolically occluded microvasculature or phlebothrombosis occurs in hamsters. Foci of necrosis surrounded by a mononuclear infiltrate occur in the viscera, along with a suppurative lymphadenitis.

Diagnosis. While necropsy signs may be suggestive of salmonellosis, culture of the mesenteric lymph nodes, feces, or cecum is the method of choice for confirming an outbreak or a carrier state of salmonellosis. The agents do not hydrolize urea and usually do not use lactose. Standard bacteriologic media are used in culture. These media may include selenite F or tetrathionate broth to selectively enhance growth of salmonella from fecal samples followed by culture on MacConkey's (MC) and brilliant green (BG) agars, XLD, and HE. A battery of additional biochemical tests is used to further identify suspect cultures. Suspected salmonella isolants must be confirmed by serology because several nonpathogenic enteric bacteria may have similar biochemical reactions. A final determination of serotype is best performed in a reference laboratory.

Treatment. Treatment of salmonellosis may suppress an epizootic to an enzootic infection, but elimination of carriers is difficult. Because of the major public health concern, colonies infected with *Salmonella* should be eliminated, premises sanitized, and clean animals used for restocking.

One therapeutic regimen that has been followed with some success is the addition of oxytetracycline to the drinking water at 10 g/L for 10 days or 250 mg/kg body weight per day (Simmons and Simpson 1980). Treated mice were isolated, the remainder of the colony killed, the room disinfected, and first litters monitored for *Salmonella*. Thus, infection in valuable rodents by minimally pathogenic strains of *Salmonella* may be controlled or eliminated by intense culture, isolation, and treatment.

Prevention. Rigid, high husbandry standards and the screening of new arrivals, existing animals, especially dogs and subhuman primates, and animal care personnel will reduce the possibility of outbreaks. Birds, wild rodents, and contaminated feed must be excluded. Elimination of the infection from conventional colonies is extremely difficult; killing, disinfection, and restocking are more practical approaches.

Public Health Significance. Salmonellosis occurs in man and can be contracted from, or given to, laboratory animals. Animal care personnel should be periodically inspected for latent infections of *Salmonella*.

REFERENCES

Corazzola, S., Zanin, E., and Bersani, G.: Food poisoning in man following an outbreak of salmonellosis in rabbits. Vet. Ital., 22:370–373, 1971.

Duthie, R.C., and Mitchell, C.A.: *Salmonella enteritidis* infection in guinea pigs and rabbits. J. Am. Vet. Med. Assoc., 78:27–41, 1931.

Habermann, R.T., and Williams, F.P.: Salmonellosis in laboratory animals. J. Natl. Cancer Inst., 20:933–947, 1958.

Innes, J.R.M., Wilson, C., and Ross, M.A.: Epizootic *Salmonella enteritidis* infection causing septic pulmonary phlebothrombosis in hamsters. J. Infect. Dis., 98:133–141, 1956.

Kent, T.H., Formal, S.B., and LaBrec, E.H.: Acute enteritis due to *Salmonella typhimurium* in opium-treated guinea pigs. Arch. Pathol., 81:501–508, 1966.

Lavergne, G.M., et al.: The guinea pig as a model for the asymptomatic human typhoid carrier. Lab. Anim. Sci., 27:806–816, 1977.

Lentsch, R.H., et al.: A report of an outbreak of *Salmonella oranienburg* in a hybrid mouse colony. Vet. Microbiol., 8:105–109, 1983.

Margard, W.L., and Litchfield, J.H.: Occurrence of unusual salmonellae in laboratory mice. J. Bacteriol., 85:1451–1452, 1963.

Margard, W.L., and Peters, A.C.: A study of gnotobiotic mice monocontaminated with *Salmonella typhimurium.* Lab. Anim. Care, 14:200–206, 1964.

Moore, B.: Observations pointing to the conjunctiva as the portal of entry in salmonella infection of guinea-pigs. J. Hyg. (Lond.), 55:414–433, 1957.

Newcomer, C.E., Ackerman, J.I., and Fox, J.G.: Laboratory rabbits as reservoirs of *Salmonella mbandaka.* J. Infect. Dis., 147:365, 1983.

Olfert, E.D., Ward, G.E., and Stevenson, D.: *Salmonella typhimurium* infection in guinea pigs: observations on monitoring and control. Lab. Anim. Sci., 26:78–80, 1976.

Olson, G.A., Shields, R.P., and Gaskin, J.M.: Salmo-nellosis in a gerbil colony. J. Am. Vet. Med. Assoc., *171*:970–972, 1977.

Rabstein, M.M.: The practical establishment and main-tenance of Salmonella-free mouse colonies. Proc. Anim. Care Panel, *8*:67–74, 1958.

Ray, J.P., and Mallick, B.B.: Public health significance of Salmonella infections in laboratory animals. In-dian Vet. J., *47*:1033–1037, 1970.

Seps, S.L., et al.: Investigations of the pathogenicity of *Salmonella enteritidis* serotype *Amsterdam* following a naturally occurring infection in rats. Lab. Anim. Sci., *37*:326–330, 1987.

Shimi, A., Keyhani, M., and Hedayati, K.: Studies on salmonellosis in the house mouse, *Mus musculus.* Lab. Anims., *13*:33–34, 1979.

Simmons, D.J.C., and Simpson, W.: *Salmonella montevideo* salmonellosis in laboratory mice: successful treat-ment of the disease by oral oxytetracycline. Lab. Anim., *14*:217–219, 1980.

Simpson, W., and Simmons, D.J.C.: *Salmonella livingstone* salmonellosis in laboratory mice: successful contain-ment and treatment of the disease. Lab. Anims., *15*:261–262, 1981.

SENDAI VIRUS INFECTION

Hosts.   Mice, rats, hamsters, guinea pigs, and swine are hosts for Sendai virus infec-tion. Suckling and weanling mice are most commonly and seriously affected. Along with the coronaviruses (MHV), Sendai virus infection is one of the most significant dis-eases of laboratory rodents.

Etiology.   Sendai virus is an RNA para-myxovirus-parainfluenza type 1 virus. Many viruses in the paramyxovirus-para-influenza group share common antigens. Clinical Sendai disease is often complicated by mycoplasmal and bacterial infections.

Transmission.   Although natural infec-tion is via the respiratory tract, the mech-anism of Sendai virus (SV) transmission is incompletely known, but passage may be by direct contact, contaminated fomites or tissues, or respiratory aerosol. The disease is extremely contagious. Acutely infected weanling mice (4 to 6 weeks of age) provide the principal reservoir for transmission in colonies that are enzootically infected. Dis-ease-free or specific pathogen-free animals are highly susceptible to infection.

Predisposing Factors.   Some inbred strains (129 and DBA/2) are very susceptible to Sendai virus infection, but other strains (C57BL/6 and SJL) and random-bred mice are moderately resistant. Concurrent *Pasteu-rella pneumotropica* and *Mycoplasma pulmonis* in-fections greatly increase morbidity and mortality. In susceptible colonies the disease is highly overt and epizootic, especially if concurrent diseases are present to produce a synergism. An inapparent enzootic form is seen when most dams have been infected and pass maternal antibodies to their young.

Clinical Signs.   In enzootically infected colonies, newborn mice are passively pro-tected by maternal antibody until they are 4 to 6 weeks of age, at which time they become infected and develop an acquired immunity that is probably lifelong. Because adults in infected colonies have active im-munity, they rarely show disease. Fre-quently, the disease is subclinical in weanling mice. Sendai virus infections are nearly always subclinical in rats, hamsters, and guinea pigs. Clinical signs of an acute infection in susceptible animals include rough hair coat, chattering, weight loss, dyspnea, decreased breeding efficiency, and variable mortality. Bacterial infections ex-acerbate the clinical disease.

Epizootic disease occurs when a previ-ously uninfected colony encounters SV in-fection. In this case, any and all animals may show clinical signs. There may be high mor-tality with many sick animals among neo-nates and sucklings. After several weeks to several months of epizootic disease, an en-zootic pattern of disease ensues and persists as long as susceptible animals are present; usually either weanlings or newly intro-duced naive or nonimmune mice are sus-ceptible. Nude mice are also highly susceptible and develop a persistent infec-tion and wasting disease.

Necropsy Signs.   The acute pneumonitis is grossly evident as multifocal reddening, swelling, and firmness of the lungs. As the disease progresses to a subacute broncho-pneumonia, consolidation occurs.

Histologically, the acute phase is char-acterized by edema, hyperemia, necrosis (day 5), and pneumonitis. Focal areas of the bronchial and bronchiolar epithelium degenerate. Beginning on day 9 there is adenomatous hyperplasia and squamous metaplasia in the alveoli and distal bronchioles, and interstitial lymphoid aggregations appear that persist for long periods of time. The bronchiolar hyperplasia may be quite pronounced and may persist for the life of the animal. The most severe lesions are seen in weanlings of susceptible strains.

Diagnosis. Consideration of the species, strain, and age of animals involved and of the clinical, necropsy, and rather characteristic histologic lesions permits a provisional diagnosis of Sendai virus infection.

The phase of Sendai virus infection determines which diagnostic test is most appropriate. In early stages of disease, during viral replication in the lungs (from 2 to 8 days post infection), viral isolation and histochemistry or immunoperoxidase procedures to detect viral antigen are effective. As lesions develop and antibody titers rise, histopathologic and serologic diagnostic tests are effective.

Serologic testing is the most commonly used diagnostic technique. Serum antibody appears 6 to 8 days after initial infection. HI (using human O or fowl erythrocytes), IFA, and ELISA tests are used commercially in the United States. All tests are useful because a high percentage of mice in an affected colony will have high titers of antibody. The ELISA and IFA tests are preferred because they are more sensitive and detect titers in a larger percentage of animals. Animals vaccinated with live attenuated or killed vaccines develop titers that must be distinguished from titers associated with the naturally occurring disease. Inoculation of suspect materials onto cell cultures can be used to diagnose Sendai virus infection prior to the time that antibodies appear and suppress active infection.

Prevention. Prevention of the highly contagious but usually subclinical Sendai virus infection involves selection of rodents from a Sendai-free source and continual seromonitoring of the colony. Because young,

recently weaned hamsters, mice, rats, and guinea pigs constitute a reservoir population through which the virus maintains itself, the elimination of such young and the exclusion of additional young for 1 to 2 months may stop an outbreak. Commercial producers successfully use this method to eliminate Sendai virus from large breeding colonies. Also, a killed vaccine of duck embryo origin is available commercially. The vaccine is prepared in 50-dose vials, and 0.1 ml of the vaccine given intraperitoneally provides approximately 7 months of protection. An attenuated live vaccine is used by at least one commercial rodent breeder.

Effect on Research. Researchers should be apprised of the fact that a wide variety of bodily functions are altered by concurrent Sendai virus infection, e.g., depression of cell-mediated immunity, blast transformation, phagocytosis, and decreased intrapulmonary antibacterial activity. The disease generally has a major impact on animals used in immunologic research, transplantation studies, and pulmonary carcinogenesis.

In animal facilities where Sendai virus is enzootic it is advisable to purchase vaccinated or older animals, i.e., 6 weeks old versus 3 weeks old, when possible. Older animals of susceptible strains have lower mortality, and clinical disease appears milder.

Public Health Significance. Little conclusive evidence exists that Sendai virus isolated from rodents affects man, although the possibility exists that this type 1 paramyovirus may cause an upper respiratory infection in humans. Antigens shared among this group of viruses, however, confuse the diagnosis.

REFERENCES

Appell, L.H., et al.: Pathogenesis of Sendai virus infection in mice. Am. J. Vet. Res., *32*:1835–1841, 1971.

Brownstein, D.: Genetics of natural resistance to Sendai virus infection in mice. Infect. Immun., *41*:308–312, 1983.

Brownstein, D.: Sendai virus. *In* Viral and Mycoplasmal Infections of Laboratory Rodents. Edited by P.N. Bhatt, et al. Orlando, Academic Press, 1986, pp. 37–61.

Brownstein, D., Smith, A., and Johnson, E.: Sendai virus infection in genetically resistant and susceptible mice. Am. J. Pathol., *105*:156–163, 1981.

Burek, J.D., et al.: A naturally occurring epizootic caused by Sendai virus in breeding and aging rodent colonies. II. Infection in the rat. Lab. Anim. Sci., *27*:963–971, 1977.

Carter, G.R.: Pasteurella infections as sequelae to respiratory viral infections. J. Am. Vet. Med. Assoc., *163*:863–864, 1973.

Carthew, P., Gannon, J., and Whisson, I.: Comparison of alkaline phosphatase and horseradish peroxidase conjugated antisera in the ELISA test for antibodies to reovirus 3, mouse hepatitis, and Sendai viruses. Lab. Anims., *15*:69–73, 1981.

Carthew, P., and Sparrow, S.: A comparison in germ-free mice of the pathogenesis of Sendai virus and mouse pneumonia virus infections. J. Pathol., *130*:153–158, 1980.

Coid, C.R., and Wardman, G.: The effect of parainfluenza type 1 (Sendai) virus infection on early pregnancy in the rat. J. Reprod. Fertil., *24*:39–43, 1971.

Eaton, G., et al.: Eradication of Sendai pneumonitis from a conventional mouse colony. Lab. Anim. Sci., *32*:384–386, 1982.

Ertl, H., Gerlich, W., and Koszinowski, U.: Detection of antibodies to Sendai virus by enzyme-linked immunosorbent assay (ELISA). J. Immunol. Meth., *28*:163–176, 1979.

Fujiwara, K., Takenaka, S., and Shumiya, S.: Carrier state of antibody and viruses in a mouse breeding colony persistently infected with Sendai and mouse hepatitis viruses. Lab. Anim. Sci., *26*:153–159, 1976.

Fukumi, H., and Takeuchi, Y.: Vaccination against parainfluenza 1 virus (typus muris) infection in order to eradicate this virus in colonies of laboratory animals. Dev. Biol. Stand., *28*:477–481, 1975.

Garlinghouse, L.E., Jr., Van Hoosier, G.L., Jr., and Giddens, W.E., Jr.: Experimental Sendai virus infection in laboratory rats I. Virus replication and immune response. Lab. Anim. Sci., *37*:437–441, 1987.

Giddens, W.E., Jr., Van Hoosier, G.L., Jr., and Garlinghouse, L.E., Jr.: Experimental Sendai virus infection in laboratory rats II. Pathology and immunohistochemistry. Lab. Anim. Sci., *37*:442–448, 1987.

Grunert, R.R.: Isolation of Sendai virus as a latent respiratory virus in mice. Lab. Anim. Care, *17*:164–171, 1967.

Hall, W., and Ward, J.: A comparison of the avidin-biotin-peroxidase complex (ABC) and peroxidase-anti-peroxidase (PAP) immunocytochemical techniques for demonstrating Sendai virus infection in fixed tissue specimens. Lab. Anim. Sci., *34*:261–263, 1984.

Ishida, N., and Homma, M.: Sendai virus. Adv. Virus Res., *23*:349–383, 1978.

Itoh, T., and Iwai, H.: Effects of Sendai virus infection on body weight, body temperature and some hematological and serobiochemical values of mice. Exp. Anim., *30*:491–495, 1981.

Jakab, G.J.: Interactions between Sendai virus and bacterial pathogens in the murine lung: a review. Lab. Anim. Sci., *31*:170–177, 1981.

Kay, M., et al.: Age-related changes in the immune system of mice of eight medium and long-lived strains and hybrids. II. Short and long-term effects of natural infection with parainfluenza type 1 virus (Sendai). Mech. Aging Devel., *11*:347–362, 1979.

Kimura, Y., Aoki, H., and Shimokata, K.: Protection of mice against virulent virus infection by a temperature-sensitive mutant derived from an HVJ (Sendai virus) carrier culture. Arch. Virol., *61*:297–304, 1979.

Lucas, C., et al.: A quantitative immunofluorescence test for detection of serum antibody to Sendai virus in mice. Lab. Anim. Sci., *37*:51–54, 1987.

Makino, S., et al.: An epizootic of Sendai virus infection in a rat colony. Exp. Anim., *22*:275–280, 1972.

Parker, J.C., and Reynolds, R.K.: Natural history of Sendai virus infection in mice. Am. J. Epidemiol., *88*:112–125, 1968.

Parker, J.C., and Richter, C.: Viral diseases of the respiratory system. In The Mouse in Biomedical Research. Volume II, Diseases. Edited by H.L. Foster, J.D. Small, and J.G. Fox. Orlando, Academic Press, 1982, pp. 110–158.

Parker, J.C., Whiteman, M.D., and Richter, C.B.: Susceptibility of inbred and outbred mouse strains to Sendai virus and prevalence of infection in laboratory rodents. Infect. Immun., *19*:123–130, 1978.

Parker, J.C., et al.: Enzootic Sendai virus infections in mouse breeder colonies within the United States. Science, *146*:936–938, 1964.

Parker, J.C., O'Beirne, A., and Collins, M.: Sensitivity of enzyme-linked immunosorbent assay, complement fixation, and hemagglutination inhibition serological tests for detection of Sendai virus antibody in laboratory mice. J. Clin. Microbiol., *9*:444–447, 1979.

Profeta, M.L., Lief, F.S., and Plotkin, S.A.: Enzootic Sendai infection in laboratory hamsters. Am. J. Epidemiol., *89*:316–324, 1968.

Rottinghaus, A.A., Gibson, S.V., and Wagner, J.E.: Comparison of serologic tests for detection of antibodies to Sendai virus in rats. Lab. Anim. Sci., *36*:496–498, 1986.

Tsukui, M., et al.: Protective effect of inactivated virus vaccine on Sendai virus infection in rats. Lab. Anim. Sci., *32*:143–146, 1982.

van der Veen, J., Poort, Y., and Birchfield, D.J.: Study of the possible persistence of Sendai virus in mice. Lab. Anim. Sci., *24*:48–50, 1974.

Ward, J.M.: Naturally occurring Sendai virus disease of mice. Lab. Anim. Sci., *24*:938–942, 1974.

Ward, J.M., et al.: Naturally occurring Sendai virus infection of athymic nude mice. Vet. Pathol., *13*:36–46, 1976.

Zurcher, C., et al.: A naturally occurring epizootic caused by Sendai virus in breeding and aging rodent colonies. I. Infection in the mouse. Lab. Anim. Sci., *27*:955–962, 1977.

SIALODACRYOADENITIS

Hosts. Rats are the natural hosts for the highly contagious and very common sialodacryoadenitis (SDA) virus. The self-limiting clinical disease commonly caused by this agent is seen primarily in young rats.

Etiology. The SDA virus is a coronavirus

(RNA) antigenically related to the mouse hepatitis virus (MHV), rat coronavirus (RCV), and the human coronavirus OC38. The virus is highly infectious and epitheliotropic and replicates in the cytoplasm, most notably in cells of the respiratory epithelium and in salivary, lacrimal, and harderian glands.

Transmission. Passage among exposed animals is rapid and most probably via respiratory aerosol or direct contact with respiratory secretions. Infected animals carry and secrete the virus for about 7 days. The disease can be endemic, but the organism does not exist in a latent, carrier state. Extension of the infection from the respiratory epithelium is via ducts of the salivary, lacrimal, and harderian glands.

Predisposing Factors. Exposed, healthy rats of both sexes and all ages are susceptible to SDA virus infection, although the severity and site of the infection vary, especially among hosts of different ages and strains. Susceptible young animals, 2 to 4 weeks of age, with no maternal antibody, develop the more serious forms of the disease.

Clinical Signs. Sialodacryoadenitis, a highly infectious enzootic or epizootic disease of rats, has high morbidity and usually low mortality. Whereas clinical signs are seen in a colony for several weeks during an epizootic, individual animals show signs for up to a week. Signs include squinting, photophobia, blinking, and eye rubbing followed by sneezing and cervical swelling within 5 to 7 days post infection. Sneezing is caused by acute rhinitis. Swelling under the neck is caused by cervical edema, enlarged cervical lymph nodes, and necrotic inflamed salivary glands. Bilateral or unilateral suborbital or periorbital swelling, prominent or bulging eyes, and often chromodacryorrhea ("red tears") and ophthalmic lesions develop secondarily to decreased lacrimation; self-mutilation may occur as a result of scratching of affected areas. Very young animals may enucleate the eye. During the infection rats usually remain active and eating, although certain behavioral activities may be depressed.

Pneumotropic strains of the rat coronaviruses, especially in such susceptible strains of rats as the Fisher 344 strain, may cause an interstitial pneumonia in young rats. This infection may be fatally exacerbated by concomitant Sendai virus infection or mycoplasmosis. During the pneumonic phase of the disease, rats may have high mortality associated with general anesthesia.

Necropsy Signs. The infection progresses rapidly from the respiratory epithelium to the lacrimal and serous or serous-mucous salivary glands, associated lymph nodes, and contiguous tissues. Affected glands are enlarged, pale, and often reddened. The thymus becomes atrophic; however, this sign and the chromodacryorrhea may be stress responses. Salivary and lacrimal glands (harderian and exorbital) may both be affected, or either can be affected without involvement of the other.

Immunofluorescent studies show viral antigen in the respiratory epithelium, lacrimal and salivary glands, and lymph nodes. Epithelial cells of the respiratory tract and ducts and acini of the salivary, lacrimal, and harderian glands undergo severe, acute necrosis. As the infectious virus is eliminated from the lesion, repair processes ensue. The ductal epithelium undergoes transient squamous metaplasia and acinar repair is characterized by regenerative hyperplasia. Harderian glands also have prominent squamous metaplasia. Accompanying changes are persistent mixed inflammatory cell infiltration of the interstitium, tubular blockage, and secondary cystic and metaplastic changes. Lymph nodes are hyperplastic with foci of necrosis.

Resolution to normal or scar requires 2 to 4 weeks. Ophthalmic lesions, associated with loss of tears, trauma, and inflammation, range from keratitis sicca to multiple inflammatory sites within and around the eyeball.

Diagnosis. Diagnosis of SDA is based on clinical signs, histologic lesions, and the detection of serum antibodies to the virus. Harderian gland lesions caused by damage associated with periorbital bleeding can be confused with lesions of SDAV infection. Conjunctivitis from other causes may mimic a prominent but secondary sign of an SDA infection. Rats without antibody titers to mouse hepatitis virus or rat coronavirus can be assumed to be free of SDA antibodies. Mouse hepatitis virus antigens cross react with SDA virus antibody in the serum and are often used in ELISA tests. Virus neutralization, IFA, and immunohistochemical tests may also be useful.

Treatment. Treatment during the rapid course of the disease can only alleviate the secondary ophthalmic bacterial infections (*Pasteurella pneumotropica* and *Staphylococcus aureus*).

Prevention. Susceptible colonies can be isolated and quarantined, but the virus is highly infectious and often manifests itself where least expected. Except in cases of eye enucleation, the disease produces essentially no disability and in many cases the rats are suitable for research. The disease confers long-lasting immunity against reinfection.

Public Health Significance. There are no reports of human susceptibility to the SDA virus.

REFERENCES

Ashe, W.K., Scherp, H.W., and Fitzgerald, R.J.: Previously unrecognized virus from submaxillary glands of gnotobiotic and conventional rats. J. Bacteriol., *90*:1719–1729, 1965.

Bhatt, P.N., Jacoby, R.O., and Jonas, A.M.: Respiratory infection in mice with sialodacryoadenitis virus, a coronavirus of rats. Infect. Immun., *18*:823–827, 1977.

Bhatt, P.N., Percy, D.H., and Jonas, A.M.: Characterization of the virus of sialodacryoadenitis of rats: a member of the coronavirus group. J. Infect. Dis., *126*:123–130, 1972.

Bhatt, P.N., and Jacoby, R.O.: Epizootiological observations of natural and experimental infection with sialodacryoadenitis virus in rats. Lab. Anim. Sci., *35*:129–134, 1985.

Carthew, P., and Slinger, R.P.: Diagnosis of sialodacryoadenitis virus infection of rats in a virulent enzootic outbreak. Lab. Anims., *15*:339–342, 1981.

Eisenbrandt, D.L., Hubbard, G.B., and Schmidt, R.E.: A subclinical epizootic of sialodacryoadenitis in rats. Lab. Anim. Sci., *32*:655–659, 1982.

Hanna, P.E., et al.: Sialodacryoadenitis in the rat: Effects of immunosuppression on the course of the disease. Am. J. Vet. Res., *45*:2077–2083, 1984.

Hunt, R.D.: Dacryoadenitis in the Sprague-Dawley rat. Am. J. Vet. Res., *24*:638–641, 1963.

Innes, J.R.M., and Stanton, M.F.: Acute disease of the submaxillary and harderian glands (sialodacryoadenitis) of rats with cytomegaly and no inclusion bodies. Am. J. Pathol., *38*:455–468, 1961.

Jacoby, R.O., Bhatt, P.N., and Jonas, A.M.: Pathogenesis of sialodacryoadenitis in gnotobiotic rats. Vet. Pathol., *12*:196–209, 1975.

Jacoby, R.O., Bhatt, P.N., and Jonas, A.M.: Viral diseases. *In* The Laboratory Rat. Vol. I. Edited by J.H. Baker, J.R. Lindsey, and S.H. Weisbroth. Orlando, Academic Press, 1979, pp. 271–306.

Jacoby, R.O.: Rat coronavirus. *In* Viral and Mycoplasmal Infections of Laboratory Rodents. Edited by P.N. Bhatt, et al. Orlando, Academic Press, 1986, pp. 625–638.

Jonas, A.M., et al.: Sialodacryoadenitis in the rat. A light and electron microscopic study. Arch. Pathol., *88*:613–622, 1969.

Lai, Y.-L., et al.: Keratoconjunctivitis associated with sialodacryoadenitis in rats. Invest. Ophthalmol., *15*:538–541, 1976.

Peters, R.L., and Collins, M.J.: Use of mouse hepatitis virus antigen in an enzyme-linked immunosorbent assay for rat coronaviruses. Lab. Anim. Sci., *31*:472–475, 1981.

Smith, A.L.: An immunofluorescence test for detection of serum antibody to rodent coronaviruses. Lab. Anim. Sci., *33*:157–160, 1983.

Utsumi, K., et al.: Infectious sialodacryoadenitis and rat breeding. Lab. Anims., *14*:303–307, 1980.

Weisbroth, S.H., and Peress, N.: Ophthalmic lesions and dacryoadenitis: a naturally occurring aspect of sialodacryoadenitis virus infection of the laboratory rat. Lab. Anim. Sci., *27*:466–473, 1977.

SPIRONUCLEOSIS (HEXAMITIASIS)

Hosts. The enteric flagellate, *Spironucleus muris,* affects mice, including inbred and athymic nude mice, rats, hamsters, and wild rodents. Generally, clinical disease is seen only in weanling mice.

Etiology. *Spironucleus (Hexamita) muris* is an opportunistic or facultative flagellate of the protozoan class Mastigacida and in the same family as Giardia. The piriform trophozoites (7 to 9 × 2 to 3 μm) possess two anterior

nuclei and eight flagella (two posterior and six anterior) and are commonly found in the upper small intestine. Small, oval cysts (5 μm) have a banded, "Easter egg" appearance and occur in both the small and large intestine. The life cycle is incompletely known, but trophozoites do not survive stomach passage as do environmentally resistant cysts. The prepatent period is 2 to 3 days, and dividing trophozoites have not been observed. *Spironucleus* feeds on bacteria, probably lactobacilli.

Transmission.   The encysted, environmentally resistant form of the protozoan is transmitted by the fecal-oral route. Young and athymic nude mice pass larger numbers of cysts than do other strains.

Predisposing Factors.   Susceptibility to *Spironucleus* infection varies with factors lowering resistance. Inbred strains C3H, NZW, DBA/2, and C57BL and the athymic nude mouse are very susceptible, as are mice of both sexes between 3 and 8 weeks of age, although clinical infection of weanlings is most often seen. Specific environmental stressors include irradiation, steroid administration, crowding, chilling, nutrition, and surgery. *Spironucleus* inhibits macrophage activity and may act synergistically with bacteria to cause disease.

Clinical Signs.   Spironucleosis may be an acute, chronic, or latent disease, depending on predisposing factors. The acute disease is characterized by a hunched posture, depression, huddling, dehydration, rough, dull hair coat, distended abdomens, and often a soft, sticky stool. Death at the peak of an outbreak occurs 3 to 4 days following onset of clinical signs, and mortality rates may approach 50%. Older mice have a marked variation in signs, with progressive wasting, listlessness, and sporadic deaths.

Necropsy Signs.   The common necropsy finding is an intestine grossly distended with a bubbly froth, giving the appearance of postmortem autolysis. The froth may be gray, red-brown, or yellow, but rarely contains blood. The duodenum and anterior jejunum are affected. Alternative or other findings include diarrhea, an enlargement and thickening of the duodenum, enlarged mesenteric nodes, ascites, and splenic and thymic atrophy.

Histopathologic signs are those of an acute catarrhal enteritis or a chronic mural hyperplasia with minimal mononuclear cell infiltration. Reduction of microvilli, epithelial desquamation, lymphocytic infiltration, villous edema, and the organisms in crypts, between villi, and in the lamina are common

findings. Paneth cells of the ileum ingest the protozoa. In chronic infections, large, debris-filled pseudocysts are seen in the duodenal wall. Villous hyperplasia is present.

Diagnosis.   Cysts and trophozoites may be seen on direct smears of intestinal content or in fixed and stained sections. The protozoans rapidly autolyze in unfixed tissues. Cysts, which may be quantitated, are better seen with iodine staining or phase contrast optics. Direct smears prepared from duodenal or jejunal contents at necropsy can be examined microscopically for trophozoites. Piriform trophozoites of *Spironucleus* have a fast, foreward, erratic movement, whereas the larger *Giardia* are halfmoon-shaped rollers. Cysts may be confused with yeast bodies.

Treatment.   The protozoastats dimetridazole and metronidazole in drinking water were reported to reduce mortality during *Spironucleus* outbreaks. Subsequently, the entire class of 5 hydroxy-imibazole compounds have been removed from the U.S. market because of a body of evidence that suggested they could be carcinogenic. An increase in the ambient temperature to 75° to 80° F has reduced mortality.

Prevention.   Cesarean derivation and cross fostering to "clean" dams reduce outbreaks, as does the provision of sufficient space, maintenance of mice in litters postweaning, culling of infected litters, and proper sanitation of caging. *Spironucleus muris* cysts are sensitive to temperatures above 45° C, 70% ethanol, 13% NaClO, and 4% formalin. They survive cold and desiccation.

Public Health Significance.   *Spironucleus muris* is not pathogenic for man.

REFERENCES

Boorman, G.A., et al.: Synergistic role of intestinal flagellates and normal intestinal bacteria in a postweaning mortality of mice. Lab. Anim. Sci., *23*:187–193, 1973.

Flatt, R.E., Halvorsen, J.A., and Kemp, R.L.: Hexamitiasis in a laboratory mouse colony. Lab. Anim. Sci., *28*:62–65, 1978.

Gruber, H.E., and Osborne, J.W.: Ultrastructural features of spironucleosis (hexamitiasis) in x-irradiated rat small intestine. Lab. Anims., *13*:199–202, 1979.

Herweg, C., and Kunstýř, I.: Effect of intestinal flagellate *Spironucleus* (*Hexamita*) *muris* and of dimetridazole on intestinal microflora in thymus-deficient (nude) mice. Zbl. Bakt. Hyg., I. Abt. Orig. A., *245*:262–269, 1979.

Keast, D., and Chesterman, F.C.: Changes in macrophage metabolism in mice heavily infected with *Hexamita muris*. Lab. Anims., *6*:33–39, 1972.

Kunstýř, I.: Infectious form of *Spironucleus* (*Hexamita*) *muris*: banded cysts. Lab. Anims., *11*:185–188, 1977.

Kunstýř, I., and Ammerpohl, E.: Resistance of faecal cysts of *Spironucleus muris* to some physical factors and chemical substances. Lab. Anims., *12*:95–97, 1978.

Lussier, G., and Loew, F.M.: An outbreak of hexamitiasis in laboratory mice. Can. J. Comp. Med., *34*:350–353, 1970.

Meshorer, A.: Hexamitiasis in laboratory mice. Lab. Anim. Care, *19*:33–37, 1969.

Sebesteny, A.: Transmission of *Spironucleus* and *Giardia* spp. and some nonpathogenic intestinal protozoa from infested hamsters to mice. Lab. Anims., *13*:189–191, 1979.

Wagner, J.E., et al.: Hexamitiasis in laboratory mice, hamsters, and rats. Lab. Anim. Sci., *24*:349–354, 1974.

STAPHYLOCOCCUS AUREUS INFECTION

Hosts. *Staphylococcus aureus* is primarily associated with humans, but infections occur uncommonly in a variety of other mammals, including cattle, dogs, cats, and laboratory animals. Subclinical nasal infections are common in guinea pigs and primates, uncommon in rabbits and dogs, and rare in rodents. Staphylococcal abscesses are common in some colonies of nude mice, and staphylococcal dermatitis occurs from time to time in rats and black mice, possibly secondary to undetermined inciting causes.

Nasal dermatitis of gerbils is frequently associated with staphylococcal infections, but the primary cause may be the irritating nature of porphyrin of harderian gland origin that accumulates around the external nares.

Etiology. *Staphylococcus aureus* is an aerobic, nonspore-forming, variably α and β hemolytic, catalase, and gram-positive coccus. Although several human phage types have been recovered from animals, various species of animals usually have host-specific staphylococcal types.

Transmission. Two important considerations in the transmission and persistence of *S. aureus* are human carriers and the resistance of the organism to drying and disinfectants. The organism may spread by direct contact between animals and man or by contaminated food, feces, cages, and bedding. Aerosol spread is also possible, as is invasion through traumatized skin or open umbilical stumps.

Predisposing Factors. Staphylococci are common microflora of the skin, mucous membranes, upper respiratory tract, and intestine. Initiation of clinical disease usually involves a cutaneous wound, constant skin wetting, or a stressor, such as crowding, dietary changes, rapid growth, low dietary protein, or high environmental temperatures. Fighting among cagemates, however, is rarely implicated as a predisposing factor, although contact with infected caretakers has been implicated.

Clinical Signs. Clinical signs of *S. aureus* infection, other than sudden death from pneumonia, septicemia, or toxemia, include fever, anorexia, depression, and death, or, more specifically and commonly, a moist dermatitis, foot swelling, subcutaneous lump, enlarged mammary glands, or a purulent discharge. Skin lesions, which are often pruritic and elicit self-mutilation, may develop scabs and heal spontaneously.

Among the dermatopathies, which range from alopecia and reddening to extensive ulceration, are cheilitis and exfoliative dermatitis in guinea pigs; pustular and ulcerative dermatitis in rats; ulcerative dermatitis and "sore nose" in gerbils; conjunctivitis in rabbits; and mastitis in several species. The lesions of a dermatitis are often exacerbated by scratching.

Internal staphylococcal lesions, which may have specific (purulent discharge) or nonspecific signs in the living animal, include acute to chronic rhinitis, gingivitis, pneumonia, and cystitis. Abscessation can occur in any tissue, including the tympanic bullae, preputial glands, brain, heart, urogenital system, lymph nodes, and spleen. Subcutaneous, muscular, or osseous abscessation has been reported in rabbits and rodents, and pododermatitis with macropodia occurs in rats and guinea pigs. Staphylococcal mastitis in rabbits, originating from scratches, bites, or nest box wounds, results in a variation of the "young doe syndrome." The resulting toxemia involves swollen glands, fever, anorexia, and death in nursing does. The young may also die.

Ulcerative pododermatitis (bumblefoot) in guinea pigs is a chronic dermatitis that progresses to arthritis. The enlarged feet with plantar scabs are 1 to 3 cm in diameter, hairless, and ulcerated. The inflammation infiltrates joints, tendon sheaths, and connective tissue and rarely produces a pus pocket that can be opened and drained; in fact, cutting the foot leads to severe bleeding. Heavy animals housed on wire or an abrasive floor are most often affected. *Staphylococcus aureus,* the usual causative agent, probably enters the foot through a cutaneous wound. Housing animals in clean-bedded, dry, or smooth-floored cages reduces the incidence of ulcerative pododermatitis. Systemic amyloidosis in the guinea pig is associated with chronic staphylococcal infection. Treatment is usually unsuccessful, as the condition rarely responds to antibiotic therapy or surgery.

"Sore nose" or facial dermatitis in gerbils is seen as a red, hairless area surrounding the nares and snout. Anxiety-producing stress in gerbils results in greatly increased activity, especially burrowing, and is associated with increased secretion of porphyrin, which may irritate the skin and initiate sore nose. Stressors include changes in cage, loss of or separation from cagemates, incompatible cagemates, overcrowding, and probably many other factors. Although the lesion may be initiated by the irritating effects of porphyrin secretions and aggravated by cutaneous abrasions incurred while burrowing, subsequent *S. aureus* infection may perpetuate a chronic dermatitis while other animals heal spontaneously.

Ulcerative dermatitis of rats is seen more often in young males. The lesions, which usually begin on the neck, shoulders, or anterior trunk, first appear as an irregularity in the hair coat. This small focus of alopecia or ulceration then enlarges over the next several days to form small to extensive ulcerations. Some lesions scab and regress, and others are pruritic and stimulate scratching and self-mutilation.

Necropsy Signs. Lesions of staphylococcal infection vary with host, site, and duration, but characteristics range from diffuse cellulitis to suppuration with abscessation, necrosis, and granuloma formation. Gram-positive clumps of large cocci may be seen on impression smears or in lesions examined microscopically.

Diagnosis. Recovery of *S. aureus* on culture confirms a diagnosis, or at least the presence of the organism at the site. Other bacteria, including *Pasteurella, Salmonella, Pseudomonas,* and *Streptococcus,* produce dermatitis or abscessation in rabbits and rodents. Trimming toenails or reducing pruritus will often provide dramatic evidence of the role of self-mutilation as a cause of the lesion.

Treatment. Treatment of staphylococcal infections, if indicated, includes cleaning, drainage, or excision of the lesion; removal of the cause of traumatic insult, if any; and selection of topical and parenteral antibiotics based on culture, sensitivity testing, and host idiosyncrasies.

Rodents are active groomers and quickly remove and consume materials applied topically. The consumed topical antibiotic may alter the intestinal flora, thus resulting in acute enterotoxemia-like conditions. Elizabethan collars, properly constructed and applied, may be useful in preventing animals from removing topical preparations.

Antibiotics that have been used include gentamicin, kanamycin, procaine penicillin G (20,000 to 40,000 units/kg for 3 days), nitrofuran topicals, oxytetracycline (55 mg/kg fresh daily in water for 4 days) (Smith 1977), tetracycline hydrochloride (0.3 g/100 ml water for 14 days) (Peckham, et al. 1974). Penicillin and other antibiotics may be fatal to rabbits and rodents. Autogenous bacterins have been used to suppress, but not eliminate, an outbreak of purulent conjunctivitis in rabbits.

Pododermatitis in guinea pigs is often impossible to cure. Ulcerative (moist) dermatopathies are cleaned, the surrounding hair is removed, and an antiseptic or antibiotic is applied. Treatment of staphylococcal dermatitis in rats and other rodents may require no more than trimming the rear toenails.

Prevention. The spread of *S. aureus* or the initiation of clinical disease in carrier animals is reduced or prevented by good sanitation, elimination of sharp or abrasive surfaces, use of clean feed and bedding, reduction of stress, and monitoring of caretakers and animals. Elimination of carriers by treatment is very difficult.

REFERENCES

Ash, G.W.: An epidemic of chronic skin ulceration in rats. Lab. Anims., 5:115–122, 1971.

Blackmore, D.K., and Francis, R.A.: The apparent transmission of staphylococci of human origin to laboratory animals. J. Comp. Pathol., 80:645–651, 1970.

Carolan, M.G.: Staphylococcosis in rabbits. Vet. Rec., 119:412, 1986.

Chew, B.P., Zamora, C.S., and Luedecke, L.O.: Effect of vitamin A deficiency on mammary gland development and susceptibility to mastitis through intramammary infusion with Staphylococcus aureus in mice. Am. J. Vet. Res., 46:287–293, 1985.

Clarke, M.C., et al.: The occurrence in mice of facial and mandibular abscesses associated with Staphylococcus aureus. Lab. Anim., 12:121–123, 1978.

Craven, N., and Anderson, J.C.: Antibiotic activity against intraleukocytic Staphylococcus aureus in vitro and in experimental mastitis in mice. Am. J. Vet. Res., 44:709–712, 1983.

Cubillos, V., Paredes, E., and Fiedler, H.H.: Estafilococosis en conejos. Arch. Med. Vet., 18:7–14, 1986.

Fox, J.G., et al.: Ulcerative dermatitis in the rat. Lab. Anim. Sci., 27:671–678, 1977.

Gupta, B.N., Conner, G.H., and Meyer, D.B.: Osteoarthritis in guinea pigs. Lab. Anim. Sci., 22:362–368, 1972.

Hagen, K.W., Jr.: Disseminated staphylococcic infection in young domestic rabbits. J. Am. Vet. Med. Assoc., 142:1421–1422, 1963.

Hinton, M.: Treatment of purulent staphylococcal conjunctivitis in rabbits with autogenous vaccine. Lab. Anims., 11:163–164, 1977.

Holliman, A., and Girvan, G.A.: Staphylococcosis in a commercial rabbitry. Lab. Anim. Sci., 119:187, 1986.

Hong, C.C., and Ediger, R.D.: Preputial gland abscess in mice. Lab. Anim. Sci., 28:153–156, 1978.

Ishihara, C.: An exfoliative skin disease in guinea pigs due to Staphylococcus aureus. Lab. Anim. Sci., 30:552–557, 1980.

Markham, N.P., and Markham, J.G.: Staphylococci in man and animals. Distribution and characteristics of strains. J. Comp. Pathol., 76:49–56, 1966.

Needham, J.R., and Cooper, J.E.: Bulbourethral gland infections in mice associated with Staphylococcus aureus. Lab. Anims., 10:311–315, 1976.

Peckham, J.C., et al.: Staphylococcal dermatitis in Mongolian gerbils (Meriones unguiculatus). Lab. Anim. Sci., 24:43–47, 1974.

Renquist, D., and Soave, O.: Staphylococcal pneumonia in a laboratory rabbit: an epidemiologic follow-up study. J. Am. Vet. Med. Assoc., 155:1221–1223, 1969.

Rountree, P.M., Freeman, B.M., and Johnston, K.G.: Nasal carriage of Staphylococcus aureus by various domestic and laboratory animals. J. Pathol. Bacteriol., 72:319–321, 1956.

Shults, F.S., Estes, P.C., and Franklin, J.A.: Staphylococcal botryomycosis in a specific-pathogen-free mouse colony. Lab. Anim. Sci., 23:36–42, 1973.

Smith, M.W.: Staphylococcal cheilitis in the guinea-pig. J. Small Anim. Pract., 18:47–50, 1977.

Snyder, S.B., et al.: Disseminated staphylococcal disease in laboratory rabbits (Oryctolagus cuniculus). Lab. Anim. Sci., 26:86–88, 1976.

Taylor, D.M., and Neal, D.L.: An infected eczematous condition in mice: methods of treatment. Lab. Anims., 14:325–328, 1980.

Taylor, J.L., et al.: Chronic pododermatitis in guinea pigs, a case report. Lab. Anim. Sci., 21:944–945, 1971.

Wagner, J.E., et al.: Self trauma and Staphylococcus aureus in ulcerative dermatitis of rats. J. Am. Vet. Med. Assoc., 171:839–841, 1977.

STREPTOCOCCUS PNEUMONIAE INFECTION

Hosts. *Streptococcus pneumoniae* affects a wide range of animals, but the young guinea pig and rat are particularly susceptible.

Etiology. *Streptococcus pneumoniae* (Genus synonyms are *Diplococcus* and *Pneumococcus*) infections in the guinea pig are caused by several serotypes including Types III, IV, and XIX. Diplococcal types reported in rats include II, III, VIII, and XVI. Diplococci are gram-positive, lancet-shaped cocci that often occur in distinctive pairs.

Transmission. Transmission of *S. pneumoniae* is by respiratory aerosol or direct contact. Clinically normal guinea pigs and rats may carry the organism in the upper respiratory passages. Contacts during shipment and subsequent stresses frequently precipitate the disease. Depending on the season of year, approximately 30% to 75% of a human population carries *S. pneumoniae* in the respiratory passages, and some of these human-associated serotypes are potential pathogens for rodents.

Predisposing Factors. Losses from *S. pneumoniae* are greater during the winter, following shipment, and in animals on marginal diets, especially in guinea pigs with hypovitaminosis C. Carrier animals frequently succumb when stressed by other diseases or experimental procedures.

Clinical Signs. Carrier animals with upper respiratory infections of *S. pneumoniae* but with few or no clinical signs of the infection are common. Signs of the clinical disease, which may be in acute episodes or in prolonged epizootics with variable morbidity and mortality, include sneezing, nasal and ocular discharges, anorexia, weight loss, coughing, dyspnea, depression, and death. Torticollis may exist if the inner ear is affected, and abortions are associated with both general and uterine infections. Similar signs, along with hematuria, are seen in affected rats. Carrier animals are prone to clinical disease due to *S. pneumoniae* following experimental manipulations, e.g., peritonitis following intraperitoneal injections, encephalitis following placement of intracranial implants, and pleuritis following cardiac puncture.

Necropsy Signs. Gross lesions of *S. pneumoniae* infection include seropurulent and fibrinopurulent pleuritis, pericarditis, epicarditis, peritonitis, meningitis, otitis media and interna, metritis, and bronchopneumonia.

Diagnosis. Diagnosis is established by observation of the gram-positive, lancet-shaped cocci in pairs or short chains on a stained, direct smear of the inflammatory exudate. Histologically, unlike many bacterial diseases, large numbers of bacterial organisms can usually be seen in tissue sections. Recovery of *S. pneumoniae* on blood agar culture in the presence of 10% CO_2 confirms the diagnosis. The organism is alpha hemolytic and bile soluble and is inhibited by ethylhydrocupreine (optochin),

an antibiotic available in impregnated paper discs. *Streptococcus pneumoniae* ferments inulin.

Treatment. Treatment in most cases is impractical because the condition may be advanced at the time of detection. Oxytetracycline at 0.1 mg/ml in the drinking water for 7 days has controlled an epizootic but not eliminated the carrier state. Such treatment, however, assumes that the animal will drink the liquid and tolerate changes in intestinal flora, and the antibiotic will reach the organisms. For rats 150 units/ g body weight of benzathine penicillin has been recommended (Kohn and Barthold 1984).

Streptococcus pneumoniae organisms are susceptible to the benzathine-based penicillins, ampicillin, bacitracin, chloramphenicol, erythromycin, lincomycin, and methacillin, but most of these antibiotics produce fatal reactions in guinea pigs. Broad-spectrum antibiotics such as chloramphenicol and Tribrissen® (Coopers, Box 167, Kansas City, MO 64141) have less imbalancing effect on intestinal flora than do narrow-spectrum antibiotics, particularly those that suppress gram-positive bacteria.

Prevention. Good husbandry, elimination of carriers, and reduction of environmental stress reduce the possibility and severity of outbreaks.

Public Health Significance. *Streptococcus pneumoniae* can cause respiratory and meningeal disease in man, especially in the elderly and in people lacking spleens, and in some cases the serotypes that affect animals may also affect humans.

REFERENCES

Ford, T.M.: An outbreak of pneumonia in laboratory rats associated with *Diplococcus pneumoniae* type 8. Lab. Anim. Care, *15*:448–451, 1965.

Homburger, F., et al.: An epizootic of pneumococcus type 19 infections in guinea pigs. Science, *102*:449–450, 1945.

Keyhani, M., and Naghshineh, R.: Spontaneous epizootic of pneumococcus infection in guinea pigs. Lab. Anims., *8*:47–49, 1974.

Kohn, D.F., and Barthold, S.W.: Biology and diseases of rats. *In* Laboratory Animal Medicine. Edited by J.G. Fox, B.J. Cohen, and F.N. Loew. Orlando, Academic Press, Inc., 1984.

Mirick, G.S., et al.: An epizootic due to Pneumococcus

type II in laboratory rats. Am. J. Hyg., *52*:48–53, 1950.

Parker, G.A., Russell, R.J., and De Paoli, A.: Extra-pulmonary lesions of *Streptococcus pneumoniae* infection in guinea pigs. Vet. Pathol., *14*:332–337, 1977.

Petrie, G.F.: Pneumococcal disease of the guinea pig. Vet. J., *89*:25–30, 1933.

Saito, M., et al.: An epizootic of pneumococcal infection occurred in inbred guinea pig colonies. Exp. Anim., *32*:29–37, 1983.

Sebesteny, A.: The isolation of carbon dioxide-requiring (carboxyphilic) type 19 pneumococcus (*Streptococcus pneumoniae*) from diseased guinea pigs. Lab. Anims., *12*:181–183, 1978.

Tucek, P.C.: Diplococcal pneumonia in the laboratory rat. Lab. Anim. Digest, *7*:32–35, 1971.

Wagner, J.E., and Owens, D.R.: Type XIX strepto-pneumonia infections in guinea pigs. Abstract #71, Proceedings 21st Annual Meeting, American Association for Laboratory Animal Science, 1970.

Weisbroth, S.H., and Freimer, E.H.: Laboratory rats from commercial breeders as carriers of pathogenic pneumococci. Lab. Anim. Care, *19*:473–478, 1969.

Zydeck, F.A., Bennett, R.R., and Langham, R.F.: Subacute pericarditis in a guinea pig caused by *Diplococcus pneumoniae*. J. Am. Vet. Med. Assoc., *157*:1945–1947, 1970.

STREPTOCOCCUS ZOOEPIDEMICUS INFECTION

Hosts. Streptococcal organisms are found in a wide variety of hosts, but among the small rodents the guinea pig is most often clinically affected.

Etiology. A beta-hemolytic *Streptococcus* (Lancefield's Group C), *S. zooepidemicus,* is commonly involved in streptococcal disease in guinea pigs. The cervical lymphadenitis caused by this agent closely resembles a much less common disease caused by *Streptobacillus moniliformis.*

Transmission. Transmission of *Streptococcus zooepidemicus* can occur through cutaneous wounds, via aerosol into the respiratory tract, or by the conjunctival or genital routes; however, in most cases the agent probably enters the host through small abrasions in the oral cavity. The oral cavity abrasions may be caused by sharp or coarse plant feeds, such as hay and oats.

Predisposing Factors. Poor husbandry and general stress factors predispose to streptococcal infections. Biting facilitates transmission, as does the traumatic effect of overgrown teeth and dietary roughage. Dietary deficiencies, especially vitamin C in guinea pigs, must be resolved.

Clinical Signs. In its chronic or enzootic form, *S. zooepidemicus* causes cervical lymphadenitis, a sporadic, pyogenic infection with abscessation and in many cases external drainage of the cervical lymph nodes. This lymphadenopathy is known as "lumps." Abscesses caused by this agent may occur in almost any organ, but cervical lymph nodes are by far the most common site. In its epizootic form, septicemias and acute pneumonias with high mortality occur.

Torticollis or "wry neck" results from streptococcal infection of the middle ear and extension to the inner ear. Respiratory involvement is manifested clinically by nasal and ocular discharge and signs of acute pneumonia, including dyspnea and cyanosis. Chronic infection can progress to a septicemia if the host is stressed or an abscess ruptures spontaneously or is opened surgically. Abortions, stillbirths, hematuria, and hemoglobinuria have been associated with septicemias. Reproductive patterns may also be disturbed by the presence of numerous, small uterine abscesses seen in some *S. zooepidemicus* infections.

Necropsy Signs. Infection with *S. zooepidemicus* ranges from an acute, fatal septicemia to chronic, suppurative processes in the lymph nodes, especially the cervical nodes, thoracic and abdominal viscera, uterus, and middle and inner ears. Extensive pulmonary consolidation, focal hepatic necrosis, and fibrinous pleuritis are additional signs.

Diagnosis. A presumptive diagnosis can be made when the cervical abscesses are seen. Chains of gram-positive cocci may be seen in direct smears of abscesses. Confirmatory diagnosis depends on *in vitro* culture from abscesses, heart blood, or affected lungs in the case of more acute disease. At 24 hours on blood agar, β-hemolytic colonies of streptococci are mucoid, 2 to 3 mm in diameter, and surrounded by a clear zone of hemolysis.

Treatment.   Concerns in therapy include the number and distribution of abscesses, presence of fistulous tracts, potential for septicemia if the abscess is opened, and the antibiotic sensitivity of the organism. If surgical drainage is indicated, the animal may be anesthetized with ketamine and xylazine and the abscess is opened, drained, and packed or flushed with an antibiotic ointment or antiseptic. With or without surgical treatment, a systemic antibiotic, such as chloramphenicol or Tribrissen® (trimethoprim and sulfadiazine—Coopers, Box 167, Kansas City, MO 64141), is given. Chloramphenicol palmitate, one of the few antibiotics relatively safe to use in guinea pigs, is given at 50 mg/kg orally for 5 to 7 days.

Prevention.   Epizootics with high mortality usually follow introduction of affected stock into a colony or mixing of newly arrived susceptible stock with resistant carrier animals. Good husbandry practices, general preventive measures, and routine palpation for enlarged cervical lymph nodes reduce the possibility of streptococcal infection in guinea pigs. Affected guinea pigs are removed from the colony or treated until the abscesses have drained and healed. In cases of enzootic *S. zooepidemicus* infection, with widespread pneumonia and septicemias, the entire colony should be eliminated.

Killed bacterins have not been effective in preventing this disease, but scratch injections into the oral cavity with American Type Culture Collection strain 12960 of a group C *Streptococcus* has stimulated immunity. The organism is grown for 24 hours in trypticase soy broth and 0.1 ml is placed onto the oral mucosa and scratched in.

Public Health Significance.   Beta-hemolytic streptococci are frequently recovered from human infections, but *S. zooepidemicus* is apparently strictly an animal pathogen.

REFERENCES

Ahern, P.F., Archer, R.K., and Sparrow, S.: An infection in rats caused by β-haemolytic streptococci of group C. Vet. Rec., *104*:507–508, 1979.

Fleming, M.P.: *Streptobacillus moniliformis* isolations from cervical abscesses of guinea-pigs. Vet. Rec., *99*:256, 1976.

Fraunfelter, F.C., et al.: Lancefield type C streptococcal infections in Strain 2 guinea pigs. Lab. Anims., *5*:1–13, 1971.

Henderson, J.D., Jr.: Cervical lymphadenitis in the guinea pig. Vet. Med. Small Anim. Clin., *71*:462–463, 1976.

Kohn, D.F.: Bacterial otitis media in the guinea pig. Lab. Anim. Sci., *24*:823–825, 1974.

Mayora, J., Soave, O., and Doak, R.: Prevention of cervical lymphadenitis in guinea pigs by vaccination. Lab. Anim. Sci., *28*:686–690, 1978.

Olson, L.D., et al.: Experimental induction of cervical lymphadenitis in guinea-pigs with group C streptococci. Lab. Anims., *10*:223–231, 1976.

Rae, M.V.: Epizootic streptococcic myocarditis in guinea pigs. J. Infect. Dis., *59*:236–243, 1937.

Seastone, C.V.: Hemolytic streptococcus lymphadenitis in guinea pigs. J. Exp. Med., *70*:347–359, 1939.

Smith, W.: Cervical abscesses of guinea-pigs. J. Pathol. Bacteriol., *53*:29–37, 1941.

Stewart, D.D., et al.: An epizootic of necrotic dermatitis in laboratory mice caused by Lancefield group C streptococci. Lab. Anim. Sci., *25*:296–302, 1975.

TRANSMISSIBLE MURINE COLONIC HYPERPLASIA

Hosts.   Transmissible colonic hyperplasia caused by *Citrobacter freundii* is a naturally occurring disease of mice (*Mus musculus*). *Citrobacter freundii* causes spontaneous disease at other sites in other hosts (e.g., septicemic cutaneous ulcerative disease—SCUD—in turtles).

Etiology.   *Citrobacter freundii*, a motile, gram-negative, aerobic enterobacterium, is an uncommon inhabitant of soil, water, and the gastrointestinal tracts of several animals, including man. *Citrobacter freundii* variant 4280 has been identified as the specific pathogen in the mouse colon. In Japan a similar disease is produced by *Escherichia coli* 0115. The relationship of the two agents deserves exploration.

Transmission.   Passage of the causative bacterium is by the fecal-oral route. Fomites may also be involved in cage-to-cage spread. Outbreaks in a colony are often prolonged over weeks or months.

Predisposing Factors.   Dietary changes, age, and genetic background of the mouse affect susceptibility to the clinical disease. Suckling mice have a higher mortality and more obvious clinical signs than do older mice. C3H mice have shown more severe clinical signs than C57BL/6 and DBA/2 mice.

Clinical Signs.   With the exception of rectal prolapse, signs of *Citrobacter freundii* 4280 infection in 2- to 4-week-old mice are nonspecific and may include depression, hunched posture, anorexia, retarded growth and dehydration, rough matted hair coat, fecal staining of the perineum, and unformed, soft feces. Mortality is variable and occurs within a week of the onset of clinical signs. Survivors may be runted. Although adult mice are susceptible to infection, they show few or no clinical signs.

Necropsy Signs.   The outstanding sign on necropsy of affected mice is a thickening (to 0.5-cm diameter) and rigidity of the terminal 2 to 3 cm of the descending colon. The thickening is due to mucosal hyperplasia, a reaction in both young and adult mice. The thickening may extend proximally to the cecum and ileum. Inflammation accompanies the hyperplasia in young mice. A sanguineous fluid may be present in the lumen.

The primary histologic sign of *Citrobacter freundii* 4280 colonic infection in mice is segmental hyperplasia of colonic epithelial cells. The cells appear crowded and pseudostratified. In suckling mice mucosal erosion, abscessation, and ulceration become superimposed on the hyperplasia. Colonies of gram-negative coccobacilli are seen in affected tissues.

Diagnosis.   Diagnosis is based on recovery of *Citrobacter freundii* 4280 from affected mice and on the characteristic colonic hyperplasia. The agent is readily cultured *in vitro* and has a characteristic biochemical reaction profile. Most isolates are lactose positive and use citrate as a sole source of carbon. Nonpathogenic *Citrobacter* biotypes are commonly isolated from the intestines and feces of rodents; thus, microbiologists working with rodent samples must be aware of *in vitro* tests necessary to distinguish the pathogenic *Citrobacter freundii* 4280 biotype from other nonpathogenic biotypes. Other enteric diseases of mice that are considered in the differential diagnosis include EDIM, mouse hepatitis, giardiasis, reovirus and rotavirus infection, salmonellosis, spironucleosis, and *Bacillus piliformis* infection.

Treatment.   A solution of sodium sulfamethazine, 0.1% in the drinking water for 60 days, has reduced clinical signs of experimental *Citrobacter* infection in a mouse colony (Ediger, Kovatch and Rabstein 1974). Tetracycline at 1000 mg/L or neomycin sulfate at 2000 mg/L will suppress the disease and eliminate the bacteria (Barthold 1980).

Prevention.   Proper husbandry and sanitation will reduce opportunities for entry of the bacterium into a colony. *Citrobacter* can occasionally be isolated from the intestinal tracts of clinically normal vertebrates, a finding which poses the problem of continuous sources of contamination.

Public Health Significance.   *Citrobacter freundii* variants have been found in human feces, but the potential pathogenicity for man of these and other *Citrobacter* is unknown.

REFERENCES

Barthold, S.W.: The microbiology of transmissible murine colonic hyperplasia. Lab. Anim. Sci., *30*:167–173, 1980.

Barthold, S.W., et al.: The etiology of transmissible murine colonic hyperplasia. Lab. Anim. Sci., *26*:889–894, 1976.

Barthold, S.W., et al.: Transmissible murine colonic hyperplasia. Vet. Pathol., *15*:223–236, 1978.

Barthold, S.W., Osbaldiston, G.W., and Jonas, A.M.: Dietary, bacterial, and host genetic interactions in the pathogenesis of transmissible murine colonic hyperplasia. Lab. Anim. Sci., *27*:938–945, 1977.

Brennan, P.C., et al.: *Citrobacter freundii* associated with diarrhea in laboratory mice. Lab. Anim. Care, *15*:266–275, 1975.

Ediger, R.D., Kovatch, R.M., and Rabstein, M.M.: Colitis in mice with a high incidence of rectal prolapse. Lab. Anim. Sci., *24*:488–494, 1974.

Muto, T., et al.: Infectious megaenteron of mice. I. Manifestation and pathological observation. Jpn. J. Med. Sci. Biol., *22*:363–374, 1969.

Nakagawa, M., et al.: Infectious megaenteron of mice. II. Detection of coliform organisms of an unusual biotype as the primary cause. Jpn. J. Med. Sci. Biol., *22*:375–382, 1969.

Silverman, J., et al.: A natural outbreak of transmissible murine colonic hyperplasia in A/J. mice. Lab. Anim. Sci., *29*:209–213, 1979.

TRICHOBEZOARS

Hosts.   Masses of hair of variable size and firmness occur commonly in mature rabbits and rarely in pet or laboratory rodents.

Etiology and Predisposing Factors.   Although hair accumulation in the stomach may sometimes be more a consequence of anorexia than a cause, reasons postulated for the excessive consumption of hair include hair loss, itching skin, boredom, lack of dietary fiber, protein, copper, or magnesium deficiency, abnormal grooming behavior, and molting or a naturally long hair coat (angora rabbits). The rabbit's inability to vomit and the small pyloric lumen predispose to hair accumulation in the stomach.

Clinical Signs.   Clinical signs include anorexia, oligodipsia, weight loss, agalactia, depression, reduction in amount of feces, and eventually death in 3 to 4 weeks from starvation and metabolic abnormalities.

Diagnosis.   Diagnosis is difficult and is based on clinical signs. Carefully performed palpation and contrast radiography may be used in diagnosis, but the results are usually inconclusive.

Treatment.   Several treatment regimens intended to facilitate breakdown and passage of the hairball have been recommended, including the use of mineral oil (20 ml), wetting agents, laxative preparations, alfalfa hay, and fresh pineapple juice (5 ml for 5 days). None of these treatments have been known to be efficacious in a significant number of cases. Gastrotomy and removal of the hairball may be attempted. Postoperative care should include fluids, antibiotics, and a preferred soft food. A mixture of high energy food supplement, ground rabbit food, 50% glucose, and yogurt given intragastrically may assist debilitated and severely anorectic patients. Anorexia following surgery may persist for 1 to 2 weeks and the rabbit may require prolonged supportive treatment, including feeding by gavage, for successful recovery. Recurrences of hairballs are common.

REFERENCES

Gillett, N.A., Brooks, D.L., and Tillman, P.L.: Medical and surgical management of gastric obstruction from a hairball in the rabbit. J. Am. Vet. Med. Assoc., *183*:1176–1178, 1983.

Lee, K.J., Johnson, W.D., and Lang, C.M.: Acute peritonitis in the rabbit (*Oryctolagus cuniculus*) resulting from a gastric trichobezoar. Lab. Anim. Sci., *28*:202–204, 1978.

Mitchell, H.S., Bradshaw, P.J., and Carlson, E.R.: Hairball formation in rats in relation to food consistency. Am. J. Physiol., *68*:203–206, 1924.

Nelson, W.B.: Fatal hairball in a long-haired hamster. Vet. Med. Small Anim. Clin., *70*:1193, 1975.

Sebesteny, A.: Acute obstruction of the duodenum of a rabbit following the apparently successful treatment of a hairball. Lab. Anims., *11*:135, 1977.

Wagner, J.L., Hackel, D.B., and Samsell, A.G.: Spontaneous deaths in rabbits resulting from gastric trichobezoars. Lab. Anim. Sci., *24*:826–830, 1974.

TULAREMIA

Hosts.   Tularemia is a relatively common, often fatal septicemic disease of wild animals, especially rodents, lagomorphs, and certain gallinaceous birds (pheasants and quail). Tularemia has a wide host spectrum, and rodents and lagomorphs are highly susceptible and have been involved in epizootics. Tularemia is rare in domestic rabbits and rodents, although animals raised outdoors and exposed to ticks are much more likely to contract infections than are similar animals kept indoors in a home or laboratory.

Etiology.   *Francisella tularensis* is a nonmotile, gram-negative, pleomorphic, bipolar coccobacillus. The agent is particularly prevalent in the South Central United States. In the past the agent has been classified in the genera *Brucella* and *Pasteurella*.

Transmission.   *Francisella tularensis* is spread by direct contact through skin and via aerosol, ingestion, or biting arthropods, especially ticks, biting flies, and mosquitoes. The tissues, blood, and feces of infected animals harbor the organism.

Predisposing Factors.   The tularemia organism is highly infectious and affects otherwise healthy animals.

Clinical Signs.   Tularemia is an acute, fatal septicemia. Clinical signs include roughened hair coat, depression, tendency to huddle, anorexia, ataxia, and death. The course in cottontails lasts approximately 1 week.

Necropsy Signs.   Signs of a septicemia include pulmonary congestion and consolidation, subpleural petechiae hemorrhage, numerous pinpoint, small bright white hepatic foci, and congestion and enlargement of the liver and spleen. The white spots on the dark background of the congested liver and spleen are said to resemble the Milky Way. Lymph nodes are enlarged, and the bacteria are widely distributed in blood vessels.

Diagnosis.   Diagnosis is based on necropsy findings of a septicemic bacterial disease. Stained impression smears of affected tissues or fluids may contain small gram-negative coccobacilli among debris and within cells and can be an aid in making a tentative diagnosis at necropsy. *In vitro* recovery of *F. tularensis* requires special attention. Special enriched media such as blood-glucose-cystine agar grow colonies that are minute and translucent. The agent is extremely hazardous to man and must be handled with extreme caution in the laboratory. Intraperitoneal injection of the suspect's blood into a guinea pig produces lymphoid necrosis, serofibrinous peritonitis, septicemia, and death in 8 to 14 days.

Treatment.   Treatment is not indicated for animals. Tetracyclines and streptomycin are the antibiotics used to treat human infections.

Prevention.   Avoid buying animals that may have been exposed to tularemia. Exclusion of wild mammals and insect vectors from the colony are preventive measures. Bacterins have been developed, but hunters should avoid "lazy" cottontails and wear gloves when skinning game.

Public Health Significance.   Humans are very susceptible and tularemia is a reportable disease. Because contamination of inapparent skin lesions results in human infections, persons handling suspect tissues should always wear gloves. Laboratory culture should be attempted only where adequate biohazard culture facilities are available. Cutaneous lesions, septicemias, and meningitis occur in cases of human tularemia. Bites of 2 and 3 host ticks are particularly dangerous, as ticks transmit the bacterium.

REFERENCES

Belding, D.L., and Merrill, B.: Tularemia in imported rabbits in Massachusetts. N. Engl. J. Med., *224*:1085–1087, 1941.

Bell, J.F., Owens, C.R., and Larson, C.L.: The virulence of *Bacterium tularense*. I. A study of the virulence of *Bacterium tularense* in mice, guinea pigs, and rabbits. J. Infect. Dis., *97*:162–166, 1955.

Eigelsbach, H.T., and Downs, C.M.: Prophylactic effectiveness of live and killed tularemia vaccines. I. Production of vaccine and evaluation in the white mouse and guinea pig. J. Immunol., *87*:415–425, 1961.

Hoff, G.L., et al.: Tularemia in Florida: *Sylvilagus palustris* as a source of human infection. J. Wildl. Dis., *11*:560–561, 1975.

Lillie, R.D., and Francis, E.: The pathology of tularemia in the guinea pig (*Cavia cobaya*). IX. Natl. Inst. Health Bull., *167*:155–176, 1936.

Moe, J.B., et al.: Pathogenesis of tularemia in immune and nonimmune rats. Am. J. Vet. Res., *36*:1505–1510, 1975.

Perman, V., and Bergeland, M.E.: A tularemia enzootic in a closed hamster breeding colony. Lab. Anim. Care, *17*:563–568, 1967.

TYZZER'S DISEASE (*BACILLUS PILIFORMIS* INFECTION)

Hosts.   The number of species susceptible to infection by *Bacillus piliformis* (Tyzzer's disease) continues to increase and presently includes mice, rats, hamsters, gerbils, rabbits, guinea pigs, horses, cattle, rhesus monkeys, marmosets, cats, dogs, several wildlife species, and others. Tyzzer's disease is probably the most common disease of gerbils. It is also common in rabbits, foals, mice, and hamsters. Many rat colonies have a high incidence of seropositive animals without evidence of clinical disease. Likewise, hamster and mouse colonies (barrier and nonbarrier sustained) may have clinically inapparent infections.

Etiology.   *Bacillus piliformis* is a gram-negative, pleomorphic, rod-shaped, PAS-positive, nonacid-fast, obligate intracellular organism of uncertain classification. It is widely distributed geographically, probably as a common and benign intestinal inhabitant, and persists in spore form for years outside a host. Recently *Bacillus piliformis* has been grown in tissue culture and it can be propagated in embryonating hens' eggs.

Transmission.   Transmission is thought to be by the fecal-oral route. Infectious spores may survive a year or more in bedding, soil, or contaminated feed.

Predisposing Factors.   Poor environmental sanitation, stress of shipping, and immunosuppressors such as radiation, corticosteroids, concurrent disease, thymectomy, and crowding contribute to the development of clinical disease. The acute, highly fatal disease is most often seen in weanling animals, but adults may also be affected. Use of sulfaquinoxaline to prevent coccidiosis and pneumonia in rabbits has been associated with an epidemic of Tyzzer's disease.

Clinical Signs.   The most common form of infection is probably subclinical with persistence of carriers and sporadic outbreaks of clinical disease. Tyzzer's disease in weanling or stressed animals is an acute, enzootic disease causing rough hair coat, lethargy, and death within 48 to 72 hours. Watery diarrhea and perineal staining may or may not accompany the disease. Chronically infected animals, in which hepatic lesions are more pronounced, exhibit weight loss, rough hair coat, and eventually death.

Necropsy Signs.   The most consistent lesion of *B. piliformis* infection is an enlarged liver with few to numerous gray, white, or yellow foci, 1 to 2 mm in diameter. The liver is not necessarily involved. The intestine or heart may be the only organ affected. Many people mistakenly believe that, when the liver is not affected, Tyzzer's disease can be ruled out. In more acute cases there may be edema, congestion, hemorrhage, and fo-

cal ulceration of the intestine, particularly around the ileocecal-colonic junction. The gut is often atonic and filled with a yellowish fluid. The small intestine of affected rats may be greatly distended. A proteinaceous exudate may be found in the abdominal cavity of guinea pigs and gerbils. Pale myocardial foci have been noted in rabbits, rats, mice, hamsters, and foals with Tyzzer's disease. There are no reports of splenic lesions.

The hepatic, intestinal, and myocardial foci which variably characterize this disease are areas of necrosis initially surrounded by a scant, mixed inflammatory cell population. These foci probably arise via an embolic shower of organisms from a primary infection in the intestine. The filamentous organisms may be seen within the cytoplasm of cells adjacent to the necrotic area, often in a "pile of sticks" arrangement.

Diagnosis.   Necropsy signs and silver, Giemsa, or PAS staining of the intracellular, filamentous *B. piliformis* organism in hepatocytes, enterocytes, or other tissues provide a definitive diagnosis of Tyzzer's disease. Indirect fluorescent antibody tests are also a diagnostic aid.

The diagnosis of Tyzzer's disease may be difficult to establish. Even with special stains, organisms in tissues may be difficult to find because of their paucity. Feces from suspect animals inoculated *per os* into known *B. piliformis*-free weanling gerbils (which are highly susceptible) is a valuable aid for detecting the presence of the Tyzzer's disease agent in a suspect colony. Little is known about the incidence of Tyzzer's disease among animals in commercial and private stock colonies because colonies are not regularly screened for Tyzzer's disease, in part because of the absence of a good screening test. A diagnostic test is badly needed for screening animals for the Tyzzer's agent before beginning research with suspect animals. A CF test is widely used for screening in Japan. The IFA test is widely used in Europe. An ELISA has been developed for the detection of anti-*Bacillus piliformis* serum antibodies in rabbits and could be applied in detection of serum antibodies in rodent species.

Treatment. The acute (1- to 4-day) course of the disease and the intracellular location of the organism reduce the effectiveness of treatment. Oxytetracycline in the drinking water at 0.1 g/L for 30 days has reportedly suppressed an outbreak. Tetracycline at 10 mg/kg body weight for 5 days "on-off-on" or at 400 mg/L for 10 days has also been used (Hunter 1971). *Bacillus piliformis* is also sensitive to penicillin, cephaloridine, chloramphenicol, and erythromycin. Sulfonamides, streptomycin, and kanamycin apparently do not affect the agent (Yokoiyama and Fujiwara 1971).

Prevention. Purchase of stock from a reputable vendor, preferably one who uses cesarean derivations and barrier rearing techniques, and good husbandry practices are the best preventive measures. Antibiotics may suppress infections, but carriers may develop. The production of resistant species of *B. piliformis* makes prevention and control of the disease more difficult. Spores survive freeze-thaw but are killed if heated at 56° C for 1 hour or 80° C for 30 minutes. Spores are resistant to ethanol and quaternary ammonia compounds. A 0.5% sodium hypochlorite (Clorox) solution and peracetic acid are effective in spore inactivation. Filter cage covers aid in reducing transmission. A formalin-killed bacterin, which is effective in preventing clinical signs in rabbits and mice, has been prepared experimentally but is not widely used. Cesarean derivation as a means of eliminating carriers in the breeding colony may be compromised if, in fact, *in utero* transmission occurs.

Public Health Significance. No public health significance is known, but the report of *B. piliformis* infection in a rhesus monkey should be noted. Antibodies to *B. piliformis* have been found in pregnant women. Because *B. piliformis* affects such a wide spectrum of animal species, man may be susceptible to clinical disease under certain circumstances.

REFERENCES

Allen, A.M., et al.: Tyzzer's disease syndrome in laboratory rabbits. Am. J. Pathol., *46*:859–882, 1965.

Boot, R., and Walvoort, H.C.: Vertical transmission of *Bacillus piliformis* infection (Tyzzer's disease) in a guinea pig: Case report. Lab. Anims., *18*:195–199, 1984.

Carter, G.R., Whitenack, D.L., and Julius, L.A.: Natural Tyzzer's disease in Mongolian gerbils (*Meriones unguiculatus*). Lab. Anim. Care, *19*:648–651, 1969.

Craigie, J.: '*Bacillus piliformis*' (Tyzzer) and Tyzzer's disease of the laboratory mouse. I. Propagation of the organism in embryonated eggs. Proc. R. Soc. (Lond.), *165*:35–60, 1966.

Cutlip, R.C., et al.: An epizootic of Tyzzer's disease in rabbits. Lab. Anim. Sci., *21*:356–361, 1971.

Fries, A.S.: Demonstration of antibodies to *Bacillus piliformis* in SPF colonies and experimental transplacental infection by *Bacillus piliformis* in mice. Lab. Anims., *12*:23–25, 1978.

Fries, A.S.: Studies on Tyzzer's disease: a long-term study of the humoral antibody response in mice, rats and rabbits. Lab. Anims., *13*:37–41, 1979.

Fries, A.S.: Studies on Tyzzer's disease: application of immunofluorescence for detection of *Bacillus piliformis* and for demonstration and determination of antibodies to it in sera from mice and rabbits. Lab. Anims., *11*:69–73, 1977.

Fries, A.S.: Studies on Tyzzer's disease: transplacental transmission of *Bacillus piliformis* in rats. Lab. Anims., *13*:43–46, 1979.

Fries, A.S., and Svendsen, O.: Studies on Tyzzer's disease in rats. Lab. Anims., *12*:1–4, 1978.

Fujiwara, K.: Tyzzer's disease. Jpn. J. Exp. Med., *48*:467–480, 1978.

Fujiwara, K., Nakayama, M., and Takahashi, K.: Serologic detection of inapparent Tyzzer's disease in rats. Jpn. J. Exp. Med., *51*:197–200, 1981.

Ganaway, J.R., Allen, A.M., and Moore, T.D.: Tyzzer's disease. Am. J. Pathol., *64*:717–732, 1971.

Ganaway, J.R., Allen, A.M., and Moore, T.D.: Tyzzer's disease of rabbits: isolation and propagation of *Bacillus piliformis* (Tyzzer) in embryonated eggs. Infect. Immun., *3*:429–437, 1971.

Ganaway, J.R., McReynolds, R.S., and Allen, A.M.: Tyzzer's disease in free-living cottontail rabbits (*Sylvilagus floridanus*) in Maryland. J. Wildl. Dis., *12*:545–549, 1976.

Ganaway, J.R.: Effect of heat and selected chemical disinfectants upon infectivity of spores of *Bacillus piliformis* (Tyzzer's disease). Lab. Anim. Sci., *30*:192–196, 1980.

Ganaway, J.R., Spencer, T.H., and Waggie, K.S.: Propagation of the etiologic agent of Tyzzer's disease (*Bacillus piliformis*) in cell culture. *In* The Contribution of Laboratory Animal Science to the Welfare of Man and Animals. Edited by J. Archibald, J. Ditchfield, and H.C. Rowsell. New York, Gustav Fischer Verlag, 1985, pp. 59–70.

Hunter, B.: Eradication of Tyzzer's disease in a colony of barrier-maintained mice. Lab. Anim., *5*:271–276, 1971.

Kawamura, S., et al.: Growth of Tyzzer's organism in primary monolayer cultures of adult mouse hepatocytes. J. Gen. Microbiol., *129*:277–283, 1983.

McLeod, C.C., et al.: Intestinal Tyzzer's disease and spirochetosis in a guinea pig. Vet. Pathol., *14*:229–235, 1977.

Nakayama, M., et al.: Transmissible enterocolitis in hamsters caused by Tyzzer's organism. Jpn. J. Exp. Med., *45*:33–41, 1975.

Ononiwu, J.C., and Julian, R.J.: An outbreak of Tyzzer's disease in an Ontario rabbitry. Can. Vet. J., *19*:107–109, 1978.

Orcutt, R.P., Pucak, G.J., and Foster, H.L.: A vaccine for Tyzzer's disease in rabbits. Abstract #5, Proceedings 31st Annual Session, American Association for Laboratory Animal Science, Indianapolis, 1980.

Peeters, J.E., et al.: Naturally occurring Tyzzer's disease (*Bacillus piliformis* infection) in commercial rabbits: A clinical and pathological study. Ann. Rech. Vet., *16*:69–79, 1985.

Port, C.D., Richter, W.R., and Morse, S.M.: Tyzzer's disease in the gerbil (*Meriones unguiculatus*). Lab. Anim. Care, *20*:109–111, 1970.

Sparrow, S., and Naylor, P.: Naturally occurring Tyzzer's disease in guinea pigs. Vet. Rec., *102*:288, 1978.

Stedham, M.A., and Bucci, T.J.: Spontaneous Tyzzer's disease in a rat. Lab. Anim. Care, *20*:743–746, 1970.

Strittmatter, J.: Elimination of Tyzzer's disease in the Mongolian gerbil (*Meriones unguiculatus*) by fostering to mice. Z. Versuchstierkd., *14*:209–214, 1972.

Takasaki, Y., et al.: Tyzzer's disease in hamsters. Jpn. J. Exp. Med., *44*:267–270, 1974.

Tsuchitani, M., et al.: Naturally occurring Tyzzer's disease in a clean mouse colony: High mortality with coincidental cardiac lesions. J. Comp. Pathol., *93*:499–507, 1983.

Tyzzer, E.E.: A fatal disease of the Japanese waltzing mouse caused by a spore-bearing bacillus (*Bacillus piliformis,* N. SP.). J. Med. Res., *37*:307–338, 1917.

Van Kruiningen, H.J., and Blodgett, S.B.: Tyzzer's disease in a Connecticut rabbitry. J. Am. Vet. Med. Assoc., *158*:1205–1212, 1971.

Waggie, K.S., et al.: A study of mouse strain susceptibility to *Bacillus piliformis* (Tyzzer's disease): the association of B-cell function and resistance. Lab. Anim. Sci., *31*:139–142, 1981.

Waggie, K.S., et al.: Experimentally induced Tyzzer's disease in Mongolian gerbils (*Meriones unguiculatus*). Lab. Anim. Sci., *34*:53–57, 1984.

Waggie, K.S., Wagner, J.E., and Kelly, S.T.: Naturally occurring *Bacillus piliformis* infection (Tyzzer's disease) in guinea pigs. Lab. Anim. Sci., *36*:504–506, 1986.

Waggie, K.S., Spencer, T.H., and Ganaway, J.R.: An enzyme-linked immunosorbent assay for detection of anti-*Bacillus piliformis* serum antibody in rabbits. Lab. Anim. Sci., *37*:176–179, 1987.

Waggie, K.S., et al.: Lesions of experimentally induced Tyzzer's disease in Syrian hamsters, guinea pigs, mice and rats. Lab. Anims., *21*:155–160, 1987.

Yokoiyama, S., and Fujiwara, K.: Effect of antibiotics on Tyzzer's disease. Jpn. J. Exp. Med., *41*:49–57, 1971.

Zook, B.C., Albert, E.N., and Rhorer, R.G.: Tyzzer's disease in the Chinese hamster (*Cricetulus griseus*). Lab. Anim. Sci., *27*:1033–1035, 1977.

Zook, B.C., Huang, K., and Rhorer, R.G.: Tyzzer's disease in Syrian hamsters. J. Am. Vet. Med. Assoc., *171*:833–836, 1977.

Zwicker, G.M., Dagle, G.E., and Adee, R.R.: Naturally occurring Tyzzer's disease and intestinal spirochetosis in guinea pigs. Lab. Anim. Sci., *28*:193–198, 1978.

UROLITHIASIS

Hosts.   Concretions of ammonium magnesium phosphate and calcium carbonate are seen sporadically in the urinary tracts of both rabbits and rodents, especially rats. Stones are more common in males because their urethra is longer and less easily distended than that of females.

Etiology and Predisposing Factors.   Although the cause of stone formation in the urinary tract in rabbits and rodents is obscure, several predisposing factors, including genetic predispostion, metabolic disorders, nutritional imbalances, and nematode and bacterial infections, have been suggested.

Clinical Signs.   Clinical signs of urolithiasis in rodents and rabbits may be absent or may include anorexia, weight loss, and listlessness. A hunched posture, anuria or hematuria, and dense abdominal masses evident on palpation or radiographs are other reported signs. Urine from affected animals is often difficult to express.

Necropsy Signs.   On necropsy, fine sand or irregularly shaped stones are found in the renal pelvis, ureters, bladder, or urethra. Structures between the nephron and concretion may be dilated and filled with fluid. Cystitis is a common sequela to urinary stasis.

Urethral proteinaceous obstructions occur in aged male guinea pigs and aged mice. These are composed primarily of congealed ejaculum that readily becomes infected; a suppurative infection ensues. A variety of bacteria may be isolated from these infections, especially *Pseudomonas aeruginosa, Escherichia coli,* and *Proteus* spp.

Diagnosis.   Diagnosis is based on clinical signs, palpation, radiography, and a chronic hematuria refractory to antibiotic therapy.

Treatment.   Treatment has included the surgical removal of stones from the bladder and the administration of chloramphenicol, tetracyclines, Tribrissen® (10 mg/kg—Coopers, Box 167, Kansas City, MO 64141), or gentamicin for the secondary cystitis. Acidification of the urine may dissolve small stones, but guinea pigs have difficulty removing an acid load.

Prevention. Because of the uncommon occurrence and uncertain etiology of uroliths, specific preventive measures other than good husbandry practices are not employed.

REFERENCES

Bauck, L.A.B., and Hagan, R.I.: Cystotomy for treatment of urolithiasis in a hamster. J. Am. Vet. Med. Assoc., *184*:99–100, 1984.

Chapman, W.H.: The incidence of a nematode *Trichosomoides crassicauda* in the bladder of laboratory rats. Treatment with nitrofurantoin and preliminary report of their influence of urinary calculi and experimental bladder tumors. Invest. Urol., *2*:52–57, 1964.

Gilmore, M.M.: Urolithiasis in a gerbil. Vet. Rec., *103*:102, 1978.

Gustafsson, B.E., and Norman, A.: Urinary calculi in germfree rats. J. Exp. Med., *116*:273–283, 1962.

Lee, K.J., et al.: Hydronephrosis caused by urinary lithiasis in a New Zealand White rabbit (*Oryctolagus cuniculus*). Vet. Pathol., *15*:676–678, 1978.

Magnusson, G., and Ramsay, C-H.: Urolithiasis in the rat. Lab. Anims., *5*:153–162, 1971.

Paterson, M.: Urolithiasis in the Sprague-Dawley rat. Lab. Anims., *13*:17–20, 1979.

Sen, A.K.: A case of calculus in the rabbit. Vet. Rec., *98*:205–206, 1976.

Spink, R.R.: Urolithiasis in a guinea pig (*Cavia porcellanus*). Vet. Med. Small Anim. Clin., *73*:501–502, 1978.

Stuppy, D.E., Douglass, P.R., and Douglass, P.J.: Urolithiasis and cystotomy in a guinea pig (*Cavia porcellanus*). Vet. Med. Small Anim. Clin., *74*:565–567, 1979.

Ungar, H., and Ungar, P.: Further studies on the pathogenesis of urate calculi in the urinary tract of white rats. Am. J. Pathol., *28*:291–301, 1952.

Wagner, J.E.: Miscellaneous disease conditions of guinea pigs. *In* The Biology of the Guinea Pig. Edited by J.E. Wagner and P.J. Manning. New York, Academic Press, 1976, pp. 228–234.

VENEREAL SPIROCHETOSIS

Hosts. Rabbits are susceptible to *Treponema cuniculi* infections. The disease is also known as treponematosis, rabbit syphilis, and vent disease.

Etiology. *Treponema cuniculi* is a fine, spiral-shaped bacterium between 10 and 30 μm in length. Antibodies to *T. cuniculi* cross react with *T. pallidum,* the agent of human syphilis.

Transmission. The spirochete is transmitted by direct contact, especially during mating. Exchange of bucks among breeders promotes dissemination of the organism. Infection of pups may occur at birth or during the nursing period.

Predisposing Factors. Rabbits in cold environments are predisposed to the clinical disease. Susceptibility and lesions vary among rabbit strains.

Clinical Signs. Serologically positive rabbits are often encountered, but the clinical, epizootic disease is uncommon. While most attention has been paid to the disease in males, because they have the more severe clinical signs, females also contract the infection. Females may develop metritis and retained placentas. Abortions, possibly unnoticed, may occur at 12 to 22 days of gestation. The herd may have a history of low conception rates and a high incidence of nest box fatalities in young under 9 days of age. Clinical signs, when they appear, begin with vesicular swelling and reddening followed by dry scaliness of the prepuce and vulva. This early stage is followed by the development of macules, papules, erosions, ulcers, and crusts on the external genitalia, perineal areas, nose, eyelids, and lips. Affected rabbits remain alert, and the condition regresses after several weeks. If the prepuce is severely affected, apparent and transient infertility may result. Healed lesions are indicated by preputial and scrotal scars.

Necropsy Signs. Many times there are no gross lesions. The perineal and genital skin and the mucous membranes of the external genitalia are the tissues most often affected. Adjacent lymph nodes may be swollen. Histologically, the thickened epidermis may be hyperkeratotic, and the dermis infiltrated by monomorphonuclear inflammatory cells, primarily plasma cells.

Diagnosis. The gross lesions of veneral

spirochetosis in rabbits resemble those caused by trauma, dermatophytes, or ectoparasites. The use of the cardiolipin antigen Wassermann-type test or the rapid plasma region (RPR) card test will provide evidence of a *T. cuniculi* infection. Absence of clinical disease in the absence of concurrent treatment with antibiotics and two negative RPR tests 30 days apart are suggested as evidence of noninfection.

A fluorescent antibody test has also been used in diagnosis. The spirochetes may be demonstrated in tissue fluids expressed from active lesions in the epidermis, dermis, uterus, or lymph nodes using darkfield microscopy of wet preparations from these tissues or by staining sections of fixed tissue with silver salts.

Treatment.   Penicillin provides a cure Penicillin is given intramuscularly at 40,000 units/kg per day for 3 to 5 days (Cunliffe-Beamer and Fox 1981). Alternately a long-acting penicillin (benzathine penicillin G) can be given intramuscularly at 84,000 units/kg 3 times at weekly intervals (DiGiacomo et al. 1984).

Prevention.   Routine screening with a serologic card test provides an indication of exposure. Periodic examination of breeding does and bucks for cutaneous lesions followed by treatment or culling will eliminate clinical carriers. Infected animals should not be bred. It may be possible to eliminate the infection from small herds by treating all animals simultaneously. In young animals or fryers bred for meat production, there may be restrictions on or specific withdrawal times from the use of certain antimicrobial agents. These must be followed. Maintaining a closed breeding herd prevents introduction of venereal spirochetosis. If new animals must be introduced, they should be clinically and serologically free of disease. It may be advisable to quarantine and treat all new arrivals intended for breeding.

Public Health Significance.   Although rhesus monkeys and a chimpanzee have been experimentally infected, man is not believed to be susceptible to *T. cuniculi* infection.

REFERENCES

Chapman, M.P.: The use of penicillin in the treatment of spirochetosis (vent disease) of domestic rabbits. North Am. Vet., *28*:740–742, 1947.

Clark, J.W., Jr.: Serological tests for syphilis in healthy rabbits. Br. J. Vener. Dis., *46*:191–197, 1970.

Cunliffe-Beamer, T.L., and Fox, R.R.: Venereal spirochetosis of rabbits: description and diagnosis. Lab. Anim. Sci., *31*:366–371, 1981.

Cunliffe-Beamer, T.L., and Fox, R.R.: Venereal spirochetosis of rabbits: epizootiology. Lab. Anim. Sci., *31*:372–378, 1981.

Cunliffe-Beamer, T.L., and Fox, R.R.: Venereal spirochetosis in rabbits: eradication. Lab. Anim. Sci., *31*:379–381, 1981.

DiGiacomo, D., et al.: Clinical course and treatment of venereal spirochetosis in New Zealand White rabbits. Br. J. Vener. Dis., *60*:214–218, 1984.

DiGiacomo, R.F., et al.: *Treponema paraluis-cuniculi* infection in a commercial rabbitry: Epidemiology and serodiagnosis. Lab. Anim. Sci., *33*:562–566, 1983.

Fried, S.M., and Orlov, S.S.: Spontaneous spirochetosis and experimental syphilis in rabbits. Arch. Dermatol. Syphilol., *25*:893–905, 1932.

Greeves, S.: Susceptibility of rabbits venereally infected with *Treponema paraluis-cuniculi* to superinfections with *Treponema pallidum*. Br. J. Vener. Dis., *56*:387–389, 1980.

Small, J.D., and Newman, B.: Venereal spirochetosis of rabbits (rabbit syphilis) due to *Treponema cuniculi:* a clinical, serological, and histopathological study. Lab. Anim. Sci., *22*:77–89, 1972.

Smith, J.L., and Pesetsky, B.R.: The current status of *Treponema cuniculi:* review of the literature. Br. J. Vener. Dis., *43*:117–127, 1967.

Warthin, A.S., Buffington, E., and Wanstrom, R.C.: A study of rabbit spirochetosis. J. Infect. Dis., *32*:315–332, 1923.

Chapter 6

Case Reports

These 53 case reports were taken, with minor modifications, from the case report files of the Research Animal Diagnostic Investigative Laboratory, University of Missouri-Columbia, and from Laboratory Animal Resources, The Pennsylvania State University. These cases reveal the complexity of disease in rabbits and rodents and demonstrate how field cases may differ considerably from textbook descriptions. Short answers are supplied at the end of the section.

RABBITS

Case 1

A colony of dwarf rabbits had sporadic deaths among the young. The closed colony was housed indoors in wire cages equipped with crocks for feed and water. Affected rabbits were first evident at 3 to 4 weeks of age, when runting, weakness, and ataxia were evident in never more than 1 member of approximately 40% of the litters. The runts, weighing one-half normal sibling weights, inevitably died at 10 to 12 weeks. Necropsy examination revealed no gross lesions or congenital abnormalities.

a. Which specific conditions in rabbits could produce ataxia?
b. Which common infectious disease of rabbits might produce abnormal neurologic signs?
c. Which tissues would be examined for histologic signs of this disease? Which stains would be used?
d. Normal-sized siblings had the same histologic lesions. Why was there no gradation of weights within the 10-week-old young?

Case 2

A drug-testing protocol required restraining rabbits for 6 hours daily in a plastic restraining box. Throughout the period the rabbits' feet were wetted with urine. Fol-

lowing 1 week of such confinement, 3 of 5 rabbits developed cutaneous ulcers on the plantar surfaces of the hocks.

a. What recommendations would you make to prevent the occurrence of this problem?
b. How would you treat affected animals?
c. Why might dampness, especially dampness caused by urine, predispose to sore hocks?

Case 3

A pet rabbit was brought to a veterinary clinic for examination. The rabbit had a firm but distended abdomen and cachexia. There were no abnormal intestinal sounds or diarrhea.

a. What diagnostic procedures would contribute to a diagnosis?
b. Eventual necropsy examination of the rabbit revealed an enlarged liver containing numerous yellow spots. How could you differentiate hepatic coccidiosis from cysticercosis (*Taenia pisiformis*) and tularemia?
c. What is the causative organism of hepatic coccidiosis?

Case 4

A first-litter doe had a long difficult delivery. Shortly after the kindling process she cannibalized several kits.

a. What causes cannibalism of newborn rabbits?
b. How can such behavior be prevented?

Case 5

A rabbit producer had two continuing problems in her rabbitry. The more serious problem was a fatal diarrhea in nursing does. The rabbits were housed in suspended wire cages, and crocks and bottles with sipper tubes were used for feeding and watering. The diarrhea was watery and green-brown. The second problem was a summer infertility.

a. What diagnostic procedures would you utilize to determine the cause of the enteropathies in the does?
b. What recommendations would you make to prevent the recurrence of the disease?
c. What is the possible cause of the herd infertility?

Case 6

A young lady, a good client with a kennel of show dogs, consults you regarding her four French Lop rabbits, which recently developed a nasal discharge, crusts in the ear canals, and ataxia.

a. Which pathogenic organism would you expect to isolate from the nasal passages? Is there a noninfectious cause of sneezing and nasal discharge in rabbits?
b. Suggest a treatment for rhinitis in rabbits.
c. What agent would you probably find in the ear canal? Why are the rabbits ataxic?

Case 7

An obese, 3-year-old, pregnant (27 days) doe that had died suddenly was submitted for necropsy examination. The rabbit had been fed table scraps and milk. Gross lesions included multifocal, small (1- to 2-mm) pits on the renal surface, 10 dead fetuses, and a light tan liver.

a. What agent is the probable cause of the renal pitting? How is this agent transmitted?
b. What is the probable cause of death? What husbandry steps might prevent or reduce the incidence of this disease in a rabbitry?
c. How long is the rabbit's gestation period?

Case 8

An owner of a herd of 60 does reported the deaths of 17 does within a 12-hour period. Affected adult and weanling animals developed dull, watery eyes, weakness, and incoordination, but the prominent sign was a fulminating, smelly diarrhea. Body temperatures were between 40° and 41° C in terminal cases.

a. Which conditions are probable differential diagnoses in this case?
b. What would you do to prevent the further devastation of this valuable breeding herd?

Case 9

A 10-month-old breeding doe in good flesh abruptly stopped eating the pelleted diet and drinking water. Despite the administration of antibiotics and multivitamins, and the provision of hay, fresh fruit, and vegetables, the rabbit remained anorectic. After 12 days no feces appeared in the pan, and 14 days later the doe died. On necropsy examination, the liver was pale, the gallbladder was distended, and the stomach and intestines were virtually empty. The cecum contained some inspissated material.

a. What causes rabbits to stop eating?
b. What additional methods might be used to coax the rabbit to resume eating?
c. Why might feces continue to appear in the pan for a week after the rabbit stops eating?

Case 10

A school class maintained a pet rabbit in a wood and wire cage in the school yard. The rabbit had an enclosed box and remained outside even during the cold winter. Although the rabbit appeared healthy during the coldest months, the urine became a bright red-orange, prompting the teacher to seek an explanation for the "bloody" urine.

a. What test could be used to determine if blood is present in the urine?
b. Why is the urine red-orange during some winter days but rarely discolored during the rest of the year?
c. Why is rabbit urine opaque?

Case 11

A veterinarian from Pennsylvania wrote, "I refer to your book *The Biology and Medicine of Rabbits and Rodents* frequently. Lab animals comprise about 15% of my clinical practice. At present I have a case which is not responding to therapy. A pet rabbit named Fred has an ulcerated area on his nose that constantly exudes serous fluid. As the fluid dries it becomes crusty, and after a week the crust looks like a small horn. The crust can be removed, but a raw, moist, ulcerated surface remains. Fred also has small scabs and crusts on the penis and scrotal pouches and several small crusty nodules on the lips. I have tried treating these areas with iodine washes, Tresaderm® (AgVet, Division of Merck & Co., Inc., Rahway, NJ), and tetracycline topically and orally, but the crusts come back. With Freds' scratching and licking, he has managed to spread similar lesions to his shoulder and left flank. I would like to biopsy an area, but the owners won't agree because of cost. Plucked hairs from the edge of the lesion do not grow on D.T.M. media and scrapings are negative for mites. Griseofulvin was tried orally without effect. I am enclosing a photograph of Fred. If you can help me with a diagnosis or treatment I would be grateful."

a. What diseases should be considered in the differential diagnosis?
b. What should be done to establish a definitive diagnosis?
c. What treatment do you recommend?

Case 12

A dead adult doe had blue-green discoloration and matting of the dewlap fur. Histologic examination of the skin revealed a severe, acute dermatitis characterized by multiple foci of epidermal ulceration and a diffuse inflammatory cell infiltration of the subcutis.

a. The blue-green discoloration of the fur suggests which bacterial organism may be present on the rabbit?
b. How can this bacterium be eliminated from a rabbitry? Suggest a treatment for this animal.
c. Which circumstances predispose to Pseudomonas infection?

Case 13

An adult doe with a history of chronic weight loss was brought to a clinic. The rabbit had lost 2 successive litters 1 week postpartum. Radiographs revealed no internal abnormalities.

a. What are causes of cachexia in rabbits?
b. Would a gastric hairball have a role in the death of the litters?
c. How are gastric hairballs diagnosed? Is there any treatment for a gastric hairball?

Case 14

A woman brought three adult non-castrated male rabbits into the veterinary clinic. All three rabbits were cagemates and had recently started chewing on their fur and biting one another over the back and around the scrotum. Examination of stool specimens revealed an absence of internal parasites.

a. What dietary deficiencies may induce fur chewing?
b. Is the housing situation a factor in the chewing and biting?

Case 15

A New Zealand White doe had delivered 10 young 4 weeks previously. All were healthy. The doe had focal areas of oily skin and alopecia over the face and shoulders. Under the skin in the inguinal region were 4 subcutaneous nodules 0.5 to 2 cm in diameter. On incision one nodule contained a caseous, inspissated material. Staphylococcus aureus was isolated from the abscess.

a. How would you treat the abscesses?
b. How would you proceed toward a definitive diagnosis of the alopecia?
c. How is acariasis treated in rabbits? Could humans be infected by a rabbit mite?

Case 16

An adult male rabbit was submitted for necropsy examination. The animal had a purulent conjunctivitis, rhinitis, and wet dewlap.

a. What are the causes of wet dewlap? How is this condition prevented and treated?
b. How would a Pasteurella multocida organism reach the orbit from the nasal passage?
c. How would you treat the conjunctivitis?

Case 17

A 2-month-old doe was submitted with a feces-soiled perineum. The attending clinician noted that the animal dragged the rear limbs.

a. What is the differential diagnosis for the paresis? How would you proceed to a definitive diagnosis?
b. What is the probable explanation for the fecal staining of the perineum?
c. In cases of spinal luxation, what factors determine a recommendation of cage rest or euthanasia?

GUINEA PIGS

Case 18

A young male guinea pig had chronic diarrhea and lethargy. He remained thin despite the owner's attempts to improve the diet and make the guinea pig gain weight. After a few weeks the guinea pig died. Necropsy revealed edema in the lamina propria of the ileum, cecum, and colon. Histopathologically, villi were blunted and no Tyzzer's bacillus could be found.

a. What diseases should be considered in the differential diagnosis of cases involving chronic diarrhea in young guinea pigs?
b. *Cryptosporidium* spp. oocysts were found in the guinea pig's feces. What does *Cryptosporidium* spp. look like? Describe the technique for finding this organism in the feces.
c. How is *Cryptosporidium* spp. transmitted? Is cryptosporidiosis a zoonotic disease?

Case 19

Within 3 days of the receipt of 30 older (2- to 3-year), mixed-sex guinea pigs, 2 animals developed crusty eyelids, rough hair coat, depression, dyspnea, and anorexia. The course of the fatal disease was 3 to 4 days. A variety of antibiotics, including gentamicin, chloramphenicol, sulfamerazine, and tetracycline, were eventually tried to stop the outbreak, but within a month only 4 animals remained alive.

a. Necropsy examination revealed pulmonary consolidation. What bacteria should be considered as etiologic agents for this disease outbreak?
b. How could the outbreak have been prevented?
c. Can other common animals serve as a source of the bacteria that are possibly involved in this case?

Case 20

A pet guinea pig was brought to a veterinary clinic. The animal exhibited the joint swelling and a history characteristic of an animal with scurvy. The veterinarian claimed he dispensed vitamin drops containing vitamin C in adequate quantity to eliminate the hypovitaminosis C. After 2 weeks the guinea pig was returned to the clinic with new problems. Despite a good appetite, the animal had become very thin and had lost much of its hair, although the signs of scurvy had receded.

a. What kind of vitamin product had the veterinarian prescribed or the client mistakenly purchased?
b. When is supplemental (in addition to that in the feed) vitamin C recommended for guinea pigs? How much is given daily?

Case 21

In a research guinea pig colony, sporadic deaths occurred until 250 of 400 animals were dead. Submandibular swellings and dyspnea were the clinical signs. Necropsy signs included conjunctivitis, enlarged cervical lymph nodes, pneumonia, pleuritis, otitis media, and metritis.

a. What pathogenic bacterium is probably involved?
b. This agent caused abortions in the colony. What other factors may produce abortions or stillbirths in guinea pigs?

Case 22

A student submitted a 600-g female guinea pig with a history of anorexia and death. Gross necropsy examination revealed atelectasis and consolidation of the entire left lung and congestion and focal abscessation of the right. The spleen was enlarged and had several white foci on its surface. The kidneys had white foci visible through the capsule.

a. The lesions in this guinea pig indicate a septicemia. The enlarged spleen with the necrotic foci is suggestive of what disease?
b. Is diarrhea a frequent finding in acute salmonellosis in research animals?
c. What prognosis can be given for colonies with endemic salmonellosis?

Case 23

An adult Strain 13 guinea pig exhibited lethargy, anorexia, emaciation, incoordination, and excessive salivation. Necropsy examination revealed a suppurative otitis media and encephalitis. A hemolytic *Streptococcus* was cultured from the middle ear.

a. What is the probable species of the *Streptococcus*?
b. How would this organism reach the middle ear and brain?
c. What are the causes of ptyalism in guinea pigs?

Case 24

An emaciated (365-g) adult female guinea pig was submitted for necropsy examination. A diagnosis of scurvy had been given by the local practitioner. Clinical and gross necropsy examination of external structures and thoracic and abdominal cavities revealed no ectoparasites or gross lesions.

a. What structures had the prosector failed to examine?
b. Anorexia and emaciation are common in guinea pigs. Suggest several causes for this problem in pet guinea pigs.

Case 25

A guinea pig with pododermatitis was noticed in a breeding colony. The animal was given ketamine, and the foot lesion was palpated and incised with a sterile scalpel. Several attempts to locate an abscess were unsuccessful. The foot, bleeding profusely, was bandaged, and the animal was held 5 days for observation. The lower portion of the affected limb was later amputated.

a. What factors predispose to pododermatitis in guinea pigs?
b. If the inflammatory response in pododermatitis is a chronic arthritis and not abscess formation, what treatment would you recommend?
c. Why might this guinea pig have become ketotic during the observation period?

Case 26

A 620-g guinea pig, which had delivered 2 young 5 days previously, was submitted for clinical examination. The right mammary gland was swollen and discolored.

a. How many mammary glands does a guinea pig have?
b. The guinea pig died 3 days after a penicillin injection. What is the probable cause of death?
c. What provisions should be made for the orphaned young?

Case 27

Ten Hartley guinea pigs were presented for evaluation from a colony with a recent history of increased stillbirths and mortality in near-term sows and neonates. There were no significant gross necropsy lesions. Histologic evaluation revealed numerous intranuclear eosinophilic inclusions in mandibular salivary gland ductular epithelial cells of two of the guinea pigs.

a. What herpes virus characteristically produces inclusions in mandibular salivary gland ductular epithelium?
b. What clinical signs are usually associated with this viral infection in guinea pigs?
c. Is this a zoonotic infection?

Case 28

A young woman had a pet breeding guinea pig sow monogamously housed with a boar. The sow delivered 2 strong, healthy guinea pigs on September 1. On September 16, 2 additional normal babies were delivered.

a. What is this phenomenon called?
b. What caused the separate births of groups of young?
c. How long is the guinea pig estrous cycle?

HAMSTERS

Case 29

A 19-month-old male hamster developed alopecia and scabs over the back and rear limbs. The hamster scratched and bit at the hairless area.

a. What disease conditions in hamsters can produce the signs seen here?
b. How would this case be treated?

Case 30

Breeding female hamsters in a colony of 150 became aggressively cannibalistic and weaned few if any young over several weeks. Necropsy examination revealed no lesions (other than bite wounds) in either dams or young. Husbandry practices regarding bedding, nesting material, and isolation were acceptable for hamster breeding colonies. Cages were polycarbonate plastic. Water bottles and slotted, flat metal hoppers were used for watering and feeding.

a. After several weeks of observation, the dams were noted to have broken incisors. How might this have happened?
b. How could malocclusion have led to the signs observed?
c. How should breeding female hamsters be fed before and during lactation?

Case 31

A 4-week-old golden hamster was submitted for clinical examination. The animal had a "wet tail," yellow diarrhea, and a prolapsed rectum.

a. At what age are hamsters weaned?
b. How is proliferative ileitis prevented in young hamsters?
c. What cestode causes an enteritis in hamsters?

Case 32

A golden hamster had not eaten for 4 days. The animal had blood at the anus and hair loss over the back and on both front legs. On necropsy examination the left adrenal gland was approximately 2 cm in diameter and had hemorrhaged into the abdominal cavity.

a. What diagnostic procedure should be undertaken to determine the agents involved in the dermatitis?
b. Do adrenal tumors occur in hamsters?

GERBILS

Case 33

An adult gerbil on a diet of sunflower seeds exclusively (no water) was submitted moribund. Three days previously the animal had jumped off a scale and fallen 3 feet to the floor. The gerbil had a paralysis of the rear limbs.

a. What dietary deficiencies may occur with a sunflower seed diet?
b. What is the probable cause of the sudden-onset paralysis?
c. How would this diet predispose to limb fractures?

Case 34

Two 4-year-old gerbils were brought to the clinic. One gerbil had a soft, subcutaneous mass (1 × 2 cm) immediately posterior to the xiphoid process. An abscess was detected by aspiration. The abscess was drained and flushed, and the gerbil was given chloramphenicol and sent home. One month later the gerbil returned to the clinic with a firm mass in the same area. The second gerbil, a female, had a rough hair coat and a slight but fixed head tilt.

a. In the first case the abscess was a secondary problem. What neoplasm might be suspected?
b. What is the probable cause of the rough hair coat?
c. Chloramphenicol and dexamethazone successfully stopped the progression of the torticollis. What is the therapeutic function of the steroid?

Case 35

Two 17-day-old gerbils were brought in for necropsy examination. Although the gerbils were cachexic, no gross lesions were present. The mother gerbil was in good health. The diet was a high quality, pelleted laboratory chow. A large water bottle was attached to the side of the cage.

a. What questions would you have asked the client about the husbandry practices?
b. At what age are gerbils weaned?
c. What are the mechanics involved in operating a sipper-tube waterer?

Case 36

A family purchased a pair of 6-week-old gerbils. The family was hoping to breed the animals and requested general information about breeding gerbils from a veterinarian. The clients were also concerned about the gerbils' rough hair coats.

a. What causes gerbils to have rough hair coats?
b. How do you verify that the gerbils are a breeding pair?
c. What type of cage, feed, bedding, feeding and watering devices, and nesting boxes should the practitioner recommend?

Case 37

A young adult female gerbil and several weanlings died suddenly with no other clinical signs. The ileocecal-colonic junction was hyperemic, and the liver contained scattered but distinct 1- to 2-mm pale foci.

a. What relatively common enteric condition of rabbits and rodents also causes focal hepatic necrosis?
b. How is the causative agent transmitted? Does it persist in the cage environment?
c. How would one obtain a specific diagnosis in this case?

Case 38

One member of a pair of male gerbils developed a copious, brown-red purulent discharge from the left eye. Adjacent skin was hairless and reddened. The swollen lids and pus covered the intact eyeball. *Staphylococcus aureus* was isolated in pure culture from the pus.

a. What is the probable site of the infection?
b. What is the source or cause of the brown-red pigment in the pus?
c. How would this gerbil be treated?

MICE

Case 39

A male mouse developed a slowly progressive posterior paresis, diarrhea, and perineal swelling. This pet mouse had been taken from a cage containing another male mouse.

a. What behavior characterizes group-housed male mice?
b. What is the probable cause of the clinical signs seen in this mouse?

Case 40

Inbred C3H female mice were inoculated with neoplastic tissue obtained from mice in another colony. Within several days of injection, two of the host mice died with no clinical signs. Over the next 3 weeks all recipient mice died. Clinical signs, when they occurred, included a rough hair coat, facial swelling, conjunctivitis, and death. As the dead mice were not found for several hours or even days following death, necropsies were not done.

a. What fatal diseases of mice should be considered in this outbreak?
b. What precautions should be taken to reduce the possibility of the spread of an infectious disease from this colony?
c. How could the disease be diagnosed?

Case 41

A black New Zealand mouse with a rough hair coat and a severe head tilt was brought to a veterinary clinic. Small, white ectoparasites were seen on the hair shafts. On gross necropsy examination the spleen was noted to be approximately 10 times normal size and the kidneys were pale brown. The tympanic bullae were grossly normal. No microbial organisms were cultured from the lungs, spleen, or middle ears. On histologic examination hematopoietic centers were observed in the liver and spleen. The glomerular basement membranes were thickened.

a. How do you account for the splenomegaly?
b. What is a possible cause for the anemia noted in this mouse?
c. How would you treat a colony of mice infested with fur mites?

Case 42

Six young male C3H mice, 4 alive and 2 dead, were emaciated and had soft, bloody feces around the anus. All mice had either a catarrhal or hemorrhagic enteropathy and thickened colons. The livers of the animals

were pale, and the spleens were enlarged. Culture of the colonic contents resulted in a heavy growth of *Proteus* spp. and coliform-like organisms.

a. Why did some mice have pale viscera?
b. What pathologic processes might account for a thickened colon?
c. What etiologic agents might be suspected in this case?
d. How would you contain an outbreak of enteritis in a large mouse colony?

Case 43

Three 4-week-old depressed and emaciated male mice had hunched postures, distended abdomens, and diarrhea. At necropsy the small intestines of the mice were pale and thickened and contained an excessive amount of mucus and gas, particularly in the anterior small intestine.

a. What diagnostic procedures would you use to obtain a definitive diagnosis?
b. What protozoal agents might be involved in this case?
c. Why, generally, are weanling animals more susceptible to acute enteropathies than are adults?
d. How can this protozoal condition be treated or controlled?

Case 44

Ten 7- to 9-g (3-week-old) white mice with stunted growth, rough hair coats, focal alopecia, and a yellowish diarrhea were submitted for necropsy examination. The large intestine contained fluid feces. Although morbidity in the colony was high, mortality was low.

a. What viral conditions in weanling mice cause diarrhea? How would you differentiate these conditions?
b. What agents or processes might cause focal alopecia in mice?
c. What murine viruses are carried subclinically in the mouse gut?

Case 45

Mice in nearly all 800 cages in a room in the research animal facility of a large medical center appeared scurfy and were scratching themselves. Many had skin lesions and hair loss and some appeared anemic. Numerous research projects costing several-hundred-thousand dollars were compromised. Filter bonnets were used on some cages, and administrators reported cages and lids were changed and washed twice weekly. Racks were washed weekly. Mice were infected with *Ornithonyssus bacoti*, the tropical rat mite.

   a. Why did some mice appear anemic?
   b. Because this mite lives off of the host, why did frequent cleaning not control the infection?
   c. How would one eliminate infections with this mite?

RATS

Case 46

Older and pregnant members of a Sprague-Dawley rat colony lost weight over several days. The animals had rough hair coats, matted eyes, labored respiration, and inappetence. Several died during a convulsion. Necropsy examination revealed pulmonary consolidation, focal abscessation, and mucopurulent material in the trachea and bronchi.

   a. Which pathogenic organisms may cause bronchopneumonia in rats?
   b. What are some gross differences that might distinguish the etiology?
   c. What diagnostic procedures are available to differentiate among the several organisms that might be involved in this case?

Case 47

A young girl brought two adult, female pet rats to the clinic. Both rats had single, 2-cm, firm, movable, subcutaneous masses in the axillary region. Fibrous connective tissue cells were seen on microscopic examination of the biopsy specimen.

   a. What is the probable origin of the masses?
   b. What histologic type of neoplasm occurs most often in the mammary gland of the rat?
   c. Which anesthetic would you use if you wished to remove the masses?

Case 48

A colony of 40 Long-Evans rats experienced an outbreak of "squinting" eyes associated with a serous conjunctivitis. The eyelids were swollen, and exudate matted the fur about the eyes. After a week the outbreak subsided, although several rats had clouded corneas. Histopathologic examination of the harderian glands (behind the globe of the eye and around the optic

nerve) revealed acute inflammation, necrosis, and squamous metaplasia of the lacrimal gland epithelium. *Staphylococcus aureus* in pure culture was recovered from the conjunctival sac of three rats. No organisms were cultured from the conjunctiva of several other affected animals.

   a. What is the hair coat pattern of the Long-Evans rat?
   b. What is the probable classification of the etiologic agent?
   c. Why would corneal damage occur?

Case 49

A junior high school student brought a 12-day-old litter of 8 rats to a clinic with the complaint that 3 had lost part of their tails and 4 others had 1 to 3 annular constrictions on the proximal portion of the tails. Except for the tail problem, the young rats and dam appeared healthy. The girl realized the affected rats could not be helped, but she wanted specific advice for preventing the disease in future litters.

   a. What is the postulated cause of the tail lesion?
   b. How can this condition be prevented?
   c. During which season is this condition most likely to occur?

Case 50

Aseptic surgical catheterization of the jugular vein was performed on 6 healthy 350-g male rats. The rats were anesthetized with doses of 300 mg/kg of chloral hydrate administered intraperitoneally. Within 4 days after surgery, the rats became depressed and anorexic and ceased defecation. Their abdomens became increasingly distended. All 6 rats died within 3 weeks after the surgery in spite of daily mineral oil gavages and penicillin injections for 4 days prior to death. At necropsy the entire small intestine was markedly dilated with fluid.

   a. What are the possible diagnoses?
   b. What is the most probable diagnosis?
   c. Describe the mechanism that resulted in the abdominal distention.

Case 51

To remove cosmetics applied in dermatologic testing, rats were shampooed and dried with a hair dryer. The animals were submitted to the laboratory because of dry scabby lesions that appeared on the tails, ears, feet, and scrotums shortly after the shampooing and drying procedures. Skin lesions were characterized by coagulative necrosis of the epidermis, dermis, and subcutaneous tissue with superficial bacterial colonization.

a. What is the most probable cause of the lesions?
b. How could the lesions be prevented?

Case 52

Investigators in urology and investigators in respiratory disease research noticed several different lesions in 3-month-old rats coming from a common institutional breeding colony. The urologists noticed small thread-like white worms in the wall of the urinary bladder. The respiratory disease researchers noticed larval forms and associated lesions in the lungs.

a. What one parasitism could cause both the respiratory and urinary tract abnormalities?
b. What laboratory tests can be done to provide confirming diagnostic evidence?
c. How can the condition be treated and eliminated from the colony?
d. Discuss the life cycle of this parasite.

Case 53

Several hours after doing injections on a group of 300-g male Sprague-Dawley rats, a research assistant had a general feeling of discomfort and itchy eyes, and rhinitis and wheals appeared on his hands. A second research assistant working on another project sometimes developed shortness of breath while working in the rat room.

a. What is the probable cause of the research assistants' ailments?
b. What preventive measure can be taken by the research assistants to avoid such reactions?
c. What environmental modifications will help to control the air allergen level?

SUGGESTED ANSWERS

These answers to the questions asked with the 53 clinical cases are intended to emphasize main points about the cases and are not detailed discussions of each case or problem.

RABBITS

Case 1

a. Traumatic spinal or limb injury, inherited defects, metabolic upset, and brain disease can lead to ataxia.
b. The common cerebral disease of rabbits is encephalitozoonosis. Cerebral nematodiasis can also cause incoordination.
c. The brain and kidneys are tissues to be examined. Periodic acid Schiff will stain the causative organism.
d. Perhaps the encephalitozoonosis was not the cause of the weight differential. Perhaps the size reduction was genetically or congenitally determined, or perhaps the effect of *Encephalitozoon* varies from rabbit to rabbit, depending on sites affected, amount of damage, and host resistance.

Case 2

a. Provisions for drainage of urine, a soft, absorptive floor, and room for movement would have reduced the causative factors involved in this case of "sore hocks."

b. The ulcerative lesions should be cleaned with an antiseptic soap, rinsed, and dried. Antiseptics or topical and systemic antibiotics may be used, if indicated. Soft, clean flooring should be placed in the box or cage.
c. Wetness reduces the thickness of the protective fur pad and promotes the growth of bacteria on skin irritated by urine or damaged from pressure necrosis.

Case 3

a. Diagnostic procedures to determine the cause of pendulous abdomen, diarrhea, and cachexia in the rabbit would involve palpation for intestinal distention, hepatomegaly, and a gastric hairball. The oral cavity should be examined for malocclusion and the feces for coccidial oocysts.
b. In hepatic coccidiosis the lesions are irregularly shaped (they follow the bile ducts) and have indistinct margins. A smear of a cut section of the liver or gallbladder would reveal oocysts on microscopic examination. *Taenia* lesions are usually rounded and have distinct margins. Larval cysts may be present in the peritoneal cavity. The numerous, small, white foci ("Milky Way") caused by *Francisella tularensis* will occur not only in the liver but also in other viscera, including the spleen. Tularemia is rare in domestic rabbits but more common in wild rabbits, especially in tick-infested areas.
c. *Eimeria stiedae* is the causative organism of hepatic coccidiosis in rabbits.

Case 4

a. Cannibalism in rabbits has been associated with prolonged delivery, poor nest building, chilling of young, hereditary nervousness, eating of placentas, disturbance of doe, presence of abnormal or dead young, water deprivation, improper nest boxes, low caloric diets, and a myriad of factors that are more difficult to define.

b. Correct the previously listed deficiencies and make the doe's environment as optimal as possible. Keep the doe calm at kindling time. Place the hutch and nesting box away from loud or unusual noises, strangers, and predators (snakes, opossums, racoons, foxes, coyotes, dogs, cats and other feral animals). Fresh drinking water and a sufficient amount of food should be provided. Mortality may decrease if the doe-kit contact is restricted to one 15- to 30-minute daily nursing session. Does that cannabilize successive litters should be culled.

Case 5

a. Clinical history, a fecal examination for coccidial oocysts, necropsy examination, and culture might provide a diagnosis, but in practice situations, a specific diagnosis is usually difficult to determine.

b. Pregnant does on restricted feeding should not be returned to full feed, especially with high energy feeds, immediately after kindling. Feed quantity should be increased 30 g per day to either full feed or 480 g maximum per day.

c. High environmental temperature (over 29° C or 85° F) for 3 days may produce infertility in bucks. The effect is both direct on the testes and indirect through decreased thyroid function.

Case 6

a. *Pasteurella multocida* is the common respiratory pathogen of rabbits. Dust and probably other allergens can cause sneezing and a watery nasal discharge in rabbits.

b. Bacterial rhinitis is rarely cured by antibiotics, but suppression of the clinical infection occurs with the administration of antibiotics such as penicillin.

c. The ear mite is *Psoroptes cuniculi*. The inflammation and pruritus within the external ear canal can cause dramatic head-shaking and scratching. The ataxia may be due to infectious middle and inner ear disease or CNS lesions associated with encephalitozoon infection.

Case 7

a. The renal lesion may be the manifestation of an interstitial nephritis caused by *Encephalitozoon cuniculi* (passed in the urine) or the arteriosclerosis-induced fibrosis associated with hypervitaminosis D in rabbits (vitamin D from cow's milk).

b. Pregnancy toxemia (acidosis) is the probable cause of death. Some causes of pregnancy toxemia or ketosis are prevented by eliminating obesity in breeding does through reduced food intake or feeding high fiber feed. The underlying defect may be inadequate nutrients reaching the fetuses because of small uterine vessels.

c. The rabbit's gestation period is between 28 and 35 days.

Case 8

a. Clostridial enterotoxemia and colibacillosis are probable diagnoses. Rotavirus infection, Tyzzer's disease, and coccidiosis should also be considered.

b. The prognosis is poor. Broad-spectrum antibiotics could be tried, along with fluid therapy, but the fatal event is probably an enterotoxemia. Symptomatic therapy may save a few animals.

Case 9

a. The causes of anorexia in rabbits are vague. Loss of olfaction, change of feed or feeder, hairballs in the stomach or fur blockage of the intestine, insufficient water, pain, illness, metabolic upset, malocclusion, and non-palatability of pelleted feed caused by rancidity or mold contamination are considerations.

b. The use of a variety of supplemental feeds, fluids administered subcutaneously or intragastrically, and anabolic steroids might have some desired effect.

c. The rabbit ingests its own feces (coprophagy); therefore, defecation would continue despite the absence of feeding.

Case 10

a. Tests for blood in urine are commonly available, especially those involving a paper strip with a color reaction.

b. The color of rabbit urine, caused by prophyrin and bilirubin derivatives, is intensified during dehydration or on certain pigmented or high calcium diets. In this case the rabbit's water was often frozen.

c. Alkaline rabbit urine contains large quantities of calcium carbonate and magnesium ammonium phosphate crystals.

Case 11

a. Differential diagnoses should include rabbit syphilis, myxomatosis, dermatophytosis, ectoparasitism, bacterial dermatitis, and neoplasia (Shope fibroma, papilloma, or squamous cell carcinoma).

b. These are classic lesions of rabbit syphilis and further testing may be unnecessary. One could culture for bacteria or fungi (D.T.M. media), do a biopsy and histopathologic examination, examine impression smears from the surface of the lesion for spirochetes, examine scrapings for ectoparasites, and do serologic tests for syphilis. The absence of growth on D.T.M. media and failure to respond to griseofulvin ruled out dermatophytosis. Myxomatosis is rare in the U.S.A. away from the West Coast, and the case was encountered in the early spring before the mosquito season.

c. Penicillin, 50,000 to 200,000 IU per 10 lb body weight, administered intramuscularly for 3 days was recommended and worked well. Rabbit syphilis may lead to reduced fertility.

Reference for Case 11

Reed, T.: Domestic Rabbits. Vol. 14(1). pp. 9–10, 1986. American Rabbit Breeders Association, Inc., P.O. Box 426, Bloomington, IL 61702.

Case 12

a. "Blue fur disease" in rabbits is caused by *Pseudomonas aeruginosa;* the blue-green discoloration is caused by the bacterial pigment pyocyanin. *P. aeruginosa* produces several toxins that may cause death of severaly affected animals. If the drinking water is contaminated, the water should be chlorinated or acidified.

b. Equipment must be thoroughly disinfected with a disinfectant effective against *Pseudomonas.* Water can be acidified. The lesion can be cleaned and treated by topical and systemic antibiotics.

c. Radiation, corticosteroid administration, and other circumstances or agents that reduce resistance may precipitate clinical *Pseudomonas* infection.

Case 13

a. Cachexia in rabbits may be caused by malocclusion, a gastric hairball, or, less commonly, a nutritional deficiency, chronic disease (coccidiosis, pasteurellosis, neoplasia), pain, advanced age, ectoparasitism, arteriosclerosis, or competition for or restricted access to feed or water.

b. Decreased food intake would lead to agalactia and dehydration, hypoglycemia, and litter death.

c. Diagnosis is based on clinical history. Palpation and radiography are not reliable aids. A gastric hairball may be removed surgically or loosened by the ingestion of hay or mineral oil; however, the prognosis for recovery or resolution remains poor, because the animal may develop ketosis.

Case 14

a. Fur chewing may indicate a fiber or salt deficiency. Adding protein, salt, fiber (hay), or MgO (5 lb per ton feed) to the diet and making certain drinking water is available may eliminate the fur chewing problem. Rabbits may ingest fur to decrease pain from enteritis.

b. Housing sexually mature male rabbits in separate cages may reduce or eliminate the fur chewing behavior, and will eliminate biting and fighting behavior. Housing adult females separately reduces the incidence of pseudopregnancies.

Case 15

a. A decision must be made to treat or not to treat an animal with multiple subcutaneous abscesses. Isolated abscesses may be treated by surgical excision and flushing with an antiseptic or antibiotic. A systemic antibiotic will protect against a septicemia resulting from surgical intervention.

b. The margins of the cutaneous lesion should be examined by hand lens for ectoparasites. When examined microscopically, a skin scraping may reveal ectoparasites or dermatophytes. A sample of broken hair from the margin of the lesion can be cultured on dermatophyte growth media. The possibility of repeated abrasion of the skin on the cage or feeder should also be considered.

c. Rabbits with ectoparasitism may be treated topically with an acaricide dust or ointment, or the rabbit may be dipped into an insecticide solution. Humans can be infected by *Cheyletiella.*

Case 16

a. Drooling or ptyalism in rabbits may be caused by malocclusion, or the dewlap may become wet and abraded if the rabbit feeds or drinks from crocks. A wet dewlap, which often is associated with a moist, bacterial dermatitis, may be prevented by correcting malocclusion or by feeding from hopper feeders and watering through sipper tubes. Superficial, moist dermatitis is treated by shaving the hair over the lesion, cleansing the skin with an antiseptic soap, rinsing, and applying a topical antibiotic powder or ointment.

b. *Pasteurella multocida* may reach the conjunctival sac through the nasolacrimal duct or by transmission on the forepaws from the nares to the eye.

c. Conjunctivitis in rabbits may be treated with an ophthalmic ointment containing an antibiotic or by systemic antibiotics. Treatment of the conjunctival infection does not necessarily eliminate the organism in the nasal passage. Flushing of the nasolacrimal duct with an antibiotic solution will reduce bacterial numbers and allow conjunctival exudate to drain to the nasal area and reduce matting around the eye.

Case 17

a. A congenital abnormality involving the spine, coxofemoral junction, or limb bones would exhibit a familial inheritance pattern. A traumatic injury would have a history of sudden onset and an association with a fall, improper restraint, or other traumatic event. A definitive diagnosis would be established by radiographic, physical, and neurologic examination.

b. The rabbit dragged the rear quarters in the dirty cage and stained the perineum and ventrum. Also, neural control of defecation and urination may be impaired.

c. The prognosis of posterior paresis in the rabbit depends on the extent of the damage to the spinal cord or nerve trunks. If the lesion is focal edema and inflammation, cage rest for 7 to 10 days may restore motor function. If the cord or nerves are torn, recovery will not occur.

GUINEA PIGS

Case 18

a. Tyzzer's disease, salmonellosis, coccidiosis, and cryptosporidiosis should be considered in cases involving chronic diarrhea in guinea pigs.

b. Cryptosporidial oocysts are 2 to 6 μm in diameter and are morphologically similar to yeasts. Diagnosis of cryptosporidial infection may be made from microscopic examination of concentrated fecal flotation samples or from acid-fast stains of fresh or formalin-fixed fecal smears.

c. *Cryptosporidium* spp. is highly infectious and may be transmitted by the fecal-oral route from various animal species to man. It is a common cause of disease in immunodeficient people.

Case 19

a. *Bordetella bronchiseptica* and *Streptococcus pneumoniae* are the bacteria that should be considered. This was an outbreak of *Bordetella bronchiseptica* infection.

b. The use of a formalin-killed bacterin might have prevented a clinical outbreak of *Bordetella* infection. Good husbandry, elimination of carriers, and stress reduction diminish streptococcal infections.

c. *Bordetella bronchiseptica* and *Streptococcus pneumoniae* can be carried by several animals, including rabbits, guinea pigs, cats, dogs, rats, and wild rodents.

Case 20

a. A multivitamin product was used in this guinea pig. The ascorbic acid was adequate, but other vitamins, especially vitamins D and A, were supplied in excessive amounts.

b. Supplemental vitamin C is given whenever the vitamin C content of the feed is suspect. Heat, length of storage, improper formulation, and dampness contribute to the breakdown of the vitamin. (See pp. 24 and 25 for suggested dosages of vitamin C.)

Case 21

a. *Streptococcus zooepidemicus* is probably the etiologic agent.

b. Dystocia, large fetal loads, salmonellosis, other streptococcal infections, nutritional deficiencies, stress, and ketoacidosis may cause abortion or stillbirths in guinea pigs.

Case 22

a. Salmonellosis in laboratory animals causes generalized visceral congestion and focal necrosis.

b. Diarrhea may occur with *Salmonella* infection, but this sign is an inconsistent finding. Affected animals may have a soft, discolored stool.

c. Because of the epizootic and fatal consequences of a *Salmonella* outbreak, the persistence of the disease in a carrier state, and the public health significance, infected colonies should be destroyed, the facility thoroughly cleaned and disinfected, animal care personnel examined, and the colony restocked with *Salmonella*-free animals.

Case 23

a. *Streptococcus zooepidemicus* is the streptococcal organism commonly affecting guinea pigs.

b. The organism may enter the guinea pig via cutaneous wounds, respiratory aerosol, or the conjunctival, oral, or genital mucosa. *Streptococcus* can reach the middle ear and brain through the circulatory system or by extension through the eustachian tube and middle and inner ear.

c. Ptyalism in guinea pigs has been associated with malocclusion, folic acid deficiency, fluorosis, scorbutus, and adrenocortical insufficiency.

Case 24

a. Emaciation in guinea pigs is often caused by malocclusion of the cheek teeth. The practitioner should also inquire about recent changes in the taste or composition of the diet.

b. Causes of weight loss in a pet guinea pig include malocclusion, metastatic calcification, vitamin C deficiency, ectoparasitism, chronic renal disease, and anorexia induced by changes in the taste or composition of the feed or changes in the feeding and watering devices.

Case 25

a. Pododermatitis in guinea pigs is most often encountered in heavy animals raised for long periods on a rough or abrasive floor, such as wire.

b. Treatment of pododermatitis is difficult because the inflammation is chronic and diffuse and relatively isolated from the vascular system. Staphylococci are frequently isolated from such lesions. Systemic antibiotics and surgical intervention might be attempted, but the prognosis for recovery is poor.

c. Ketosis in guinea pigs is a common sequela to anorexia. If for any reason (pain, septicemia) a guinea pig with a foot lesion should stop eating, ketosis could result.

Case 26

a. A guinea pig has two (inguinal) mammary glands.

b. Penicillin, and most other antibiotics, used in the guinea pig or hamster may induce an alteration of the intestinal flora that results in enteritis or enterotoxemia.

c. Newborn guinea pigs are precocious and can eat softened, pelleted feed within a few hours of birth.

Case 27

a. Cytomegalovirus (CMV) infection produces eosinophilic intranuclear inclusions in mandibular salivary gland ductular epithelial cells. These inclusions are markedly enlarged in infected guinea pigs. Smaller intranuclear inclusion may be found in kidney tubular cells.

b. Cytomegalovirus usually causes subclinical infection in guinea pigs. The incidence of infection in affected colonies is high.

c. Although the guinea pig is an animal model for CMV infections of man, there is no evidence that virus can be transmitted between guinea pigs and man. In fact, attempts to transmit guinea pig CMV to other species have failed.

References for Case 27

Griffith, B.P., and Hsiung, G.D.: Cytomegalovirus infection in guinea pigs. IV. Maternal infection at different stages of gestation. J. Infect. Dis., *141*:787–793, 1980.

Griffith, B.P., Lucia, H.L., Tillbrook, J.L., and Hsiung, B.D.: Enhancement of cytomegalovirus infection during pregnancy in guinea pigs. J. Infect. Dis., *147*:990–998, 1983.

Case 28

a. This phenomenon is referred to as superfetation.

b. Superfetation is associated with fertile matings at two different, usually successive, estrous cycles.

This phenomenon also occurs in man. Hormonal and placental similarities exist between guinea pigs and man.

c. The estrous cycle of the guinea pig is 16 days.

HAMSTERS

Case 29

a. Dermatophytosis, bite wounds, bacterial dermatopathies, and acariasis (*Demodex*) might cause lesions similar to those seen.
b. The cause of the problem should be determined (in this case, demodectic mange) and an appropriate treatment, such as Amitraz (Mitaban ® Liquid Concentrate—Upjohn, Kalamazoo, MI). Exercise caution in the use of Amitraz on debilitated or young animals.

Case 30

a. The female hamsters had worn and broken their incisor teeth on the feeding apparatus. The flat noses of the hamsters precluded them from sticking their faces through the narrow openings, so the dams bit over the metal in an attempt to gnaw the pellet.
b. Malocclusion and difficulty in reaching and gnawing the pellet led to decreased food intake, time away from the nest, and stress to the dam. These factors probably led to maternal neglect, agalactia, litter weakness and death, and cannibalism.
c. Feed should be placed onto the cage floor several days before parturition, and the cage then should be left alone until a week or more postpartum. Food near the nest, which simulates the natural hoarding behavior of hamsters, allows the dam ready access to ample feed.

Case 31

a. Hamsters are weaned at approximately 3 to 3.5 weeks of age.
b. Proliferative ileitis probably caused by a Campylobacter agent may be prevented by use of good sanitary measures and avoiding contact with and introduction of diseased animals. *Campylobacter* spp. may respond to treatment with erythromycin.
c. *Hymenolepis nana* and possibly *H. diminuta* are cestodes that may be associated with catarrhal enteritis in hamsters although asymptomatic infections are far more common.

Case 32

a. A skin scraping from the margin of the skin lesion should be examined for *Demodex* spp. and dermatophytes, although dermatophytosis has not been reported in the golden hamster. Alopecia in hamsters is also associated with endocrinologic imbalances.
b. Pheochromocytomas, cortical adenomas, and adenocarcinomas have been reported in golden hamsters. They are relatively common among the neoplasms of hamsters. Neoplasia may predispose an animal to demodicosis.

GERBILS

Case 33

a. Sunflower seeds contain low levels of calcium and high levels of fat; despite these deficiencies, when provided free choice gerbils may prefer sunflower seeds to other diets.
b. The sudden-onset paralysis probably resulted from a traumatic injury to the spine or limbs.
c. If the diet was in fact calcium deficient, the bones may have been more susceptible to fracture.

Case 34

a. Cutaneous neoplasms have been associated with the ventral, midline sebaceous scent gland in the gerbil. These neoplasms are usually basal cell or squamous cell carcinomas.
b. Rough or dull hair coats in sick animals may be associated with anorexia, dehydration, elevated body temperature, wetting of the fur with saliva, and pruritus. The fur of healthy gerbils becomes matted in humid environments.
c. Torticollis indicates an inflammation of the inner ear. If an inflammatory process damages the vestibular apparatus of the inner ear, the animal may never regain a normal posture. Administration of a steroid may reduce the damage done by inflammation. If the torticollis is not severe, an affected animal may compensate and continue to eat and drink despite the head tilt.

Case 35

a. Were the young gerbils able to reach and operate the water bottle? Were the young nursing the mother? Could the young animals gnaw the large, hard pellets? Did the young exhibit signs of clinical disease?
b. Gerbils are weaned between 20 and 26 days of age. The gerbils in this case were preweanlings.
c. Water bottles with small, curved, improperly angled, or partially blocked sipper tubes may release water until the partial vacuum above the water prevents further release of water. The bottle may appear full, but the water is not available to small animals.

Case 36

a. Rough hair coats in gerbils occur when the relative humidity is 50% or higher. Gerbils may also have rough hair coats if they are febrile, the water bottle leaks, or if the bedding is damp or contains resin.
b. Gerbils are sexed by noting the anogenital distance, which in adults is about 5 mm for females and 10 mm for males. The female has a vulvular opening at the base of the urogenital papilla, whereas the male has a larger papilla. If the gerbils have identical arrangements of the anal and urogenital structures, they are not a male-female pair.
c. Gerbils should be provided with deep bedding and concealed nesting and hiding places within the cage. The bedding should not be abrasive or irritating. Gerbils should be fed a quality, pelleted rodent feed from a hopper feeder and watered from a water bottle. The well-known water conservation

mechanism of the gerbil is a mechanism for survival in the desert and should not be used as a rationale for excluding water from the pet or laboratory gerbil. Nesting boxes may be simple metal boxes or cans 10 cm on a side.

Case 37

a. Tyzzer's disease often involves focal hepatic necrosis, but many other enteric conditions, including coliform infection, can pass via the portal circulation to the liver.
b. *Bacillus piliformis* is primarily transmitted through the ingestion of contaminated feces. Infectious, sporelike bodies from the filamentous organism remain in the environment for months.
c. Necropsy signs and Giemsa, PAS, or silver staining of histologic sections are the common methods for identifying the disease and the intracellular organism.

Case 38

a. The infection is probably located within the lacrimal glands and other retrobulbar structures.
b. The pigment is either blood or the porphyrin of the harderian (lacrimal) gland.
c. Sucrose-sweetened tetracycline in the drinking water for 8 days effected a cure for about 8 months, at which time the clinical signs reappeared.

MICE

Case 39

a. Adult male mice housed in groups will fight and bite one another. The fighting wounds, usually seen over the back, rump, and tail head, include alopecia, scabbing, and ulceration. In some cases abscessation occurs. The social hierarchy existing within the group influences the relative severity of the lesions.
b. This mouse had a staphylococcal abscess extending from beside the anus through the posterior abdomen and into the spinal cord and canal. Differential diagnoses include traumatic injury and neoplasia.

Case 40

a. Mousepox should be suspected in all cases of continuing mortality in a mouse colony, especially when that mortality is high and when the introduction of new mice or mouse tissues precedes the outbreak.
b. Recommendations should include evaluation of the colony providing the tissue, inoculation of the tissue into test mice maintained under strict quarantine, use of sentinal mice with the experimental mice, necropsy, and repeated serologic testing.
c. Serologic testing, virus inoculation, mouse inoculation, and histopathologic examination are used to diagnose mousepox.

Case 41

a. New Zealand mice over 12 months of age have been reported to have autoantibodies against erythrocytes and thrombocytes, although the prevalence varies with population and substrain. New Zealand mice also develop antinuclear antibodies, glomerulosclerosis, and a lupus-like glomerulonephritis. The splenomegaly may be associated with active, extramedullary erythropoiesis. The ectoparasites should be identified. In this case, they were debris-feeding mites.
b. The autoimmune hemolytic disease explains the anemia. The mouse should be examined for hemorrhage into the intestinal tract and also for ectoparasites. *Polyplax serrata,* the house mouse louse, is a blood sucker. The resulting anemia and extramedullary hematopoiesis could account for the splenic enlargement.
c. The placement of small pieces of a dichlorvos-impregnated resin strip onto the mouse cages or of dichlorvos pellets into the bedding will reduce the ectoparasite population. The strips should be placed in the cages for 48-hour periods every week or so until the parasites are gone. Caution is required because organophosphate insecticides affect breeding and the cholinesterase level in mice. Dusting with silica dust may be effective. Treatment with a synthetic pyrethroid, 0.25% Permethrin dust (Anchor Inc., St. Joseph, MO 64502), is also effective and has minimal effects on the metabolic processes of the animal.

Case 42

a. Pale viscera indicates an anemia. As some of these mice had a hemorrhagic enteritis, this may account for the anemia.
b. Inflammation of the intestinal wall, with edema and inflammatory cell infiltration, will result in a thickened intestinal wall, as will hyperplasia of the intestinal epithelium. Hyperplastic colitis of mice is caused by *Citrobacter freundi.*
c. Enteric organisms that may be involved in enteropathies in mice include *Giaridia, Hexamita, Citrobacter, Salmonella,* and *Pseudomonas.*
d. Rooms with affected mice should be isolated; the affected mice should be quarantined and treated or killed. Food, water, bedding, and animal handlers should be examined for the causative agent. The facility should be thoroughly disinfected and restocked with known disease-free mice. Filter cage covers used in a mouse colony will reduce the transmission of the highly infectious disease agents.

Case 43

a. The intestinal content can be cultured on a specific medium intended for the isolation of intestinal pathogens, and a direct smear of duodenal content and the mucosa can be examined for protozoa.
b. The causative agent in this case is probably *Spironucleus muris.*
c. Weanling animals lack acquired immunity. It can, therefore, not be very easily controlled.
d. Dimetridazole has been used on weanling mice to reduce mortality caused by protozoal infections but has been removed from the U.S. market. Flagyl® (metronidazole—Searle & Co., San Juan, Puerto Rico 00936), used in the treatment of trichomoniasis of humans, may be tried.

Case 44

a. Diarrhea in mice may be caused by a wide variety of organisms, including the viruses of epizootic diarrhea of infant mice (EDIM), mouse hepatitis virus, and reovirus 3 (Reo 3). EDIM affects weanling mice under 21 days of age. Mortality remains low, and the feces may be pasty and yellow. Reovirus 3 infections are characterized by an oily or fatty diarrhea occurring in preweanling mice. On necropsy examination mice affected with Reo 3 may have pale foci in the liver and other viscera. Serologic or fluorescent antibody tests may be used to establish a definitive diagnosis.

b. Dermatophytoses, excessive grooming or barbering, biting, and abrasion on a cage surface will cause a focal alopecia in mice.

c. Murine viruses that may be shed in the feces include TMEV, mouse hepatitis virus, reovirus 3, EDIM virus ectromelia virus, K virus, and the minute virus of mice.

Case 45

a. The mite is a rapid blood sucker.

b. The filter bonnets were not being cleaned or fumigated when cages, lids, and racks were cleaned. Mites may have resided in the filter bonnet from week to week. Nests with neonates were transferred from dirty to clean cages, possibly transferring the mites. Contrary to what administrators reported, close observation of husbandry practices over several weeks revealed that many cages with low cage densities were not changed twice weekly. Bedding was changed on a haphazard schedule when it appeared dirty. Many times all bedding was dumped out and new bedding was added without washing and sanitizing the cages. If bedding appeared clean it was not dumped and additional clean bedding was added.

c. Institute and enforce regular thorough cage, lid, and rack cleaning and wash, autoclave, or fumigate filter bonnets. Provide clean bedding and nesting material after each cage wash, decontaminate the environment, and prevent reintroduction of mites.

RATS

Case 46

a. Bacteria commonly associated with pneumonia in rats include *Corynebacterium kutscheri, Streptococcus pneumoniae,* and *Pasteurella pneumotropica. Mycoplasma pulmonis* is a common respiratory pathogen in rats. Sendai and corona viruses and the CAR bacillus may also contribute to respiratory disease of rats.

b. Suppuration, from which a stained direct impression smear might be made, might indicate a bacterial etiology. *Corynebacterium* causes a distinctive multifocal abscessation against a hyperemic parenchyma. *Mycoplasma pulmonis* infection in advanced stages causes bronchiectasis. Sendai virus may cause acute, focal necrotizing pneumonia and bronchitis. Corona virus primarily affects the upper respiratory tract and glands of the head and neck.

c. Direct smear, gross and microscopic signs, culture, and serologic testing will provide a diagnosis.

Case 47

a. Subcutaneous masses in rats are usually mammary neoplasms.

b. Mammary neoplasms in rats are most often fibroadenomas.

c. Ketamine-xylazine and methoxyflurane are satisfactory anesthetics for rats.

Case 48

a. Long-Evans rats have dark hair over the head and dorsoanterior trunk and are often described as "hooded" rats.

b. The causative agent of sialodacryoadenitis in rats is a corona virus.

c. Corneal damage would result from the decreased or absent lacrimal secretion and from traumatization of the eyeball by scratching the irritated area with rear feet.

Case 49

a. The specific causes of ringtail are unknown; however, ringtail in young rats results when the ambient humidity is low (20% or less). Temperature may also be involved. Because the rat's tail serves as a temperature control device, the ischemic lesion may involve an aberration of microcirculation in the tail, which functions as a temperature regulatory mechanism.

b. Ringtail can be prevented by ensuring an elevated ambient humidity in the rat breeding cage. Humidity can be raised by placing the young rats in a nest box supplied with adequate bedding and nesting material and by controlling the relative humidity of the room.

c. Ringtail usually occurs during cold seasons, when humidity is low in artificially heated rooms.

Case 50

a. Possible diagnoses include Tyzzer's disease, sepsis, adynamic ileus, bacterial enteritis, or protozoan enteritis (*Spironucleus muris* infection).

b. The likely diagnosis is adynamic ileus caused by an intraperitoneal injection of a high concentration of chloral hydrate. Fluid that accumulates in the adynamic distended intestine is a good place for proliferation of pathogenic protozoa, such as *S. muris.*

c. Chloral hydrate given intraperitoneally reduces intestinal motility either by its irritating chemical nature or by stimulating adrenergic reflexes that inhibit smooth muscle contractability. Chloral hydrate should not be used as the sole general anesthetic in rats. It is a poor analgesic and severely depresses respiration at anesthetic doses.

References for Case 50

Fleischman, R.P., McCracken, D., and Forbes, W.: Adynamic ileus in the rat induced by chloral hydrate. Lab. Anim. Sci., *27*:238–243, 1977.

Giel R.G., Davis C.L., and Thompson S.W.: Spontaneous ileitis in rats. A report of 64 cases. Am. J. Vet. Res., *22*:932–936, 1961.

Hottendorf, G.H., Hirth, R.S., and Peer, R.L.: Megaloileitis in rats. J. Am. Vet. Med. Assoc., *155*:1131–1135, 1969.

Leary, S.L., and Manning, P.J.: Fat rats. Lab. Anim., *11*(1):17–18, 1982.

Case 51

a. Judging by the location of the lesions on non-haired parts of the body and the histologic appearance, the lesions were probably thermal burns caused by heat from the hair dryer. Heat lamps cause similar burns.

b. Shampooed rats should be towel dried. If a dryer must be used, care should be taken to maintain it on a low heat setting and to remove the heat as soon as the animals are dry. Anesthetized animals are also susceptible to burns from hair dryers, heat lamps, or overheating from heating pads because rats cannot sense the heat and move away from the heat source.

Case 52

a. *Trichosomoides crassicauda* infection.

b. Examine urine for parasite ova. To do this, collect urine (use of metabolism cages and 2% glucose in 0.09% saline for drinking water to cause diuresis may be helpful); filter (use of a syringe to force urine through a 22-mm diameter, coarse Millipore prefilter, which retains the eggs, may be advantageous); and examine the filter pad microscopically at about 40X. The dark, brownish-yellow, doubly operculated eggs are easily seen. Histologically, one can see profiles of adult parasites and eggs in the wall of the urinary bladder, ureter, or pelvis of the kidney and migrating larval forms in the lungs.

c. Administer a single 200 µg/kg dose of ivermectin (Ivomec® Injectable—Merck and Co., Rahway, NJ) subcutaneously. Institute an effective cage cleaning routine associated with repeated transfer of recently treated animals to clean cages and elimination of chances for fomite transmission of eggs.

d. Adult female worms live partially embedded in the mucosa of the urinary tract and pass eggs intermittently in the urine. The small males reside inside the reproductive tract of female worms. Eggs ingested by a rat hatch in the stomach, and the resulting larvae migrate, via the lungs, to the kidneys and urinary bladder where the worms mature.

References for Case 52

Findon, G., and Miller, T.E.: Treatment of *Trichosomoides crassicauda* in laboratory rats using ivermectin. Lab. Anim. Sci., *37*:496–499, 1987.

Weisbroth, S.H., and Scher, S.: *Trichosomoides crassicauda* infection of a commercial rat breeding colony. I. Observations on the life cycle and propagation. Lab. Anim. Sci., *21*:54–61, 1971.

Case 53

a. The research assistants are probably having an allergic response to rat serum protein, salivary protein, or dander. The older the rat, the higher the concentration of serum protein in the urine.

Ammonia is produced by certain urea-splitting bacteria. Ammonia production increases as cages become more dirty (beginning about 3 days after changing). Serum proteins present in the urine become airborn with the ammonia and serve as allergens when contacted or inhaled by hypersensitive individuals. Direct contact with blood, feces, urine, or saliva can cause urticaria. In the case of the second research assistant, dypsnea and pulmonary reactions can occur almost immediately on inhalation of the allergen and can be severe, even fatal.

b. Delegating animal care duties to non-hypersensitive workers and wearing protective clothing (i.e., mask, gloves, gown) reduce exposure to allergens. Frequent cage changing (every 2 or 3 days) reduces ammonia concentrations.

c. Maintaining clean cages and proper ventilation is necessary to control allergen levels in the air. The airflow in the environment should be directed from ceiling to floor. Cage-rack units with special exhaust systems effectively reduce allergen exposure.

Index

Page numbers in *italics* indicate figures; numbers followed by *t* indicate tables.

Abscess(es), drainage of, 69
Acariasis, 111-115
Acepromazine, 63-64
  as preanesthetic, 62
  dosage table for, 60*t*
  in rabbits, 65
Acetaminophen, for analgesia, 70
Acetylsalicylic acid, for analgesia, 70
Acquired immune deficiency syndrome (AIDS), cryp-
      tosporidiosis and, 130
Aggression, in gerbils, 97
  in hamsters, 95
Alcide, as disinfectant, 5
Allergy(ies), to guinea pigs, 26
  to laboratory animals, 115-116
  to mice, 43, 47
  to rats, 53
Alopecia, in gerbils, 95
  in guinea pigs, 90-91
  in hamsters, 93
  in mice, 98
  in rabbits, 86
  in rats, 100
Ampicillin, fatal reactions to, 56
  for *Pasteurella pneumotropica*, 174
Analgesia, 69-70
Anaphylaxis, in guinea pigs, 21
Anatomy, of gerbils, 34-36
  of guinea pigs, 20-21
  of hamsters, 27-28
  of mice, 42-43
  of rabbits, 10
  of rats, 48-50
Anemia, in mice, 100
  in rabbits, 90
  in rats, 103
Anesthesia, 61-68
  administration of, 62-63
  depth of, 64
  drugs for, 63-64. *See also* Anesthetics
  in gerbils, 66
  in guinea pigs, 65-66
  in mice, 66-67
  in rabbits, 64-65
  in rats, 67
  overdose, 67-68
  postoperative care, 68

preoperative procedures, 61-62
  secretion removal during, 62
  topical, 64
Anesthetics, administration of, 62-63
  dosage table for, 57*t*
  See also Anesthesia
Anorexia, 117
  in rabbits, 87
Antagonists, dosage table for, 57*t*
Anthelmintics, dosage table for, 58*t*
Antibiotics, dosage table for, 58*t*-59*t*
  reactions to, fatal, 28, 56-57
Antimicrobials. *See* Antibitotics; specific agents
Apodemus species, Korean hemorrhagic fever and, 141
Appendage amputation, in mice, 98
Appendage inflammation, in mice, 98
Artificial insemination, of rabbits, 15-16
Artificial respiration, 67-68
Ascorbic acid, in guinea pigs, 24-25, 142-143
Aspiculuris tetraptera infestation, 168-170
Atropine sulfate, 63-64
  as preanesthetic, 62
  dosage table for, 58*t*, 60*t*

Bacillus piliformis infection, 198-200
Bacitracin, fatal reactions to, 57
Bactrin(s). *See* Vaccination
Bedding, cedar and pine, anesthetic degradation and,
      61
  for gerbils, 37
  for mice, 44
  for rats, 51
  See also Housing; Nest(s)
Behavior, of gerbils, 36
  of guinea pigs, 22-23
  of hamsters, 29-30
  of mice, 43
  of rabbits, 10, 12
  of rats, 50
Bittner agent, 165
Blood collection, 73, 76
Blood values. *See* Physiologic characteristics
"Blue breasts," in rabbits, 146
"Blue-fur" disease, 148
Bordetella, in guinea pigs, 26
Bordetella bronchiseptica, infection with, 117-119

Breeding program(s), for gerbils, 39
  for guinea pigs, 26
  for hamsters, 32-33
  for mice, 45
  for rabbits, 16
  for rats, 52
Brown fat, in rats, 48
Bruce effect, 46
Buffalo rat, 48
Bumblefoot, 191
Buphthalmia, in rabbits, 90
Butorphanol tartrate, for analgesia, 70

Cage(s), shoe box, *31*
    for guinea pigs, 23
    for mice, 43
  See also Housing
Cage cards, 3
"Caked udder," in rabbits, 146
Campylobacter, nonspecific enteropathy and, 135
  proliferative ileitis and, 178-179
Cannibalism, among gerbils, 39
  among hamsters, 33
  among mice, 46
  in rabbits, 17
  See also Litter desertion or death
Carbaryl powder, dosage table for, 59*t*
Carbon dioxide, for euthanasia, 71-72
Carbon monoxide, for euthanasia, 72
Cardiac puncture, 73
Castration, 69
Cavia porcellus, 19
Cavy. *See* Guinea pig
Cerebral ischemia, in gerbils, 34
Cervical dislocation, for euthanasia, 72
Cesarean delivery, of guinea pigs, 20
Cestodiasis, infestation with, 119-121
Cheek pouch(es), in hamsters, 28
Cheyletiella parasitivorax, in rabbits, 111-115
Chirodiscoides caviae, in guinea pigs, 111-115
Chloramphenicol, dosage table for, 58*t*
    for *Bordetella* infection, 118
    for mastitis, 146
    for mycoplasmosis, 158
    for *Pasteurella pneumotropica*, 174
    for *Streptococcus zooepidemicus* infection, 195
Chloroform, for euthanasia, 71
Chlorpromazine, 63-64
  dosage table for, 60*t*
  in guinea pigs, 66
  in mice, 67
  in rabbits, 65
Chlortetracycline, for nonspecific enteropathy, 136
Circling, by guinea pigs, 93
Citrobacter freundii, in transmissible colonic hyperplasia,
    195-196
Cleaning, of facility, 4-5
  of gerbils, 38
  of guinea pigs, 24
  of hamsters, 31
  of mice, 44
  of rabbits, 13
  of rats, 51
  See also Disinfectant(s)
Clindamycin, clostridial enterotoxemia and, 137

Clinical procedures, 55-82
  analgesia, 69-70
  anesthesia, 61-68
  blood collection, 73, 76
  drugs, 55-61, 57*t*-60*t*
  euthanasia, 71-73
  radiography, 70-71
  serologic testing, 74*t*-75*t*, 76
  surgical, 68-69
  See also specific topics
Clinical signs and differential diagnoses, 85-109
  in gerbils, 95-97
  in guinea pigs, 90-93
  in hamsters, 93-95
  in mice, 98-100
  in rabbits, 86-90
  in rats, 100-103
Clostridial enterotoxemia, antibiotic therapy and, 56,
    137-139
Clostridium, enterotoxemia and, 56, 137-139
  nonspecific enteropathy and, 135
Coccidiosis, hepatic, 122-124
  intestinal, 124-125
Colibacillosis, 126-127
Coliform enterotoxemia, antibiotic therapy and, 56
Colimycin, for nonspecific enteropathy, 136
Combiotic, 55
Conofite, 131
Constipation, in hamsters, 94
  in rabbits, 87
Convulsions, in gerbils, 34, 97, 139-140
  in guinea pigs, 92
  in hamsters, 94
  in mice, 99
  in rabbits, 89
  in rats, 102
  See also Seizures
Copper sulfate, for clostridial enterotoxemia, 138
Corynebacterium kutscheri infection, 127-129
Cottontail(s), 10
Cricetus cricetus, 27
Crictulus griseus, 27
Cryptosporidiosis, 129-131
Cryptosporidium, 129-131
Cutaneous or subcutaneous swelling, in gerbils, 96
  in guinea pigs, 91
  in hamsters, 93
  in mice, 98
  in rabbits, 86
  in rats, 100
Cuterebra fly larvae, in rabbits, 86

Death, in gerbils, 97
  in guinea pigs, 93
  in hamsters, 95
  in mice, 100
  in rabbits, 90
  in rats, 102-103
Decapitation, for euthanasia, 72
Demodex, infestations with, 111-115
Dermatitis, in gerbils, 96
  in guinea pigs, 91
  in hamsters, 93
  in mice, 98
  in rabbits, 86-87

in rats, 100
  moist, 148-149
Dermatopathy(ies), *Staphylococcus aureus* infection and, 190-192
Dermatophytosis, 131-132
Diarrhea, in gerbils, 96
  in guinea pigs, 91
  in hamsters, 93
  in mice, 99
  in rabbits, 87
  in rats, 101
Diazepam, 63-64
  as preanesthetic, 62
  dosage table for, 60t
  in gerbils, 66
  in guinea pigs, 66
  in rabbits, 65
  in rats, 67
Dichlorvos, dosage table for, 58t, 59t
  for acariasis, 114
Diet(s), disease prevention and, 4
  in guinea pigs, 25
    hypovitaminosis C and, 143
  in mice, 44
  in rabbits, clostridial enterotoxemia and, 137-138
    prenatal mortality and, 88
  renal disease and, 166-167
  See also Feeding and watering; Nutrition
Differential diagnoses, clinical signs and, 85-109
Dimethyl sulfoxide, for analgesia, 70
Diplococcus. See Streptococcus pneumoniae infection
Disease(s), factors predisposing to, 3-4
Disinfectant(s), 4-5
  for mousepox, 153
  for rabbit housing, 13
Disinfection, of facility, 4-5
Distrycillin A.S., 55
Drugs, antibiotic, 56-57, 58t-59t
  approved, 55-56
  dosing procedures for, 60t
  in feed and water, 61, 61t
  injection procedures for, 60t
  table of dosages, 57t-60t
  See also specific drugs
Dyrex, 56
Dyspnea, in gerbils, 96
  in guinea pigs, 91
  in hamsters, 94
  in mice, 99
  in rabbits, 87
  in rats, 101

Ear mite, in rabbits, 111-114
Ear notch-punch code, 3, *3*
Ectromelia, 151-154
Eimeria species, in intestinal coccidiosis, 124-125
Eimeria stiedae, in hepatic coccidiosis, 122-124
Encephalitis, mouse hepatitis virus infection and, 149-150
Encephalitozoon cuniculi, 132-134
Encephalitozoonosis, 132-134
Endotracheal intubation, for anesthetics, 62-63
Enflurane, in rats, 67
Enteropathy, mucoid, 154-155
  nonspecific, 134-137

Enterotoxemia, antibiotic therapy and, 56
  clostridial, 137-139
  milk, 138
  neonatal, 138
Environmental factors, disease prevention and, 3-4
Epilepsy, in gerbils, 139-140
Epizootic diarrhea of infant mice, 180-181
Erythromycin, fatal reactions to, 57
  for proliferative ileitis, 179
Escherichia coli, enteropathy and, 126-127
  proliferative ileitis and, 178-179
Estrous cycle, of gerbils, 38-39
  of guinea pigs, 25-26
  of mice, 45
  of rabbits, 14-15
  of rats, 52
Ether, 63-64
  for euthanasia, 71
  in guinea pigs, 65
  in hamsters, 66
  in mice, 67
  in rabbits, 65
Ethyl chloride spray, 64
Euthanasia, 71-73
Experimental procedures, disease prevention and, 4
Exsanguination, for euthanasia, 72
Eye enucleation, in rats, 102
Eye lesions, in rats, 103

Facility(ies), sanitation of, 4-5
Feeding and watering, daily intakes of adult animals, 61t
  of gerbils, 38
  of guinea pigs, 24-25
  of hamsters, 31
  of mice, 44
  of rabbits, 13-14
    equipment for, 13
  of rats, 51
  See also Diet(s); Nutrition
Fenbendazole, for pinworms, 170
Fentanyl-droperidol, 63-64
  in guinea pigs, 65
  in hamsters, 66
  in rabbits, 65
  in rats, 67
Fentanyl-fluanisone, 63-64
  in guinea pigs, 65
  in rats, 67
Foot stomping, in gerbils, 97
Formaldehyde gas fumigation, 5
Formalin test, for analgesia, 70
Fracture(s), reduction and fixation of, 68-69
Francisella tularensis, 197-198
Fumigation, of facility, 5

Genetic factors, disease prevention and, 4
Gentamicin, for *Staphylococcus* infection, 191
Gerbil(s), aggression in, 97
  alopecia in, 95
  anatomic and physiologic characteristics of, 34-36, 35t-36t
  anesthesia in, 66
  as pets, 36-37, *37*
  behavior of, 36

Gerbil(s) (*continued*)
  biology and husbandry of, 34-40
  breeding of, 38-39, *38-39*
    programs for, 39
  case reports in, 210, 217-218
  cutaneous or subcutaneous swelling in, 96
  cutaneous signs in, 95-96
  death in, 97
  dermatitis in, 96
  diarrhea in, 96
  disease prevention in, 40
  dyspnea in, 96
  epileptiform seizures in, 34, 97, 139-140
  estrous cycle of, 38-39
  feeding and watering of, 38
  foot stomping in, 97
  gastrointestinal signs in, 96
  housing for, 37-38
  incoordination in, 97
  infertility in, 96
  life span of, 36-37
  litter desertion and death in, 96-97
  miscellaneous signs in, 97
  mite infestations in, 111-115
  nasal discharge in, 96
  neoplasia in, 164
  neuromuscular signs in, 97
  ocular discharge in, 96
  origin and description of, 34
  pregnancy and rearing of, 39
  prenatal mortality in, 96
  public health concerns about, 40
  reproductive signs in, 96
  research uses of, 40
  respiratory signs in, 96
  restraint of, 37, *37*
  rough hair coat in, 95-96
  sexing of, 38, *38-39*
  sore nose in, 95, 148-149, 191
  sources of information about, 40
  Staphylococcus aureus infection in, 190-191
  tail skin loss in, 97
  torticollis in, 97
  Tyzzer's disease in, 198-200
  weight loss in, 97
Gliricola porcelli infestation, 175-176
Golden hamsters. *See* Hamster(s)
Griseofulvin, dosage table for, 58*t*
  for dermatophytosis, 131
Guinea pig(s), alopecia in, 90
  anatomic and physiologic characteristic of, 20-21, 21*t*-22*t*
  anesthesia in, 65-66
    pedal reflex and, 64
  as pets, 22-23, *23*
  behavior of, 22
  biology and husbandry of, 19-27
  breeding of, 25-26, *25*
  case reports in, 208-209, 215-217
  castration of, 69
  clostridial enterotoxemia in, 137-139
  convulsions in, 92
  cryptosporidiosis in, 129-131
  cutaneous or subcutaneous swelling in, 91
  cutaneous signs in, 90-91

death in, 93
  dermatitis in, 91
  diarrhea in, 91
  disease prevention in, 26
  dyspnea in, 91
  estrous cycle of, 25-26
  feeding and watering of, 24-25
  gastrointestinal signs in, 91
  heat stroke in, 140-141
  housing for, 23-24
  hypovitaminosis C in, 92
  incoordination in, 92
  infertility in, 92
  lice in, 175-176
  life span of, 22
  litter desertion or death in, 92
  malnutrition in, metastatic calcification and, 2-3
  malocclusion in, 145-146
    premolar trimming and, 68
  nasal discharge in, 91
  neoplasia in, 163
  nephrosis in, 166-168
  neuromuscular signs in, 92
  ocular discharge in, 91-92
  origin and description of, 19-20
  ovariectomy of, 69
  paralysis in, 92
  pregnancy and rearing, 26
  pregnancy toxemia in, 147-148
  prenatal mortality in, 92
  ptyalism in, 91
  public health concerns and, 26
  reactions to antibiotics, 28, 56-57
  reproductive signs in, 92
  research uses of, 26-27
  respiratory signs in, 91-92
  restraint of, 22, *23*
  rough hair coat in, 91
  salmonellosis in, 182-183
  sexing of, 25, *25*
  sources of information about, 27
  stampeding or circling by, 93
  Streptococcus pneumoniae infection in, 191-194
  Streptococcus zooepidemicus infection in, 194-195
  torticollis in, 92
  weight loss in, 92-93
Gyropus ovalis infestation, 175-176

Haemodipsus ventricosis, 175-176
  anemia and, 90
Halogen-bearing disinfectants, 5
Halothane, 63-64
  for euthanasia, 71
  in guinea pigs, 65, 66
  in hamsters, 66
  in mice, 67
  in rabbits, 65
  in rats, 67
Hamster(s), aggression in, 95
  alopecia in, 93
  anatomic and physiologic characteristics of, 27-28, 28*t*-29*t*
  anesthesia in, 66
  as pets, 29-30, *30*
  behavior of, 29

biology and husbandry of, 27-34
breeding of, 31-33, *32*
  programs for, 32-33
case reports in, 209-210, 217
clostridial enterotoxemia in, 137-139
constipation in, 94
convulsions in, 94
cutaneous or subcutaneous swelling in, 93
cutaneous signs in, 93
death in, 95
dermatitis in, 93
diarrhea in, 93
disease prevention in, 33
dyspnea in, 94
estrous cycle of, 32
feeding and watering of, 31
gastrointestinal signs in, 93-94
hibernation in, 95
housing of, 30-31, *31*
incoordination in, 94
infertility in, 94
life span of, 29-30
litter desertion or death in, 94
lymphocytic choriomeningitis virus in, 144-145
miscellaneous signs in, 94-95
mite infestations in, 111-115
neoplasia in, 163-164
nephrosis in, 166-168
neuromuscular signs in, 94
ocular discharge in, 94
origin and description of, 27
pregnancy and rearing of, 32
prenatal mortality in, 94
proliferative ileitis in, 178-179
public health concerns in, 33
reactions to antibiotics, 28, 56-57
rectal prolapse in, 94
reproductive signs in, 94
research uses in, 33
respiratory signs in, 94
restraint of, 30, *30*
rough hair coat in, 93
sexing of, 31-32, *32*
sources of information about, 33-34
torticollis in, 94
urogenital exudate in, 95
weight loss in, 94-95
Hantavirus infection, 141
Harderian glands, in gerbils, sore nose and, 95, 148-149, 191
  in rats, 48, 50
  oribital sinus bleeding and, 73
  sialodacryoadenitis and, 187
Hare(s), 10
Health, general husbandry concerns in, 2
Heat stroke, 140-141
Hemorrhagic fever with renal syndrome, 141-142
Hexamitiasis, 188-190
Hibernation, in hamsters, 95
History protocol, 85-86
Housing, for gerbils, 37-38
  for guinea pigs, 23-24
  for hamsters, 30-31, *31*
  for mice, 43-44
  for rabbits, 12-13

for rats, 50-51
  general husbandry concerns in, 1-2
Humidity, general husbandry concerns in, 2
  ringtail and, 100-101
Husbandry, facility sanitation, 4-5
  factors predisposing to disease, 3-4
  general considerations of, 1-7
  health, 2
  housing, 1-2
  major concerns in, 1-3
  nutrition, 2-3
  of rabbits and rodents in research, 5-6
  physical comfort, 2
  publications about, 1
"Hutch burn," in rabbits, 148
Hydroxyzine hydrochloride, for analgesia, 70
Hymenolepis diminuta, infestation with, 119-121
Hymenolepis nana, infestation with, 119-121
Hypnosis, in guinea pigs, 66
  in rabbits, 65
Hypochlorite(s), as disinfectants, 5
Hypothermia, postoperative, 68
Hypovitaminosis C, 142-143

Ibuprofen, for analgesia, 70
Identification, general husbandry concerns in, 3, *3*
Ileitis, proliferative, in hamsters, 178-179
Illumination, general husbandry concerns in, 2
Immune reaction, depression of, in mice, 46
Inbreeding, of mice, 42
Incoordination, in gerbils, 97
  in guinea pigs, 92
  in hamsters, 94
  in mice, 99
  in rabbits, 89
  in rats, 102
Infectious diarrhea of infant rats, 180-181
Infertility, in gerbils, 96
  in guinea pigs, 92
  in hamsters, 94
  in mice, 99
  in rabbits, 88
  in rats, 102
Inhalation anesthetics, delivery of, 62
Injection procedures, 60*t*
Innovar-Vet, for analgesia, 70
Insecticide(s), dosage table for, 59*t*
Intramedullary pins, 68
Intraperitoneal injection, of anesthetics, 62
Intravenous injection, of anesthetics, 63
Iodophor compounds, as disinfectants, 5
Ivermectin, dosage table for, 58*t*, 59*t*
  for acariasis, 113

Ketamine, 63-64
  dosage table for, 57*t*
  in gerbils, 66
  in guinea pigs, 65-66
  in hamsters, 66
  in mice, 67
  in rabbits, 64, 65
  in rats, 67
Korean hemorrhagic fever, 141-142
Kurloff's bodies, in guinea pigs, 20

Laryngoscope(s), for endotracheal intubation, 62–63
Lepus, 10
Life span(s), of gerbils, 36
  of guinea pigs, 22
  of hamsters, 29–30
  of mice, 43
  of rabbits, 12
  of rats, 50
Light:dark cycle, for gerbils, 37
  for guinea pigs, 24
  for hamsters, 31
  for rabbits, 13
  for rats, 51
  general husbandry concerns in, 2
  See also Illumination
Limb necrosis, in rats, 100–101
Lincomycin, clostridial enterotoxemia and, 137
  fatal reactions to, 56
Lindane, for acariasis, 113
Lipemia, in gerbils, 34, 36
Liponyssus bacoti, in rats, 53
  infestations with, 111–115
Liponyssus sylviarum, in rats, 53
Listrophorus gibbus, in rabbits, 111–115
Litter characteristics, of gerbils, 39
  of guinea pigs, 26
  of hamsters, 32, 33
  of mice, 46
  of rabbits, 17
  of rats, 53
Litter desertion or death, in gerbils, 96–97
  in guinea pigs, 92
  in hamsters, 94
  in mice, 99
  in rabbits, 17, 88–89
  in rats, 102
Liver spots, in rabbits, 90
Long-Evans rat, 48
Lymphocytic choriomeningitis, 144–145
  in mice, 46

Malathion, for acariasis, 113
Malocclusion, 145–146
Marchal bodies, in mousepox, 152
Mastitis, 146–147
Mebendazole, for pinworms, 169
Meperidine, for analgesia, 69–70
Meriones unguiculatus, 34. *See also* Gerbil(s)
Mesocricetus auratus, 27. *See also* Hamster(s)
Metabolic factors, disease prevention and, 4
Methoxyflurane, 63–64
  for euthanasia, 71
  in gerbils, 66
  in guinea pigs, 66
  in hamsters, 66
  in mice, 67
  in rabbits, 65
  in rats, 67
Microsporum infection, 131–132
Microtus, 42
Mitaban, for acariasis, 114
Mites, infections with, 111–115
Moist dermatitis, 148–149
  malocclusion and, 146
Mongolian gerbil. *See* Gerbil(s)
Morphine, for analgesia, 69

Mouse(mice), alopecia in, 98
  anatomic and physiologic characteristics of, 41*t*–42*t*, 42–43
  anemia in, 100
  anesthesia in, 66–67
  appendage inflammation or amputation, 98
  as pets, 43
  behavior of, 43
  biology and husbandry of, 40–47
  breeding of, 44–45
    programs for, 45
  case reports in, 210–212, 218–219
  convulsions in, 99
  Corynebacterium kutscheri infection in, 127–129
  cutaneous or subcutaneous swelling in, 98
  cutaneous signs in, 98
  death in, 100
  dermatitis in, 98
  diarrhea in, 99
  disease prevention in, 46
  dyspnea in, 99
  ectromelia in, 151–154
  encephalomyelitis in, 155–156
  epizootic diarrhea of infant, 180–181
  estrous cycle of, 45
  feeding and watering of, 44
  gastrointestinal signs in, 99
  hepatitis virus infection in, 149–151
  housing of, 43–44
  incoordination in, 99
  infertility in, 99
  lice in, 175–176
  life span of, 43
  litter desertion or death, 99
  lymphocytic choriomeningitis virus in, 144–145
  mammary tumors in, surgery for, 69
  miscellaneous signs in, 99–100
  mite infestations in, 111–115
  mycoplasmosis in, 156–159
  nasal discharge in, 99
  neoplasia in, 164–165
  nephrosis in, 166–168
  neuromuscular signs in, 99
  ocular discharge in, 99
  orchidectomy of, 69
  origin and description of, 42
  ovariectomy of, 69
  Pasteurella pneumotropica infection in, 174–175
  pendulous abdomen in, 99
  pinworms in, 168–171
  pregnancy and rearing of, 46
  prolapsed rectum in, 99
  prolapsed uterus in, 99
  public health concerns about, 46
  reovirus type 3 infection of, 179–180
  reproductive signs in, 99
  research uses of, 47
  respiratory signs in, 99
  restraint of, 43, *43, 44*
  rotavirus infection of, 180–181
  rough hair coat in, 98
  salmonellosis in, 182–183
  Sendai virus infection in, 184–186
  sexing of, 44, *45*
  sources of information in, 47
  spironucleosis in, 188–190

Staphylococcus aureus infection in, 190-191
transmissible colonic hyperplasia in, 195-196
weight loss in, 99-100
Mouse hepatitis virus infection, 149-151
Mouse leukemia virus, 165
Mouse mammary tumor virus, 165
Mouse polio, 155-156
Mouse sarcoma virus, 165
Mousepox, 151-154
Mucoid enteropathy, 154-155
Multiceps serialis, infestation with, 119-121
Murine encephalomyelitis, 155-156
Murine mycoplasmosis, 156-159
Mus musculus, 42. *See also* Mouse(mice)
Mycoplasma pulmonis infection, 156-159
Myobia musculi, in mice, 111-115
Myocoptes musculinus, in mice, 111-115
Myxomatosis, 160-161
vaccine for, for rabbits, 18

Nalbuphine hydrochloride, for analgesia, 70
Nalorphine, dosage table for, 58*t*
for anesthetic overdoes, 67
Nasal discharge, in gerbils, 96
in guinea pigs, 91
in mice, 99
in rabbits, 87
in rats, 101
Neomycin, for nonspecific enteropathy, 136
for transmissible colonic hyperplasia, 196
Neoplasia, 161-166
in gerbils, 164
in guinea pigs, 163
in hamsters, 163-164
in mice, 164-165
in rabbits, 161-163
in rats, 165-166
Nephrosis, 166-168
Nest(s), for gerbils, 37, 39
for mice, 46
for rabbits, 17
See also Housing
Neutrophil(s), in rabbit, 10
Niclosamide, for cestodiasis, 121
Night stool, in rabbits, 14
Nitrofurantoin, for nonspecific enteropathy, 136
Nitrogen gas, for euthanasia, 72
Norway rat, 47. *See also* Rat(s)
Nosematosis, 132-134
Notoedres muris, in rats, 111-115
Nutrition, for rabbits, 14-15
general husbandry concerns in, 2-3
See also Diet(s); Feeding and watering

Obesity, pregnancy toxemia and, 147-148
Ocular discharge, in gerbils, 96
in guinea pigs, 91
in hamsters, 94
in mice, 99
in rabbits, 88
in rats, 101-102
Onychomys, 42
Orbital sinus bleeding, 73
Orphan care, for gerbils, 39

for guinea pigs, 26
for hamsters, 33
for rabbits, 17
Oryctolagus, 9-10. *See also* Rabbit(s)
Osborne-Mendel rat, 48
Oxymorphone hydrochloride, for analgesia, 70
Oxytetracycline, for nonspecific enteropathy, 136
for *Pasteurella pneumotropica*, 174
for salmonellosis, 183
for *Staphylococcus* infection, 191
in drinking water, 57
Oxytocin, dosage table for, 59*t*
Oxyuriasis, 168-171

Paraldehyde, 63-64
in rabbits, 65
Paralysis, in guinea pigs, 92
in rabbits, 89
in rats, 102
Paresis, in guinea pigs, 92
in rats, 102
Passulurus ambiguus infestation, 168-170
Pasteurella multocida, in rabbits, 18, 171-174
Pasteurella pneumotropica, infection with, 174-175
Pediculosis, 175-176
Pendulous abdomen, in mice, 99
in rabbits, 87
in rats, 101
Penicillin, dosage table for, 59*t*
enterotoxemia and, 56
for pasteurellosis, in rabbits, 172
for *Staphylococcus* infection, 191
for venereal spirochetosis, 203
Pentazocine lactate, for analgesia, 70
Permethrin, for acariasis, 113
Peromyscus, 42
Phenacetin, for analgesia, 70
Phenol derivative disinfectants, 5
Physical comfort, general husbandry concerns in, 2
Physiologic characteristics, of gerbils, 34-36, 35*t*-36*t*
of guinea pigs, 20-21, 21*t*-22*t*
of hamsters, 27-28, 28*t*-29*t*
of mice, 41*t*, 42-43
of rabbits, 10, 10*t*-11*t*
of rats, 48-50, 49*t*-50*t*
Pinworm(s), 168-171
Piperazine, dosage table for, 58*t*
for pinworms, 169
Pituitary gland neoplasms, in rats, 166
Placentoma, in guinea pigs, 163
Pneumococcus. See Streptococcus pneumoniae infection
Pneumocystis carinii infection, 176-178
Pneumocystosis, 176-178
Pneumonia, in rabbits, 171-174
Pododermatitis, ulcerative, in guinea pigs, 191
Polydipsia, in rabbits, 87
Polyplax infestation, 175-176
Porphyrin(s), secretion of. *See* Red tears
Poxvirus-induced tumors, in rabbits, 162
Praziquantel, for cestodiasis, 121
Preanesthetics, 61-62
dosage table for, 60*t*
Pregnancy, in gerbils, 39
in guinea pigs, 26
in hamsters, 32-33
in mice, 46

Pregnancy (*continued*)
  in rabbits, 16-17
  in rats, 52-53
  metabolic toxemias of, 147-148
Prenatal mortality, in gerbils, 96
  in guinea pigs, 92
  in hamsters, 94
  in rabbits, 88
Presis, in rabbits, 89
Procaine, fatal reactions to, 56
Procaine penicillin, dosage table for, 59*t*
Prolapsed rectum, in mice, 99
Prolapsed uterus, in mice, 99
Proliferative ileitis, 278-179
Pseudomonas, in mice, prevention of, 46
Pseudopregnancy, in mice, 46
  in rabbits, 17-18
Psorergates simplex, in mice, 111-115
Psoroptes cuniculi, in rabbits, 111-114
Ptyalism, in guinea pigs, 91
  in rabbits, 87
  in rats, 101
Pyrvinium pamoate, for pinworms, 169

Quaternary ammonium disinfectants, 5

Rabbit(s), alopecia in, 86
  anatomic characteristics of, 10
  anemia in, 90
  anesthesia in, 64-65
  anorexia in, 87
  artificial insemination of, 15-16
  as pets, 10, 12
  atherosclerosis in, malnutrition and, 2
  atropinesterase activity in, 62
  behavior of, 10, 12
  biology and husbandry of, 9-19
  bloody urine in, 90
  breeding programs for, 16
  carrying methods for, 12, *12*
  case reports in, 205-208, 213-215
  castration of, 69
  clostridial enterotoxemia in, 137-139
  constipation in, 87
  convulsions in, 89
  cutaneous or subcutaneous swelling in, 86
  cutaneous signs in, 86-87
  death in, causes of, 90
  dermatitis in, 86-87
  diarrhea in, 87
  disease prevention in, 18
  dyspnea in, 87-88
  encephalitozoonosis in, 132-134
  estrous cycle of, 14-15
  eye enlargement in, 90
  feeding and watering of, 13-14
  gastrointestinal signs in, 87
  heat stroke in, 140-141
  hepatic coccidiosis in, 122-124
  housing of, 12-13
  hypnosis in, 65
  hysterectomy in, 69
  incoordination in, 89
  infertility in, 88
  intestinal coccidiosis in, 124-125
  lice in, 175-176

life span of, 12
litter desertion or death in, 88-89
liver spots in, 90
lumbar spine fracture in, 89
malocclusion in, 145-146
  incisor trimming and, 68
mastitis in, 146-147
miscellaneous signs in, 89-90
mites in, 86, 111-115
mucoid enteropathy in, 154-155
myxomatosis in, 160-161
nasal discharge in, 87
neoplasia in, 161-163
neuromuscular signs in, 89
nonspecific enteropathy in, 134-137
obesity and anorexia in, 117
ocular discharge in, 88
origin and description of, 9-10
paresis and paralysis in, 89
Pasteurella multocida infection in, 171-174
pendulous abdomen in, 87
physiologic characteristics of, 10, 10*t*-11*t*
pinworms in, 168-171
polydipsia in, 87
pregnancy and rearing of, 16-18
pregnancy toxemia in, 147-148
prenatal mortality in, 88
ptyalism in, 87
public health concerns for, 18
reactions to antibiotics, 56-57
reproductive signs in, 88
research uses of, 5-6, 18
respiratory signs in, 87
restraint of, 12, *12*
rotavirus infection in, 180-181
sexing, 14, *15*
sore hocks in, 86
sources of information about, 18
splay leg in, 89
Staphylococcus aureus infection in, 190-191
stomach tube for, 61
temperature for, room, 13
torticollis in, 89
trichobezoars in, 197
tularemia in, 197-198
urine of, 13
urolithiasis in, 201-202
vaginal discharge in, 89
venereal spirochetosis in, 202-203
weight loss in, 89-90
yellow fat in, 90
Rabbit enteropathy complex, 137
Rabbit hutch(es), 12-13
Rabbit syphilis, 202
Rabies, in mice, 46
  in rabbits, 18
  in rats, 53
Radfordia affinia, in mice, 111-115
Radfordia ensifera, in rats, 111-115
Radiography, 70-71
Rat(s), alopecia in, 100
  anatomic and physiologic characteristics of, 48-50,
      49*t*-50*t*
  anemia in, 103
  anesthesia in, 67
  as pets, 50

behavior of, 50
biology and husbandry of, 47-54
breeding of, 51-52
  programs for, 52
case reports in, 212-213, 219-220
convulsions in, 102
Corynebacterium kutscheri infection in, 127-129
cutaneous or subcutaneous swelling in, 100
cutaneous signs in, 100-101
death in, 102-103
dermatitis in, 100
diarrhea in, 101
disease prevention of, 53
dyspnea in, 101
estrous cycle in, 52
eye enucleation in, 102
eye lesions in, 103
feeding and watering of, 51
gastrointestinal signs in, 101
general illness in, 102
germ-free, production of, 53
housing of, 50-51
incoordination in, 102
infectious diarrhea of infant, 180-181
infertility in, 102
lice in, 175-176
life span of, 50
limb or tail necrosis in, 100-101
litter desertion or death in, 102
mammary tumors in, surgery for, 69
miscellaneous signs in, 102-103
mite infestations in, 111-115
mycoplasmosis in, 156-159
nasal discharge in, 101
neoplasia in, 165-166
nephrosis in, 166-168
neuromuscular signs in, 102
ocular discharge in, 101-102
origin and description of, 47-48
paresis or paralysis in, 102
Pasteurella pneumotropica infection in, 174-175
pendulous abdomen in, 101
pinworms in, 168-171
pregnancy and rearing in, 52-52
ptyalism in, 101
public health concerns about, 53
red tears in, 48, 50, 101-102
reproductive signs in, 102
research uses of, 53
respiratory signs in, 101-102
restraint of, 50, *50*
ringtail in, 100-101
rotavirus infection of, 180-181
rough hair coat in, 100
salmonellosis in, 182-183
sexing of, 51-52, *51*
sialodacryoadenitis in, 186-188
sources of information about, 54
Streptococcus pneumoniae infection in, 192-194
torticollis in, 102
weight loss in, 102
"Rat mesh," for guinea pigs, 24
  for rats, 50
Rat-bite fever, 53
Rattus norvegicus, 47
Rattus rattus, 48

Rearing, of gerbils, 39
  of guinea pigs, 26
  of hamsters, 32
  of mice, 46
  of rabbits, 17-18
  of rats, 52-53
Rectal prolapse, in hamsters, 94
Red tears, in rats, 48, 50, 101-102
  sialodacryoadenitis and, 187
Renal disease, chronic, 166-168
Reovirus type 3 infection, of mice, 179-180
Reserpine, in guinea pigs, 66
Restraint, for radiography, 70-71
  of gerbils, 37, *37*
  of guinea pigs, 22-23, *23*
  of hamsters, 30, *30*
  of mice, 43, *43-44*
  of rabbits, 12, *12*
  of rats, 50-51, *50*
Rhabdomyomatosis, in guinea pigs, 163
Ringtail, in rats, 100-101
Ringworm, 131-132
Rodent(s), in research, 5-6
Rotavirus infection, 180-182
Rough hair coat, in gerbils, 95-96
  in guinea pigs, 91
  in hamsters, 93
  in mice, 98
  in rats, 100

Salmonella enteritidis, 182
Salmonella typhimurium, 182
Salmonellosis, 182-184
  in mice, 46
Scurvy, in guinea pigs, 142-143
Seizures, in gerbils, 34, 97, 139-140
Sendai virus infection, 184-186
Serologic test(s), 74*t*-75*t*, 76
Sexing, of gerbils, 38, *38-39*
  of guinea pigs, 25, *25*
  of hamsters, 31-32, *32*
  of mice, 44, *45*
  of rabbits, 14, *15*
  of rats, 51-52, *51*
Sialodacryoadenitis, 186-188
"Slobbers," in rabbits, 148
Sodium pentobarbital, 63-64
  dosage table for, 57*t*
  for euthanasia, 72
  in gerbils, 66
  in guinea pigs, 66
  in hamsters, 66
  in mice, 67
  in rabbits, 64, 65
  in rats, 67
"Sore hocks," in rabbits, 86
"Sore nose," in gerbils, 95, 148-149, 191
Spironucleosis, 188-190
Spironucleus muris infection, 188-190
Splay leg, in rabbits, 89
Splints, 68
Sprague-Dawley rat, 48
St. Louis encephalitis, in rats, 53
Stampeding, by guinea pigs, 93
Staphylococcus aureus infection, 190-192
Streptobacillus moniliformis, in rats, 53

Streptococcus pneumoniae infection, 192-194
Streptococcus zooepidemicus infection, 194-195
Streptomycin, toxic reactions to, 56
"Stroke" syndrome, in gerbils, 34
Sulfamerazine, for mycoplasmosis, 158
Sulfamethazine, dosage table for, 59*t*
   for *Bordetella* infection, 118
   for coccidiosis, 124
   for transmissible colonic hyperplasia, 196
Sulfaquinoxaline, 55
   dosage table for, 59*t*
   for coccidiosis, 123
   for nonspecific enteropathy, 136
   for pasteurellosis, 172
Surgical procedures, 68-69
Swiss albino mouse. *See* Mouse(mice)
Sylvatic plague, rats and, 53
Sylvilagus, 10
Syphacia infestation, 168-170
Syrian hamster. *See* Hamster(s)

T-61, 55-56
   for euthanasia, 72
Taenia pisiformis, infestation with, 119-121
Taenia taeniaformis, infestation with, 119-121
Tail amputation, 69
Tail bleeding, 73
Tail necrosis, in rats, 100-101
Tail skin loss, in gerbils, 97
Teeth, trimming of, 68
Temperature(s), environmental, for gerbils, 37
   for guinea pigs, 24
   for hamsters, 31
   for mice, 44
   for rabbits, 13
   for rats, 51
   general husbandry concerns about, 2
   heat stroke and, 140-141
Tetracycline, dosage table for, 59*t*
   fatal reactions to, 56
   for murine mycoplasmosis, 157-158
   for nonspecific enteropathy, 136
   for proliferative ileitis, 179
   for *Staphylococcus* infection, 191
   for transmissible colonic hyperplasia, 196
   in drinking water, 57
Theiler's murine encephalomyelitis viruses, 155
Thiabendazole, dosage table for, 58*t*
   for cestodiasis, 121
   for pinworms, 170
Thiamylal sodium, dosage table for, 57*t*
TM-10, 55
Toenail trimming, 68
Torticollis, in gerbils, 97
   in guinea pigs, 92
   in hamsters, 94
   in rabbits, 89
   in rats, 102
"Transmissible ileal hyperplasia," in hamsters, 178
Transmissible murine colonic hyperplasia, 195-196
Treponema cuniculi infection, 202-203
Tribrissen, for *Bordetella* infection, 118
   for *Streptococcus zooepidemicus* infection, 195

Trichlorfon, for pinworms, 169
Trichobezoars, 197
Trichophyton mentagrophytes infection, 131-132
Trixacarus caviae, in guinea pigs, 111-115
Tularemia, 197-198
   Haemodipsus ventricosis infestation and, 176
Tumor(s), lymphocytic choriomeningitis virus in, 144
   See also Neoplasia
Tylosin, dosage table for, 59*t*
   for mycoplasmosis, 158
Tyzzer's disease, 198-200

Ulcerative dermatitis of rats, 191
Uredofos, for cestodiasis, 121
   for pinworms, 169
Urine, bloody, in rabbits, 90
Urine characteristics, in rabbits, 13
   of gerbils, 38
Urogenital exudate, in hamsters, 95
Urolithiasis, 201-202

Vaccination, for mousepox, 152-153
   for mycoplasmosis, 158
   for myxomatosis, 160
   for pasteurellosis, 173
   for Sendai virus, 185
Vaccine. *See* Vaccination
Vaginal discharge, in rabbits, 89
Venereal spirochetosis, 202-203
Vent disease, 202
Ventilation, general husbandry concerns in, 2
Vitamin A, in rabbits, 14
Vitamin C. *See* Ascorbic acid
Vitamin D, in rabbits, 14
Vitamin K, dosage table for, 58*t*

Water, for mice, acidification of, 46
   for rabbits, 15
   See also Feeding and watering
Weight loss, in gerbils, 97
   in guinea pigs, 92-93
   in hamsters, 94-95
   in mice, 99-100
   in rabbits, 89
   in rats, 102
"Wet dewlap," in rabbits, 148
"Wet tail," in hamsters, 178
Whitten effect, 45
Wire cage floors, for guinea pigs, 22, 23-24
   for rabbits, 16, 18
   for rats, 50
Worms, in rabbits, 13

Xylazine, 63-64
   dosage table for, 60*t*
   in guinea pigs, 65-66
   in hamsters, 66
   in mice, 67
   in rabbits, 64, 65
   in rats, 67

Yellow fat, in rabbits, 90
Yohimbine hydrochloride, 64
Yomesan, 56